Sparse Matrices and Their Uses

The Institute of Mathematics and its Applications Conference Series

---

RECENT TITLES

The Mathematical Theory of the Dynamics of Biological Populations, *edited by* M. S. Bartlett and R. W. Hiorns
Recent Mathematical Developments in Control, *edited by* D. J. Bell
Numerical Methods for Constrained Optimization, *edited by* P. E. Gill and W. Murray
Computational Methods and Problems in Aeronautical Fluid Dynamics, *edited by* B. L. Hewitt, C. R. Illingworth, R. C. Lock, K. W. Mangler, J. H. McDonnell, Catherine Richards and F. Walkden
Optimization in Action, *edited by* L. C. W. Dixon
The Mathematics of Finite Elements and Applications II, *edited by* J. R. Whiteman
The State of the Art in Numerical Analysis, *edited by* D. A. H. Jacobs
Fisheries Mathematics, *edited by* J. H. Steele
Numerical Software—Needs and Availability, *edited by* D. A. H. Jacobs
Recent Theoretical Developments in Control, *edited by* M. J. Gregson
The Mathematics of Hydrology and Water Resources, *edited by* E. H. Lloyd, T. O'Donnell and J. C. Wilkinson
Mathematical Aspects of Marine Traffic, *edited by* S. H. Hollingdale
Mathematical Modelling of Turbulent Diffusion in the Environment, *edited by* C. J. Harris
Mathematical Methods in Computer Graphics and Design, *edited by* K. W. Brodlie
Computational Techniques for Ordinary Differential Equations, *edited by* I. Gladwell and D. K. Sayers
Stochastic Programming, *edited by* M. A. H. Dempster
Analysis and Optimisation of Stochastic Systems, *edited by* O. L. R. Jacobs, M. H. A. Davis, M. A. H. Dempster, C. J. Harris and P. C. Parks
Numerical Methods in Applied Fluid Dynamics, *edited by* B. Hunt
Recent Developments in Markov Decision Processes, *edited by* R. Hartley, L. C. Thomas and D. J. White
Power from Sea Waves, *edited by* B. Count
Sparse Matrices and their Uses, *edited by* I. S. Duff

IN PREPARATION

The Mathematical Theory of the Dynamics of Biological Populations II, *edited by* R. W. Hiorns and D. Cooke
Third IMA Conference on Control Theory, *edited by* J. E. Marshall, W. D. Collins, C. J. Harris and D. H. Owens

ACADEMIC PRESS INC. (LONDON) LTD.
24/28 Oval Road,
London NW1

United States Edition published by
ACADEMIC PRESS INC.
111 Fifth Avenue
New York, New York 10003

British Library Cataloguing in Publication Data
Sparse Matrices and their Uses.
1. Matrices – Congresses
I. Duff, Iain S.
II. Institute of Mathematics and its Applications
512.9'43 QA188 80-42355

ISBN 0-12-223280-1

Printed in Great Britain by
Whitstable Litho Ltd., Whitstable, Kent

Sparse Matrices and their Uses

Based on the proceedings of the IMA Numerical Analysis Group Conference, organised by The Institute of Mathematics and its Applications and held at the University of Reading, 9th-11th July, 1980

Edited by

IAIN S. DUFF

AERE Harwell
Didcot, Oxfordshire
England.

1981

ACADEMIC PRESS

A Subsidiary of Harcourt Brace Jovanovich, Publishers

London New York Toronto Sydney San Francisco

CONTRIBUTORS

P.J. BERESFORD; *Atkins Research and Development, Woodcote Grove, Ashley Road, Epsom, Surrey KT18 5BW.*

R.E. BORLAND, DNACS; *National Physical Laboratories, Teddington, Middlesex.*

N.G. BROWN; *Computational and Statistical Science Department, University of Liverpool, Liverpool.*

D. CATLEY, AFIMA, *The British Ship Research Assn., Wallsend, Tyne and Wear.*

E. DETYNA, FIMA, *Department of Computer Science, University of Reading, Whiteknights Park, Reading RG6 2AX.*

I.S. DUFF, AFIMA; *Computer Science and Systems Division, Building 8.9, AERE Harwell, Didcot, Oxon OX11 0RA.*

A.M. ERISMAN, *Boeing Comp. Service Ltd., Energy Tech, App. Division, 565 Andover Park West, Tukwila, Washington 98188, USA.*

A. GEORGE, *Faculty of Mathematics, Department of Computer Science, University of Waterloo, Ontario, Canada.*

R.G. GRIMES, *Boeing Computer Services Co., 565 Andover Park West, Tukwila, Washington 98188, USA.*

U. HARMS, *Wunstorferstr, 14118, D-3000 Hannover 91, Germany.*

N. HOUBAK, *Technical University of Denmark, Numerical Institute, 2800 Lyngby, Denmark.*

D.A.H. JACOBS, FIMA; *Central Electricity Research Labs., Research Division, Kelvin Avenue, Leatherhead, Surrey KT22 7SE.*

A. JENNINGS, FIMA; *Department of Civil Engineering, The Queen's University of Belfast, David Keir Building, Stranmillis Road, Belfast BT7 1NN*

J.G. LEWIS; *Energy Tech. Division, Boeing Computer Services Co., 565 Andover Park West, Tukwila, Washington 98188, USA.*

H.M. MARKOWITZ; *IBM T.J. Watson Research Centre, Yorktown Heights, New York.*

R.K. NESBETT; *IBM Research Lab, K34-281, 5600 Cottle Road, San Jose, California 95193, USA.*

A.F. PEROLD; *Graduate School of Business Administration, Harvard University, Boston, Massachusetts 02163, USA.*

J. RAE; *Oil Reservoir Modelling Group, AERE Harwell, Didcot, Oxon.*

J.K. REID, AFIMA; *Computer Science and Systems Division, AERE Harwell, Oxfordshire OX11 ORA.*

D.S. SCOTT; *Mathematics and Statistics Research Department, Union Carbide Corporation, Nuclear Division, P.O. Box Y, Bldg. 9704-1, Oak Ridge TN37830 USA.*

D.B. SZYLD; *New York University, Inst. for Economic Analysis, 251 Mercer Street, New York, NY 10012, USA.*

P.G. THOMSEN; *Institute for Numerical Analysis, Building 303, The Technical University of Denmark, DK-2800 Lyngby, Denmark.*

Ph. L. TOINT, GradIMA; *Department of Mathematics, F.N.D.P., 64 Rue de Bruxelles, B-5000 Namur, Belgium.*

R. WAIT, GradIMA; *Computational and Statistical Science Department, University of Liverpool, Liverpool.*

Z. ZLATEV; *Hagens Alle 28, 2900 Hellerug, Denmark.*

PREFACE

The papers in this book were presented at a conference held at the University of Reading from July 9th to July 11th, 1980. The conference was organised by the Institute of Mathematics and its Applications whose Numerical Analysis Group (IMANA) was responsible for the initial proposal.

It was originally intended that the conference would be principally expository with several surveys and with other contributions discussing the application of sparse matrix techniques and software to various problem areas. In the event, many authors saw fit to introduce new research or new lines for research in their papers and so the resulting volume has, hopefully, something for everyone and should appeal equally to sparse matrix researchers, users of sparse matrix technology, and scientists and engineers who wish to learn more of this exciting and expanding field.

It was planned that each of the invited talks should last for fifty minutes to be followed by a ten minute questions and discussion period. Partly because this division was explicit in the programme, it was rather successful. My very free precis of these discussions follows each paper. Although the authors have checked my summaries, I am responsible for any misrepresentations contained therein.

Since it was the organisers' intention to structure the conference so that the topics on the three days were, respectively, the solution of linear systems and related areas, the eigenproblem, and systems arising from the solution of partial differential equations, this order has been adhered to in these proceedings. Since each paper is preceded by a short abstract, we do not discuss them at length in this preface. The first day's contributions consist of papers by Duff and Erisman on the solution of linear equations, by Toint on minimization and by Perold (jointly with Markowitz of ordering fame) on portfolio optimization. The eigenproblem papers were presented by Jennings (a survey with particular attention given to problems in structural vibration), Scott (Lanczos' algorithm), Nesbet (eigensystem techniques in quantum chemistry and atomic physics) and Beresford (engineering applications). Surveys on methods for the solution of sparse symmetric linear systems were given by Jacobs (iterative methods) and George (direct methods) while Reid considered frontal methods. Topics more specialized to the solution of partial differential equations were presented by Detyna (rapid elliptic solvers) and Rae (oil reservoir modelling).

A poster session was held at the conference and the level of interest in the posters was so high that it was decided to ask the authors to contribute short papers to this volume. Although there has been no refereeing process, I have tried hard as an editor to establish some conformity and consistency among the quite varied submissions. These papers follow the invited talks and are arranged in alphabetical order by author. They give a further flavour of current sparse matrix research and form a valuable addition to this volume.

It is interesting to compare this conference with the proceedings of the last conference held in England which was specifically on sparse matrices. That conference, also organised by the Institute of Mathematics and its Applications, was held in Oxford in April 1970 and was on the solution of linear equations. It is evident that, in addition to the spread of sparsity techniques to other areas, the scale of problems now being solved is considerably greater than ten years ago. It is also clear that the exploitation of sparsity is now widely appreciated to be a necessary element in the solution of large problems and techniques for this exploitation are now gaining acceptance in many different and varied fields. What is also clear, however, is that there are many areas in which apparently little progress has been made. We still do not fully understand the underlying structure of sparse problems in any but the simplest cases and the relationship between computing architecture, modelling and algorithms is far from completely mastered. Certainly, there has been an enormous growth in good and efficient general purpose software for sparse matrix problems. This is probably a significant contributory factor to the acceptance of such methods in so many different fields.

Finally, I would like to thank all of the authors for their patience and understanding when the editorial axe flew in their direction. Some cuts and alterations were very substantial and I sincerely hope that they have helped to make this book easier and more pleasing to read in addition to assisting the reader to a better comprehension of the concepts discussed. I would also like to thank the girls in the IMA office who have done an excellent job of retyping the whole book in a uniform format prior to photographic reproduction by Academic Press.

Iain S. Duff
AERE, Harwell

April 1981

ACKNOWLEDGEMENTS

The Institute thanks the authors of the papers, the editor, Dr. I.S. Duff (AERE, Harwell) and also Miss J. Fulkes, Mrs. S. Hockett and Miss D. Wright of the IMA for typing the papers for publication.

CONTENTS

CONTRIBUTED PAPERS

A SPARSE FUTURE

I.S. Duff

(AERE, Harwell)

ABSTRACT

We illustrate the present healthy state of sparse matrix research and indicate growth areas for future work by looking at a pot-pourri of current developments. In particular, we stress some recent results on the solution of linear equations by direct methods which call into question apparently self-evident observations of yesteryear and discuss their influence on future research. We describe embryonic attempts to develop general purpose algorithms which can take advantage of underlying structure, the impact of new machine architectures, interactive use of sparse matrix packages, and the concept of superaugmentation.

1. INTRODUCTION

In 1970, Ralph Willoughby of IBM, one of the founding fathers of sparse matrix technology, forecast that the field would be ripe for research and development for about ten more years. The purpose of this paper is not only to illustrate the health of sparse matrix research in 1980 but also to point to developments which indicate that the "sparse future" will be rosy and fruitful. We will make no attempt at a comprehensive "state of the art" review, since these can be found elsewhere in, for example, Duff (1977b) or Reid (1977). Instead we concentrate on one or two areas of particular interest and promise. Throughout, most of our discussion will be concerned with direct methods for the solution of sparse linear equations although many of our remarks have relevance in other contexts.

Sections 2 and 3 are concerned with methods for ordering variables and equations for the direct solution of sparse linear equations although the preordering and substructuring techniques discussed in section 3 are useful in many other applications and are integrally connected to the exploitation of many types of structure. The contrast between the complexity of sparse code on the one hand and the simplicity of full code on the other is highlighted in sections 4 and 5 and in the latter section we examine ways in which advantage can be taken of the simpler full code in the solution of sparse systems. Although we are concerned throughout with the impact of new machine architectures, we devote the whole of section 6 to looking at their influence on algorithms and software. We return to the theme of underlying structure in section 7 stressing the vast differences in sparse matrices from varying application areas and, in section 8, we discuss the important issue of the interface between sparse algorithms and the user community. Finally, in section 9, we comment on what joys the "sparse future" may hold in store for us.

Throughout this paper, the letter n will denote the order of the matrix and τ will be used to indicate the number of non-zeros. All times are in milliseconds on an IBM 370/168 unless stated otherwise.

2. A CALL TO ORDER

Much of the research into direct methods for the solution of sparse linear systems lies in determining the order in which pivots are chosen in the Gaussian elimination process. This choice can be made with a view to preserving sparsity, optimizing data structures, or maintaining stability. Often our pivoting strategy will seek to satisfy all three requirements but there is a large body of problems where we do not have to fear numerical instability, symmetric positive definite systems being the commonest and largest subclass. We first discuss ordering techniques for this class before considering the more general case.

Enormous savings can be made if we can avoid the necessity to pivot on numerical grounds. Not only do we not need to store or reference the non-zero values during the pivot selection stage but we do not necessarily have to access or update the full structure of subsequent reduced matrices and, by restricting our pivoting to the diagonal, we decrease greatly the complexity of our pivot selection process. Indeed, one of the beauties of diagonal pivoting on symmetric systems is that graph theory can be used in its analysis and implementation (Parter (1961), Rose (1970)). To any symmetric matrix, of order n say, there can be associated a labelled graph on n vertices such that there is an edge between vertex i and vertex j (edge (i,j)) if and only if entry a_{ij} (and, by symmetry, a_{ji}) of the matrix is non-zero. We give an example of a matrix and its associated graph in Figure 2.1.

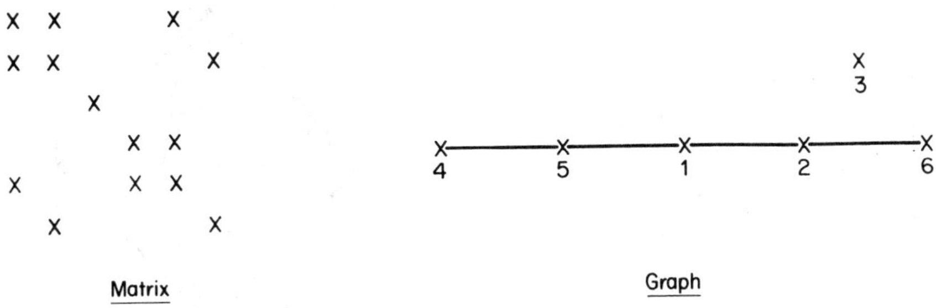

Figure 2.1 Symmetric matrix and associated graph

The main benefits of using such a correspondence can be summarised as

(i) The structure of the graph is invariant under symmetric permutations of the matrix (they correspond merely to a relabelling of the vertices).

(ii) For mesh problems, there is usually an equivalence between the mesh and the graph associated with the resulting matrix. We can thus work more directly with the underlying structure (see sections 3 and 7 also).

(iii) We can represent cliques (fully connected subgraphs) in a graph by listing the vertices in a clique without storing all the interconnecting edges.

To illustrate the importance of this clique concept in Gaussian elimination, we show in Figure 2.2 a matrix and its associated graph (also the underlying mesh if, for example, the matrix represents a five point discretization of the Laplacian operator). If the circled diagonal entry in the matrix were chosen as pivot (admittedly not a very sensible choice on sparsity grounds) then the resulting reduced matrix would have the dashed (pivot) row and column removed and have additional nonzeros (fill-ins) in the checked positions and their symmetric counterparts. The corresponding changes to the graph cause the removal of the circled vertex and its adjacent edges and the addition of all the dashed edges shown.

Thus the vertices 3,6,8,11 form a clique and, indeed, it is an easily proven property of Gaussian elimination that all vertices of the graph of the reduced matrix which were connected to the vertex associated with the pivot will become pairwise connected after that step of the elimination process. Although this clique formation (and, later in the process, clique amalgamation) has been observed for some time (for example, Rose (1970)) it is only very recently that techniques making use of (iii) above have been used in ordering algorithms. For our small clique above, there are 4 vertices and 6 interconnecting

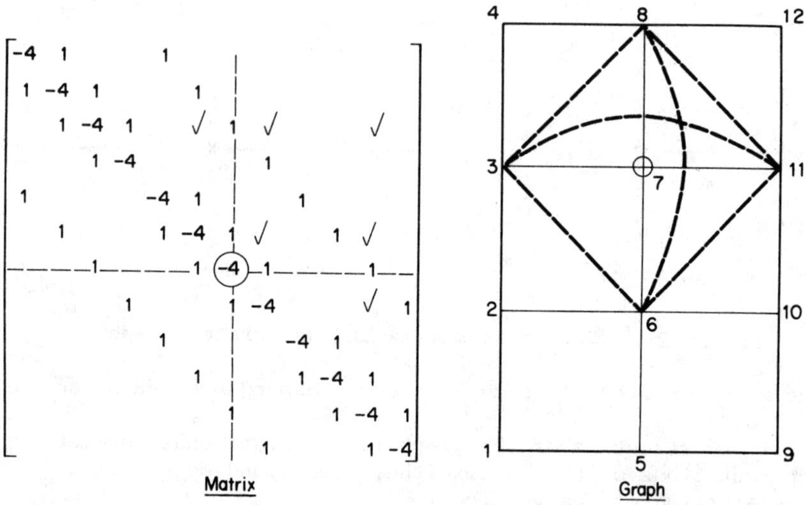

Figure 2.2 Use of cliques in Gaussian elimination

edges but as the elimination progresses this difference will be more marked since a clique on n vertices has n(n-1)/2 edges corresponding to the off-diagonal non-zeros in the corresponding full submatrix. Since any subsequent clique amalgamation can be performed in a time proportional to the number of edges in the cliques concerned, both work and storage are linear rather than quadratic in the number of nodes in the cliques.

We will not now discuss algorithms based on this observation. They are considered further in the contributions of Reid (1980) and George (1980) to these proceedings. However, we do wish to stress two very important aspects of such orderings which make this development one of the most exciting in this decade. Firstly, because we do not have to mimic the Gaussian elimination operations during this ordering phase, the ordering will generally be much faster than the actual elimination itself. This is in sharp contrast to the situation for unsymmetric systems or indeed to the situation for symmetric systems a decade ago. The second and in many ways more important aspect is that of storage where the ordering subroutine(s) can be implemented so that they require only slightly more storage (a small multiple of n) than that for the matrix itself. This means that even for very large problems whose decomposition must necessarily be out-of-core it may be possible to do an in-core ordering. Additionally, since we know in advance how much storage we need for this phase we know a priori that we can find the ordering and in the process we can forecast the storage required for symbolic and numerical factorization and the work for numerical decomposition (the symbolic factorization can be performed in time proportional to the storage, see for example George (1980)). This ability could be crucially important if we want to know whether the in-core solution by a direct method is feasible or whether an out-of-core solver or iterative method must be used. To illustrate the remarkable power of this approach, we show, in Table 2.1, ordering times for various algorithms on the graded-L problems of George and Liu (1978). The last two implementations use the clique amalgamation approach although many features of this technique are also incorporated in the 1978 subroutine.

TABLE 2.1

Ordering times on graded-L (IBM 370/168 millisecs)

Order Non-zeros	265 1009	1009 3937	3466 13681
1970 subroutine	1565	29876	-
1973 subroutine	683	6857	62437
1978 subroutine	244	1270	6047
Early implementation	267	1113	4043
Recent implementation	147	577	2047

We have not yet mentioned the strategy used to choose pivots. A very common technique is called the minimum degree strategy. Here we choose, at each stage, that diagonal element as pivot in the row with fewest non-zeros. One difficulty in using this approach is that we must keep track of the position of the non-zero elements, thus necessitating storage for integer indexing information in addition to that for the non-zeros themselves. Thus, if storage is at a premium, there may be some attraction in methods which do not take as great advantage of sparsity but take more account of data organisation. A classic example of such methods is the variable band technique (see for example, Jennings (1966)) where the only overhead is an array to indicate the length or position of the first non-zero in each row. The longer vectors and contiguous directly addressed storage employed suggest that such methods will adapt well to vector machines. While variable band schemes can be very competitive on problems in structural analysis where the sparsity pattern is very regular, they can perform very badly on the more irregular problems arising in networks (Erisman (1980)). We will discuss this phenomenon a little more in section 7. For the moment, it is interesting to see the effect of these two orderings on the matrix of Figure 2.3(a) from the Harwell collection of sparse matrices (Duff and Reid (1979)). This matrix seems, and indeed is, a good candidate for the variable band scheme pivoting down the diagonal in the natural order. This gives the essentially complete fill-in within the band shown in Figure 2.3(b) whereas the minimum degree ordering disperses the non-zeros as in Figure 2.3(c) necessitating the extra indexing information but perhaps surprisingly causing much less fill-in in the overall matrix (Figure 2.3(d)). The relevant counts are shown in Table 2.2.

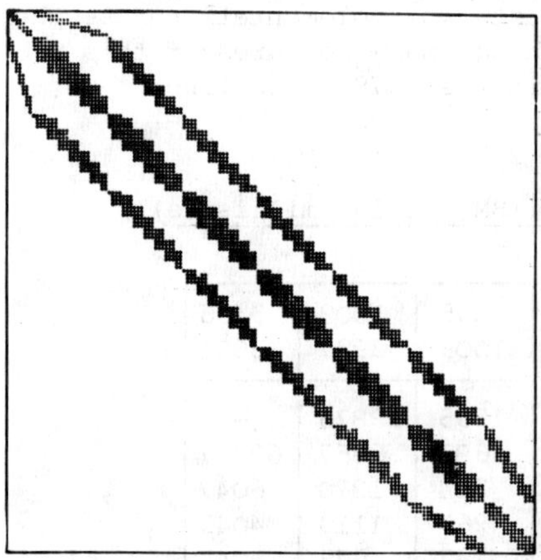

Figure 2.3(a) Original matrix

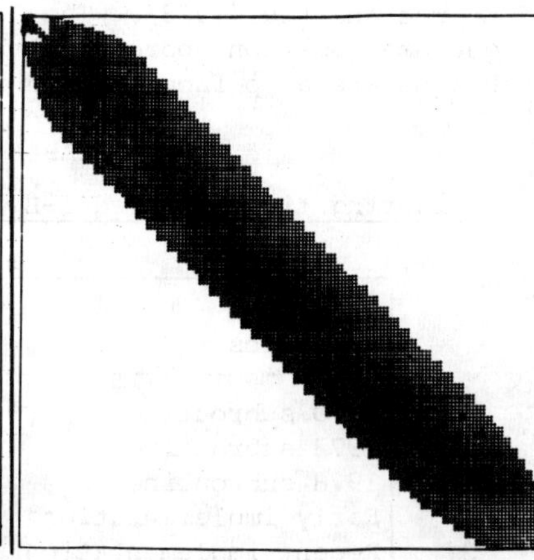

Figure 2.3(b) Pattern after pivoting down the diagonal in order

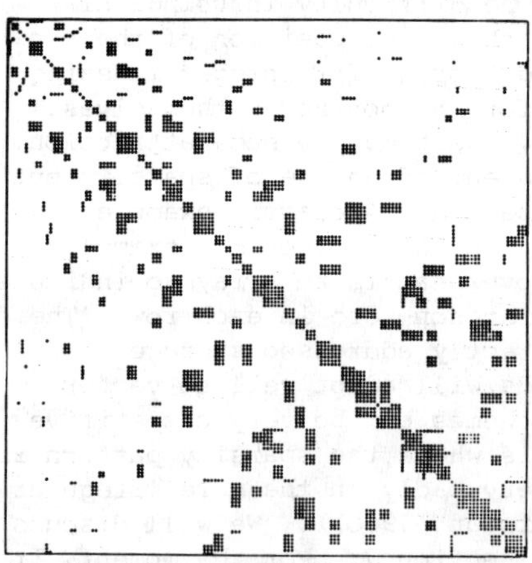

Figure 2.3(c) Reordered according to minimum degree strategy

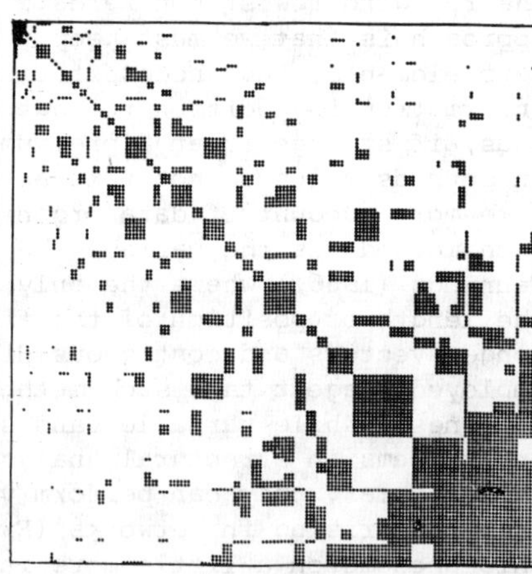

Figure 2.3(d) Pattern after pivoting down the diagonal of 2.3(c) in order

Figure 2.3 Pivot orderings on symmetric matrix of order 147

TABLE 2.2

Comparison of variable band and minimum degree algorithms on matrix from Figure 2.3 of order 147 with 1298 non-zeros. Integers and reals are assumed to occupy the same storage.

	Non-zeros in factors	Total storage required
Minimum degree	2353	3611
Variable band	3017	3164

In the general unsymmetric case, account must be taken of the value of the non-zeros to prevent possible numerical instability resulting from loss of information when a large number is added to a small one. This necessitates performing the decomposition at the same time as the ordering, thus, in general, precluding the techniques used above. The normal strategy used in such instances is the Markowitz (1957) criterion for preserving sparsity and a form of threshold pivoting to control stability. The Markowitz strategy is to choose that non-zero for which the product of the number of other non-zeros in its row with the number of other non-zeros in its column is minimized while the typical threshold criterion ensures that any pivot, a_{ij} say, satisfies the inequality.

$$|a_{ij}| \geq u.\max_{\ell} |a_{i\ell}| \qquad (2.1)$$

where u is a scalar in the range 0 to 1. We illustrate in Figures 2.4(a) and 2.4(b) how this criterion will tend to destroy any structure originally present in the matrix scattering the non-zeros apparently randomly and necessitating an even higher integer overhead than with the minimum degree ordering in the symmetric case.

It thus may be thought to be self-evident that we could never contemplate using such an ordering on a symmetric system since not only do we lose any gains from techniques mentioned earlier but we also destroy symmetry. To indicate that things are not always what they seem, we take as our example the solution of the sparse linear least squares problem

$$\min_{\underline{x}} \|\underline{b}-A\underline{x}\|_2$$

using the augmented system

$$\begin{pmatrix} I & A \\ A^T & 0 \end{pmatrix} \begin{pmatrix} \underline{r} \\ \underline{x} \end{pmatrix} = \begin{pmatrix} \underline{b} \\ \underline{0} \end{pmatrix} \qquad (2.2)$$

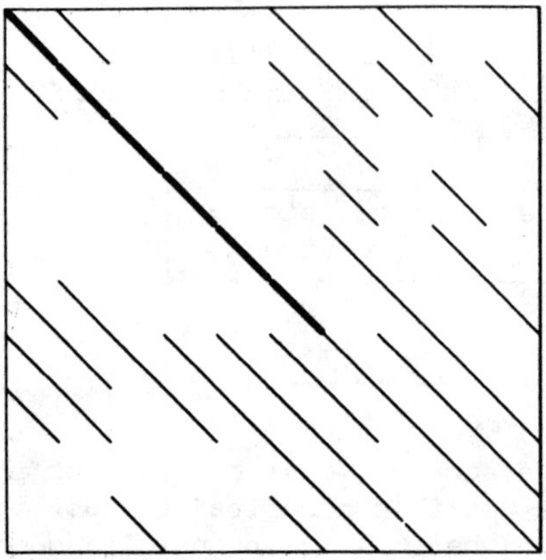

Figure 2.4a Original Matrix

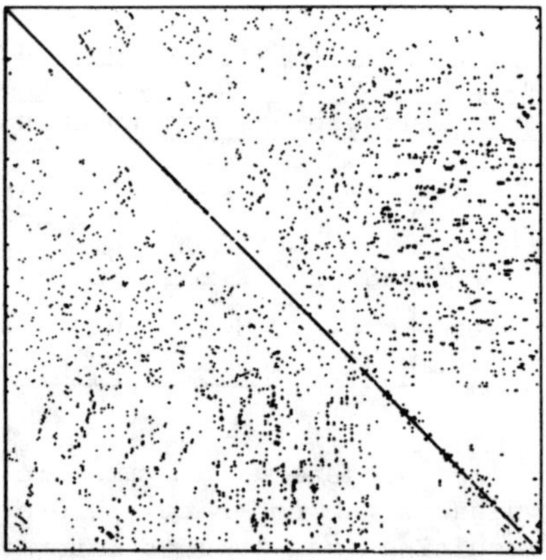

Figure 2.4b Matrix reordered so that pivots are on the diagonal in order

Figure 2.4 Markowitz ordering on unsymmetric matrix of order 541

as suggested by Hachtel (1974). An important feature of equation (2.2) is that, if our coefficient matrix A is of order mxn, then the last n components of our right hand side in (2.2) are always zero and we need only solve for the last n components $\underline{x}$ (the residual, if required, being obtained from the calculation $\underline{b}-A\underline{x}$). Duff and Reid (1976) show that, by modifying the Markowitz count to take advantage of this, an unsymmetric algorithm can beat a symmetric one with respect to operations in decomposition and storage for the factors. In a different context, George (1980) shows that an unsymmetric way of looking at the factors of a symmetric matrix can sometimes be more efficient than treating the factors symmetrically.

Much effort has been expended in attempting to develop general purpose heuristic ordering algorithms which will perform better than the Markowitz criterion. It is amazing that, after all those 23 long years, no technique can lay claim to having done so. It is very ironic, particularly in view of the remarks in the previous paragraph, that one of the strengths of Markowitz's ordering lies in its symmetry. That is, except for effects of tie-breaking and the effect of the threshold criterion (2.1), the Markowitz criterion will give the same set of pivots from the matrix or from its transpose. We may consider a strategy of first selecting the column with least number of non-zeros and then, within that column, selecting as pivot (subject to a threshold test of the form (2.1)) that non-zero in the column lying in the row with the fewest number of non-zeros (among all those rows with non-zeros in the column). We call this strategy r_i in c_j. Clearly this criterion is not symmetric and when we applied it to a matrix of order 363 with 3157 non-zeros we obtained the results shown in Table 2.3.

TABLE 2.3

The symmetry of the Markowitz ordering

	Non-zeros in factors		Multiplications for decomposition	
	Markowitz	r_i in c_j	Markowitz	r_i in c_j
Matrix	3886	3966	6440	7811
Matrix transpose	3904	8046	7865	51375

Finally, there is a fairly large class of unsymmetric matrices for which pivoting down the diagonal in any order is stable and, if they are also nearly symmetric in structure, an ordering technique like those discussed at the beginning of this section may be used on, for example, the structure of $A+A^T$. Although this can work well on certain classes of near-symmetric matrices, the results of Duff (1979) show that it is extremely dangerous to employ such a strategy on more general systems. Furthermore the fast symmetric orderings may not be able to yield a good estimate of the work or storage required for subsequent factorization. However, it would be a worthwhile and interesting task to discover what classes of matrices are amenable to such pivoting techniques. For example, it would appear that matrices arising from the discretization of second order non-self-adjoint elliptic partial differential equations with constant coefficients viz.

$$u_{xx} + u_{yy} + bu_x + bu_y + cu = f$$

fall into this category. For such matrices, Curtis (1980) has shown that, if upwind or central differencing is used on a square grid for the first derivative discretization, their eigenvalues are either all real and positive or all lie on a line parallel to the imaginary axis, the latter case being only possible if central differencing is used. We believe that some of the symmetric techniques discussed in this section might be used for unsymmetric systems in a more robust way than this (see section 3).

3. BEFORE THE ORDERING BEGINS

One problem with the ordering techniques of section 2 is that they take little account of the global structure of the underlying problem and a simple use of the variable band orderings works well only on rather simple geometries. In this section, we study techniques which attempt to use the underlying structure of the problem. As we shall stress again in section 7, we believe this to be a very exciting development in sparse matrix research which is still very much in its infancy.

The analogy to the variable band orderings of section 2 are the frontal schemes (for example, Irons (1970)) long loved by engineers and finite element practitioners. These schemes need not be restricted to finite element problems but will tend to do better if there is an underlying mesh present rather than if one is created artificially. Nor need the matrix be symmetric (see, for example, Hood (1976)). Indeed, this method of solving unsymmetric problems has much to commend it since advantage can be taken of the structure of the underlying problem while at the same time guarding against numerical instability. Indeed all we do is to make local perturbations to a ordering based globally on the structure (so that, for example, pivots satisfy the threshold criterion (2.1)). A similar technique could be used in the symmetric but indefinite case where local 2×2 pivoting could be used (for example, Bunch and Kaufman (1977)). There are two great benefits in using this approach. The first is that only the active front matrix need be kept in-core and so primary storage requirements can be kept low even when the whole problem is very large. The second lies in the coding of the inner loop. Normally, no advantage is taken of sparsity at this level and code for full matrices is used (see section 5). This has the advantage of both being simple and easily vectorizable (see section 6).

Of course, the simple frontal scheme can perform very badly on a multi-lobed structure like that of a hub with propeller blades and for this we need a more general multi-frontal approach like that discussed by Reid in these proceedings (Reid (1980)). Here the basic idea is similar to that described above but now the entire front matrix can be output to backing store and another front started in the structure. Later in the elimination process the output fronts are combined and further eliminations take place. This notion of working independently on different parts of the structure is what engineers have for many years called substructuring.

Global orderings of this type are not, of course, applicable to all problems and there are many in linear programming, network analysis, and chemical engineering for example, whose underlying structure is too unsymmetric to be amenable to such techniques. Here again we can often preorder (sometimes called sorting) the matrix to a suitable form either automatically or by a knowledge of the structure of the underlying problem. One form which can be proven essentially unique (Duff (1977c)) and for which good algorithms exist (Gustavson (1976), Duff (1978), Duff and Reid (1978)) is that of block triangular form. However, we do not know of as efficient algorithms for obtaining a bordered triangular form. This, for example, occurs slightly disguised in the "cheminée" form of economic modelling (for example, Nepomiastchy et al (1978)) illustrated in Figure 3.1. A similar form and algorithms for obtaining it also occur in linear programming (for example, Hellerman and Rarick (1971)).

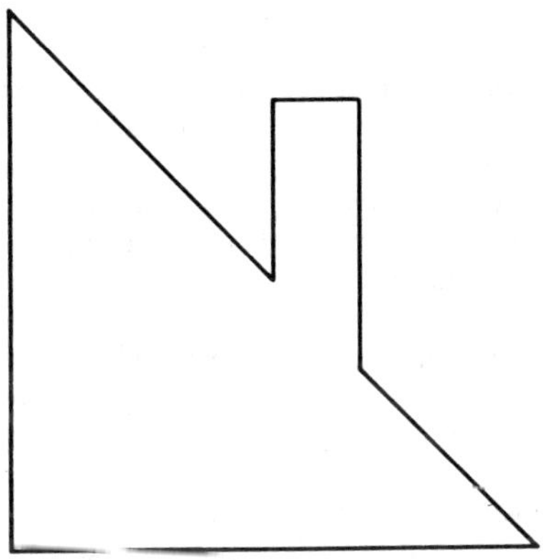

Figure 3.1 The Cheminée Form

For a large class of problems this form is only attainable at the cost of a relatively wide "cheminée" but removal of only a few columns can make it possible to order the rest into block triangular form. These bordered block triangular forms occur naturally in chemical process engineering where there are only a few global feedback variables and in operations research problems where a basically one way hierarchical structure has two way communication between one part of the structure and the rest. Although complicated heuristics have been suggested for finding such a structure (for example, Lin and Mah (1977)), there is again no very good algorithm or software and indeed one could surmise that the trial removal of successive columns followed by an attempt to block triangularize using the highly efficient algorithms mentioned earlier might prove as effective as any more complex technique.

Pottle (1979) inter alios has developed efficient techniques for the solution of doubly bordered block diagonal matrices on multi-processor machine environments and awaits good algorithms for obtaining this form when the formulation of the problem conceals this underlying structure.

4. THE COMPLEXITY OF SPARSITY

In this short section, which is as much aimed at the casual user of sparse packages as at the sparse matrix engineer, we consider briefly some implications of the complexity in the design of sparse code and indicate some pitfalls which should be avoided.

It is important to realise that sparse code is vastly more complicated than full code for performing the same function. We illustrate this in Figure 4.1 where we show a segment of Fortran code for performing the innermost loop in sparse Gaussian elimination (actually extracted directly from the Harwell MA28 code, Duff(1977a)) together with its full counterpart. Gustavson (1978) also illustrates well the complexity of performing, in an efficient way, the relatively simple operations of matrix multiplication and transposition on sparse data structures. Thus it is easy to see why the pivot selection and factorization phase in the Harwell code for sparse unsymmetric linear equations (Duff (1977a)) can run to 678 Fortran statements rather than the 122 statements of a corresponding routine for full systems. It is evident that there is a vast quantitative as well as qualitative difference between full and sparse codes making sparse codes very difficult to write well in addition to sometimes causing problems for the user because of their sheer bulk. Of course, that is one reason why we would like to use full code as much as possible during sparse computations (see section 5) and another reason why simpler codes like variable band schemes should be used where applicable.

The reason for such complicated codes is not that their authors like to exalt in the beauties and intricacies of their code (although, of course, they do) nor because they are paid for each line of code generated (would this were the case) but rather so that the run time efficiency can be enhanced. By this, we do not mean nitpicking changes to gain the odd 10% in speed but rather orders of magnitude savings in certain situations. For example, if a vector of length n is scanned at each stage in Gaussian elimination, the resultant $O(n^2)$ loop will totally dominate the execution time unless the system fills in rapidly. This is one reason for the elaborate code in Figure 4.1, the others are to preserve storage and to avoid having to keep sparse vectors ordered. Indeed, for very sparse systems, any loops of a complexity greater than $O(n)+O(\tau)$ should be looked at warily. It is not my experience that prospective authors of papers on sparse matrix codes are as aware of this as they should be.

```
C     ARRAY IQ IS AN INTEGER ARRAY ALL OF WHOSE ENTRIES
C     ARE POSITIVE EXCEPT THOSE ENTRIES IQ(I) FOR WHICH COLUMN I OF
C     THE PIVOT ROW IS NON-ZERO.  THIS NON-ZERO WILL BE FOUND IN POSITION
C     IJPOS-IQ(I) OF ARRAY A.  A PREVIOUS PASS THROUGH THE PIVOT ROW
C     HAS SO INITIALIZED IQ.  THE COLUMN INDEX OF NON-ZERO A(I) IS
C     HELD IN ICN(I).   FOR EACH NON-PIVOT ROW WITH A NON-ZERO IN THE
C     PIVOT COLUMN WE DO THE FOLLOWING
C
      DO 630 JJ=ROWI,IEND
      J=ICN(JJ)
      IF(IQ(J).GT.0) GO TO 630
      PIVROW=IJPOS-IQ(J)
      A(JJ)=A(JJ)+AU*A(PIVROW)
      ICN(PIVROW)=-ICN(PIVROW)
  630 CONTINUE
      DO 810 JJ=IJP1,PIVEND
      J=ICN(JJ)
      IF(J.LT.0) GO TO 800
C
C     FILL-IN PERFORMED.  IN MA28 THERE IS EXTRA CODE HERE TO ENSURE
C     THAT POSITIONS IEND,... ARE FREE TO HOLD FILL-IN.   THIS MAY HAVE
C     NECESSITATED MOVING THE NON-PIVOT ROW.
C
      A(IEND)=AU*A(JJ)
      ICN(IEND)=J
      IEND=IEND+1
C
C     IN MA28 WE NOW ALTER COLUMN ORIENTED REPRESENTATION TO ALLOW FOR
C     FILL-IN.
C
      GO TO 810
C
C    RESET ICN FOR NEXT PIVOT ROW
 800  ICN(JJ)=-J
C
 810  CONTINUE
```

<u>Sparse code</u>

```
      DO 630 J=1,N
      A(I,J)=A(I,J)+AU*A(IPIV,J)
630   CONTINUE
```

<u>Full code</u>

Figure 4.1 A fragment of sparse code and its full counterpart

Of course such complexities also create difficulties in assessing the merits of various competing codes. This is why variable band schemes can do well in comparison with their much more sophisticated counterparts. It also stresses the need for not just using fill-in or multiplication count as a yardstick for comparing codes since, in some instances, there can be more work in the book-keeping concerned with the sparse data structures. It also means that timing comparisons can be misleading unless we use a battery of test cases with very varied structure (for example, the differences in performance reported by Erisman

(1980)). Even then, because some machines (for example, the IBM/3033) are very much more efficient than others (for example, the CDC 7600) at fixed point calculations, any conclusions must necessarily be quite machine dependent and any rigid ranking, even for a single type of structure, may not be possible or meaningful.

5. IN THE FULLNESS THEREOF

We would be turning the clock back more than twenty years if we recommended the use of full matrix code on sparse problems. Nevertheless, in this section, we wish to look constructively at some genuine benefits of using full code and how these benefits can be reaped when solving sparse problems.

We can summarise the principal benefits of using full code as follows:

(i) it is easy to implement
(ii) there are no integer overheads
(iii) it will vectorize easily
(iv) it can occur naturally in some formulations of sparse matrix calculations and is, in particular, very applicable towards the end of the elimination when the reduced matrix becomes very dense.

We now look at each of these points in turn. The first two require no elaboration; Figure 4.1 makes the point quite clearly.

The vectorization of the inner-loop of Gaussian elimination is relatively easy to arrange. On most vector machines standard code will vectorize without change and after a small amount of manipulation and perhaps a little assembly level coding the super-vector speeds obtained from pipelined machines with chaining of pipes can be attained. We examine this further in the next section.

In a sense all sparse matrix algorithms involve full code since each Gaussian elimination operation involves operating on a full submatrix whose columns are those columns with non-zeros in the pivot row and whose rows are those with non-zeros in the pivot column. Methods like the frontal methods discussed in section 3 take advantage of this but it is difficult for general software to do so. Of course, as we saw in Figure 2.3(d), the reduced matrix becomes full before the elimination is complete, and so it would be quite appropriate to use full matrix code at that stage. Indeed, it may be beneficial not to wait until the reduced matrix is completely full but rather to switch to full code after it becomes relatively dense. We thus may lose a little on storage and operation count but we will gain on simplicity and speed of decomposition. The IBM SLMATH code DMOOP (IBM(1976)) based on work of Fred Gustavson, employs such an option although the switch over is set within the code so that full code is only used when the reduced matrix is full. The Harwell code MA31A (Munksgaard (1980)) also uses such an option where, because the matrices are symmetric and positive definite, the ordering problem after switching to full code is completely trivial since the order of elimination within the full block is totally immaterial. We illustrate the gains in using such a technique in Table 5.1. This shows that substantial savings can be obtained by switching long before the reduced matrix is full and indeed the point at which the increase in non-zeros held begins to outweigh the benefits of using full code is at a surprisingly low density. This effect would, of course, not be so marked for subsequent factorizations (where the pivot order is kept the same) or solutions where indeed the faster times would be obtained at the higher densities. However, the benefits of using full code may be even more substantial on a vector computer.

TABLE 5.1

The use of full matrix code in sparse elimination. Times are in milliseconds on an IBM 3033.

Density of active matrix when switch to full code occurs	Order Non-zeros	147 1294	363 1962	400 1199
	Non-zeros in factors			
30		3616	7808	7094
40		3043	6656	6673
60		2638	5785	6179
80		2446	5174	5914
100		2348	4725	5768
no switch		2348	4725	5768
	Time			
30		222	491	410
40		214	432	394
60		216	432	408
80		229	475	423
100		260	616	521
no switch		296	820	1281

6. THERE'S A NEW GENERATION

When we speak loosely of vector machines or parallel hardware, we will here mean computers which employ high degrees of pipelining or vector registers or simultaneous instruction execution or genuine parallel processors or indeed have a combination of these features. Several, so called third generation machines do use some of these devices, for example pipelining on the CDC 7600, but the new generation of machines are characterized by their ability to perform long vector operations much more quickly than corresponding computations involving scalars. Thus, we include in this class machines as diverse as the CRAY-1, CYBER 205, ICL DAP and AP 120 B. As of June 1980, typical speeds for vector computation are shown in Table 6.1.

TABLE 6.1

Speeds in megaflops* for new machines

Best third generation machines	2
Fourth generation machines	
Essentially scalar mode	5
Vector mode	25
Super-vector mode	120

Although it is most important to ask how existing software will run on such machines and to bear possible vectorization in mind when designing new codes, it is also important to realise that most scientific computation is presently performed on sequential machines and much will be for many years to come.

Obviously such general purpose algorithms as minimum degree need to be re-evaluated in the light of such machines. A balance must be sought between squeezing the utmost out of sparsity and organising the computation so that more long-vector operations are used. The inner-loop of many sparse calculations shown in Figure 6.1 will, because of indirect addressing and the use of short vectors (which is the aim of the minimum degree criterion), perform badly on most vector machines (for example, Dembart and Neves (1977)). Clearly, the problem of indirect addressing is avoided if the variable band techniques of section 2 are used although, even there, one normally tries to keep vectors short causing significant overheads on many pipelined machines. The frontal methods can be better organised and a balance set between maintenance of sparsity and use of long vectors by allowing the front to grow beyond the minimum required.

*A megaflop is one million floating point operations (of type + * - ÷) per second

Some authors (Dembart and Neves (1977), Pottle (1979)) have even suggested artificially increasing vector length by adding judiciously placed zeros!

```
      DO 100 JJ=J1,J2
      J=ICN(JJ)
      W(J)=W(J)+AMULT*A(JJ)
  100 CONTINUE
```

Figure 6.1 Inner-loop in a sparse matrix calculation

Many of the preordering techniques of section 3 lend themselves very well to developing algorithms for machines capable of parallel processing. For instance, the decomposition of any diagonal block of a block triangular form can be performed quite independently of the decomposition of any other one. Pottle (1979) has examined the solution of problems in doubly banded block diagonal form on such architectures.

With the cost of VLSI chips coming down even faster than inflation is going up, it is very likely that we will see the appearance of machines specifically designed to solve one particular problem (for example, the implementation of Gaussian elimination on band (and full) systems by Kung and Leiserson (1979)) Perhaps a self-adjoint elliptic partial differential equation solving machine will be available from Tandy* within the next decade.

7. LET STRUCTURE REIGN

Many fortunes (and reputations) have been won and lost in the quest of yet better solution techniques to discretizations of elliptic partial differential equations and, in particular, to 5 or 9 point discretizations of the Laplacian operator. For some people, the only type of sparse matrix is of the form shown in Figure 2.2. However, most of us would agree that such matrices have a very special structure and very special properties and indeed it is by the very exploitation of these that fast and efficient solution techniques have been developed.

It would be quite unfair to say that the story ends there since the profile or variable band techniques knowingly or unknowingly make use of the peculiar connectivity found in models in, for example, structural engineering, and the preordering techniques of reordering to block triangular form or bordered block triangular form are very effective in many areas of linear programming although they are inapplicable in the general

---

*Radio Shack in the United States

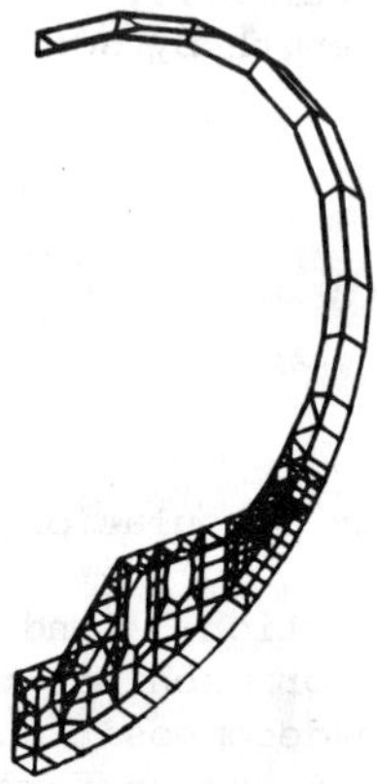

Figure 7(a) Hull-tank-region

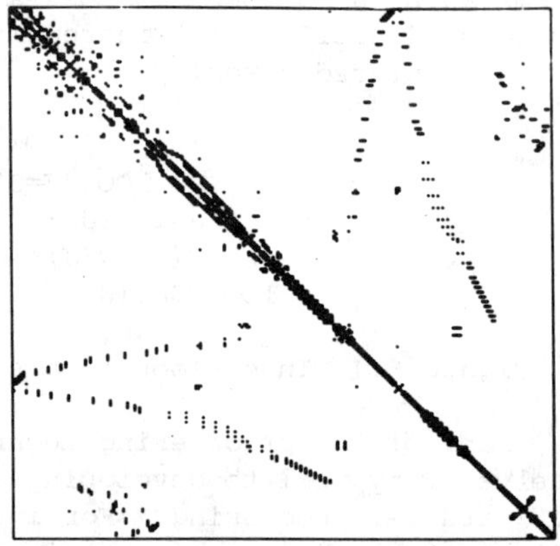

Figure 7(b) Matrix of order 221

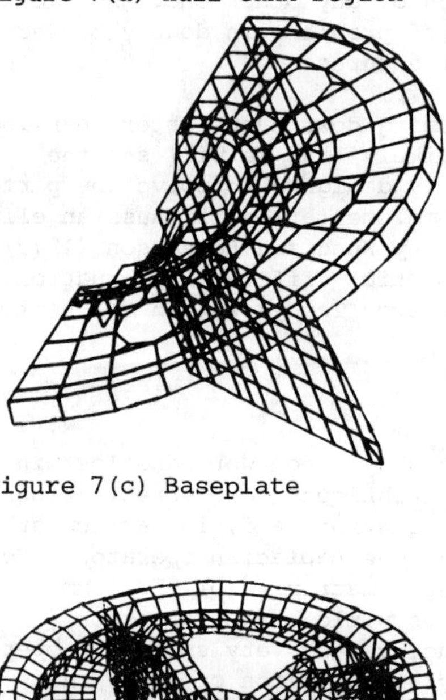

Figure 7(c) Baseplate

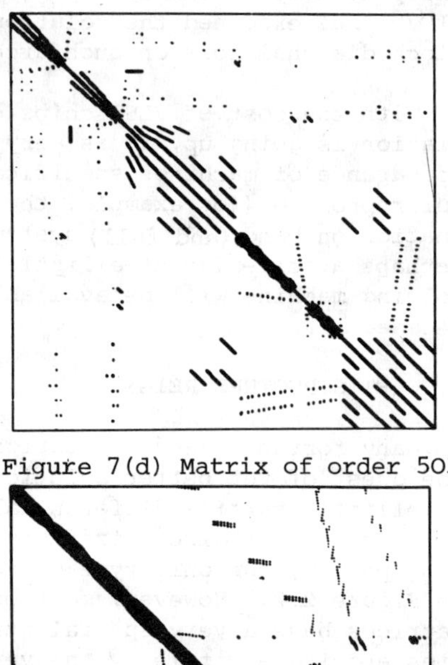

Figure 7(d) Matrix of order 503

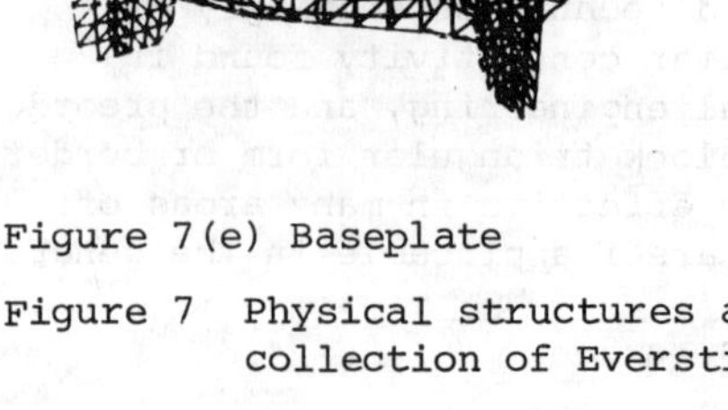

Figure 7(e) Baseplate

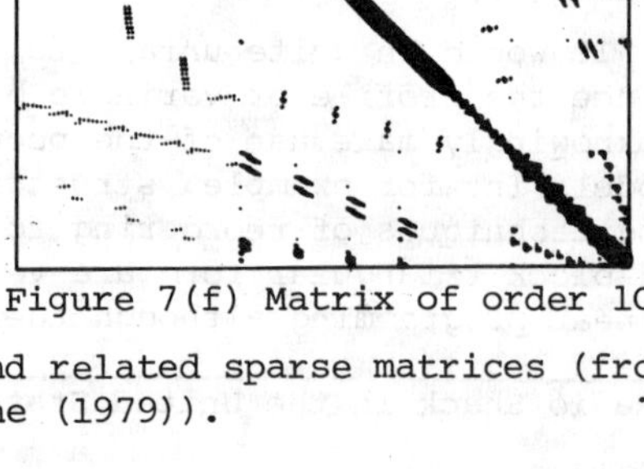

Figure 7(f) Matrix of order 1005

Figure 7 Physical structures and related sparse matrices (from collection of Everstine (1979)).

structural analysis model. To illustrate the relationship between physical structures and sparse matrices we show three structures from Everstine (1979) in Figures 7.1 (a), (c) and (e) and their corresponding matrices in Figures 7.1(b), (d) and (f). It is interesting that, while a variable band scheme does expectedly well on reordered versions of matrices 7.1(b) and (d) it is outperformed by minimum degree on example 7.1(f), (Lewis, private communication). We show, in Figure 7.2, a matrix from a completely different application area, that of statistical modelling (see Duff and Reid (1979) for fuller details). Lewis and Poole (1980) (see also Erisman (1980)) have run several of the accredited finite element matrix solvers on examples from electric power networks and found that the minimum degree ordering is up to 100 times better in most respects than the variable band schemes.

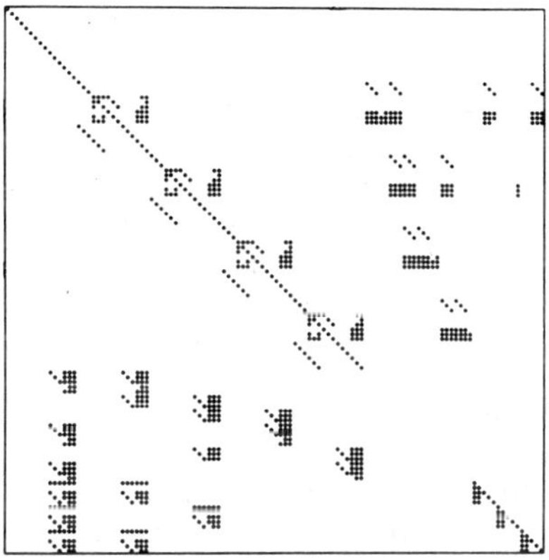

Figure 7.2 A matrix of order 113 from statistical modelling

The purpose of the foregoing is to stress the effect of structure on the performance of sparse matrix algorithms, and to warn against making hasty conclusions on insufficient data. Process flow engineering, power systems and power networks, scheduling problems, partial differential equations, stochastic modelling, economic modelling, and linear programming all produce sparse matrices with vastly different characteristics. One might ask oneself whether it is feasible or worthwhile to attempt to develop general purpose methods, or whether it is preferable to concentrate on just one problem area (which may itself be very diverse) and develop highly efficient techniques which are almost inapplicable elsewhere. As an author of software for a multi-purpose library, I believe the former to be difficult but feasible and, if one can use general codes to obtain a solution, albeit without optimum efficiency, then at least a benchmark has been set for subsequent optimization to problem characteristics. Whether one is developing such general software or whether one is working in a particular application area it is of vital importance to have a wide set of test problems both to aid in the development of the algorithms and in their eventual evaluation. A start has been made to this collec- and dissemination of test problems, for example Duff and Reid (1979) and Everstine (1979), and, although it may not seem a very academic pursuit, such collections will grow and more efforts will be made to identify particular problem classes with particular matrices and with appropriate algorithms. In some areas, for example partial differential equations, it is possible to parametrize the problem and obtain a sequence of similar problems ideal for tuning software. The use of parametrized models will probably extend but that can only happen after the underlying structure has been identified. In the future, we will think of a sparse matrix not in terms of its application area but in terms of its identifiable and useful structural characteristics.

Obviously the model formulation greatly influences the structure of the resulting matrix and indeed historically the formulation has often been designed to produce a small denser matrix because at the time of the original formulation there were few, if any, good sparse algorithms around and sparse matrices were to be avoided as much as possible. Erisman (1980) gives one example of this in node reduction in power system analysis. In the future, there may be much remodelling to enhance sparsity even though the model size may be much larger.

A simple example of this is given by the solution of sparse linear least squares problems where Hachtel (1974) has suggested using the augmented system

$$\begin{pmatrix} I & A \\ A^T & O \end{pmatrix} \begin{pmatrix} \underline{r} \\ \underline{x} \end{pmatrix} = \begin{pmatrix} \underline{b} \\ \underline{O} \end{pmatrix} \tag{7.1}$$

to solve the problem

$$\min_{\underline{x}} \|\underline{b}-A\underline{x}\|_2 \tag{7.2}$$

which formulation we have seen already in section 2. However, this can be extended so that, if we have a weighted constrained linear least squares problem $\min_{\underline{x}} \|\underline{b}-A\underline{x}\|_2$ subject to $C\underline{x} = \underline{d}$ we can write it in the form

$$\begin{pmatrix} O & O & I & A \\ O & O & O & C \\ I & O & W & O \\ A^T & C^T & O & O \end{pmatrix} \begin{pmatrix} \underline{y} \\ \underline{\lambda} \\ \underline{r} \\ \underline{x} \end{pmatrix} = \begin{pmatrix} \underline{b} \\ \underline{d} \\ \underline{O} \\ \underline{O} \end{pmatrix} \tag{7.3}$$

where, as in section 2,

(i) we need only solve for $\underline{x}$
(ii) we can take advantage of the zeros in the right hand side and
(iii) the coefficient matrix is symmetric if W is.

Notice, from (7.3), that we can reformulate problem (7.2) in the form

$$\begin{pmatrix} O & I & A \\ I & I & O \\ A^T & O & O \end{pmatrix} \begin{pmatrix} \underline{y} \\ \underline{r} \\ \underline{x} \end{pmatrix} = \begin{pmatrix} \underline{b} \\ \underline{O} \\ \underline{O} \end{pmatrix}$$

which is even larger, sparser and more augmented than system (7.1). Clearly, there is a limit to the advantages derived from continued superaugmentation but I don't believe such limits are yet recognized.

Finally, it is my contention that really not much is known about the effect of underlying structure on sparse matrix problems and that this will be one of the major growth areas in the future. This path ahead is admittedly not easy.

8. INTERFACES

For the previous generation of algorithm designers, the work stopped as soon as the method was proven to be efficient on some class of problems and the paper (or two) was published. It was left to some first year graduate student to extend its domain to cover some genuinely interesting problems and then to release it to the remarkably uncomplaining user.

Only recently has the art of mathematical software design been appreciated as a worthwhile pursuit for qualified researchers. There has been a growing awareness amongst the numerical analysis community that much of the real usefulness of their research, in all bar the most esoteric reaches of the subject, is lost unless it can be packaged in a form suitable for the customer, perhaps a first rate physicist who may be very naive about the inner-workings of numerical computation. As algorithm designers, we have to decide whether we merely wish to receive the adulation of our peers, or whether we wish to benefit a much larger scientific community.

These general comments on numerical software are so much more true in the present context. Most of sparse matrix research is motivated by the need to solve practical problems or to provide tools so that others may do so. Yet even here, it is not fully appreciated how much work is involved or the benefits derived from building a convenient, versatile and robust user interface. All the principal routines for the Harwell sparse unsymmetric equation solver were written and debugged over a period of two months (August 1976). However, it was eight months later (admittedly not with full time working) before the MA28A package was released to the Harwell subroutine library.

One of the main benefits of dressing our algorithmic servant in good clothes is that it enhances the confidence the user will have in the software and often determines whether he will use it at all. If some problem in his presentation of data to the software arises this will normally be detected and diagnosed by the software package and if the error cannot be detected and the user's program fails he will be less likely to accuse the "black box" of causing the failure and thus manage to isolate the source of the trouble that much faster. We might even dare to suggest that the user, perhaps dangerously, is attracted by the gloss on the outside irrespective of what evils that gloss may be covering. Certainly some well-used and well-trusted packages in statistics and structural analysis have had software of dubious quality embedded within.

There are costs, of course, involved in providing such an interface. Not only the manpower required to create it but also the overheads in run-time and the length of the code. However, applications where 10% savings in time are critical are few and far between and the savings in human effort when good robust software is readily available far outweigh these expenses.

It is particularly important in sparse matrix software to shield the user from the complexity of the data structures involved and in which he is very unlikely to have any interest. The screening in SPARSPAK (George and Liu (1979)) is more or less complete since the user must merely declare one array of storage, entry of data being by one of several subroutines. In MA28, (Duff (1977a)), the input is demanded in a way meaningful to the user where he is asked to supply the row and column indices of each non-zero.

Another important aspect of a good user interface is to provide good diagnostics in the event of failure and in some instances take further action to provide the user with more information on the error. In addition to providing error diagnostics, SPARSPAK has a subroutine to enable the user to extract useful information about the run and a parameter to allow various levels of detail on the progress of the solution to be output. In MA32, the Harwell unsymmetric frontal solver, computation can continue symbolically even if insufficient space has been allocated to the in-core frontal matrix, the user finally receiving a lower bound on the size required for subsequent runs (a similar facility for the workspace array in MA28 has proved very popular). Additionally, if insufficient space has been allocated for the factors (either in-core or out-of-core), then the amount required for subsequent success will be returned at the end of the run.

Indeed these facilities and the ability to perform orderings in a small predetermined amount of storage from which an accurate forecast of time and space requirements for subsequent solution can be obtained (see section 2) open the door to an exciting new prospect: that of interactive computing with sparse matrices. This could well be another important growth area in the next decade.

Thus, the reasons for the long time-lag between algorithms and good software is evident and as a postscript to this section, when we observe the large gap between algorithms and software and the almost total dearth of good software in some areas, we can clearly see where many advances will be made in the near future.

9. SPARSE FRONTIERS

In looking at where sparse matrix research is heading and what the sparse future holds, we will first summarize some of the main points made in the previous sections and then add a few comments on areas not discussed in the main body of the paper.

The development of decomposition tools akin to block triangular ordering will undoubtedly be refined and extended to a fairly wide range of decompositions. Much of the impetus for this will come from a desire to use parallel computing environments more efficiently as well as enabling very large problems to be solved on small mainframes and mini-computers. More efficient ordering strategies will take advantage of this global decomposition by using local subordering perturbing this overlying or underlying ordering to make best use of sparsity or vector machines or to maintain stability.

Certainly there will, particularly in the near future, be much work on developing algorithms which are efficient on fourth generation computers but I feel this could be overrated since judicious application of the work discussed in the last paragraph should give extremely good results in many machine environments with only a minimal amount of tuning. Of course, machines of the twenty first century will extend the frontiers of computation even further but, if we have by then understood fully the nature of sparse matrix structure, we will be in a much better position to exploit these new technologies.

Although it is far from obvious how our knowledge of underlying structure will develop, what is certain is that there will be an enormous growth in the availability of high quality numerical software for sparse matrix applications. Initially this will be in the form of subroutines implementing current algorithms but later we will see the development of whole packages for the solution of a variety of problems on a variety of computing environments. At least some of these packages will be interactive to the extent that a user can perform his own tuning as his solution develops. Of course, a larger range of sparse capabilities will be added to existing packages like NASTRAN or ASAS in structural engineering or SPSS in statistics. Versions of these packages will be developed for use on mini-computers and in real-time applications.

On the algorithmic side, better methods will be developed for unsymmetric problems, for example more robust iterative methods (Concus and Golub (1975), Vinsome (1976), Axelsson (1980), Jacobs (1980))and local subordering techniques for direct methods. Already work is being done on extending the highly successful Lanczos algorithm (see Scott (1980), for example) and variants of it to unsymmetric problems (Parlett and Saad, private communication). The whole concept of preconditioning (for example, Meijerinck and van der Vorst (1977)) is still in its infancy, and new methods will be developed as well as theory to explain old ones. Finally, many more problem areas like sparse non-linear least squares and general non-linear systems and optimization are only now being explored for the first time (Toint (1980)). Also that old grandaddy of sparse matrix research stimuli, linear programming, can still provide us with some challenge (Markowitz and Perold (1980)).

ACKNOWLEDGEMENTS

I would like to thank Al Erisman and John Reid for reading an early draft of this paper and making some useful remarks.

REFERENCES

Axelsson, O. (1980) Conjugate gradient type methods for unsymmetric and inconsistent systems of linear equations, Lin. Alg. and its Applics. 29, pp. 1-16.

Bunch, J.R. and Kaufman, L. (1977) Some stable methods for calculating inertia and solving symmetric linear systems, Math. Comp. 31, pp. 163-179.

Concus, P. and Golub, G.H. (1975) A generalized conjugate gradient method for nonsymmetric systems of linear equations, Comput. Sci. Dept. Stanford Report STAN-CS-75-535, Presented at IRIA meeting, Paris, Dec. 1975.

Curtis, A.R. (1980) On a property of some test equations for finite difference methods, Harwell Report CSS 84.

Dembart, B. and Neves, K.W. (1977) Sparse triangular factorization on vector computers, in Special Report EL-566-SR, Electric Power Research Institute.

Duff, I.S. (1977a) MA28 - a set of Fortran subroutines for sparse unsymmetric linear equations, Harwell Report R.8730, HMSO, London.

Duff, I.S. (1977b) A survey of sparse matrix research, Proc. IEEE 65, pp. 500-535.

Duff, I.S. (1977c) On permutations to block triangular form, J. Inst. Maths. Applics. 19, pp. 339-342.

Duff, I.S. (1978) On algorithms for obtaining a maximum transversal, Harwell Report CSS 49, to appear in TOMS.

Duff, I.S. (1979) Practical comparisons of codes for the solution of sparse linear systems, in Duff and Stewart (1979), pp. 107-134.

Duff, I.S. and Reid, J.K. (1976) A comparison of some methods for the solution of sparse overdetermined systems of linear equations, J. Inst. Maths. Applics. 17, pp. 267-280.

Duff, I.S. and Reid, J.K. (1978) An implementation of Tarjan's algorithm for the block triangularization of a matrix, TOMS 4, pp. 137-147.

Duff, I.S. and Reid, J.K. (1979) Performance evaluations of codes for sparse matrix problems, in Performance Evaluation of Numerical Software, (L. Fosdick ed), North Holland, pp. 121-135.

Duff, I.S. and Stewart, G.W. (eds.) (1979) Sparse Matrix Proceedings 1978, SIAM Press.

Erisman, A.M. (1980) Sparse matrix problems in electric power systems analysis, these proceedings.

Everstine, C.G. (1979) A comparison of three resequencing algorithms for the reduction of matrix profile and wavefront, Int. J. Num. Meth. Engng. 14, pp. 837-853.

George, A. (1980) Direct solution of sparse positive definite systems: some basic ideas and open problems, these proceedings.

George, A. and Liu, J.W.H. (1978) An automatic nested dissection algorithm for irregular finite element problems, SIAM J. Numer. Anal. 15, pp. 1053-1069.

George, A. and Liu, J.W.H. (1979) The design of a user interface for a sparse matrix package, TOMS 5, pp. 139-162.

Gustavson, F.G. (1976) Finding the block lower triangular form of a matrix,in Sparse Matrix Computations, (Bunch and Rose eds.) Academic Press, pp. 275-289.

Gustavson, F.G. (1978) Two fast algorithms for sparse matrices: multiplication and permuted transposition, TOMS 4, pp. 250-269.

Hachtel, G.D. (1974) Extended application of the sparse tableau approach - finite elements and least squares, Elec. Sci. and Eng. Dept. and Comput. Sci. Dept. UCLA.Tech. Report.

Hellerman, E. and Rarick, D. (1971) Reinversion with the preassigned pivot procedure, Math. Prog. 1, pp. 195-216.

Hood, P. (1976) Frontal solution program for unsymmetric matrices, Int. J. Num. Meth. Engng. 10, pp. 379-399.

IBM (1976) IBM System/360 and System/370 IBM 1130 and IBM 1800 Subroutine Library - Mathematics, User's Guide, Program Product 5736-XM7, IBM catalogue SH12-5300-1.

Irons, B.M. (1970) A frontal solution program for finite element analysis, Int. J. Num. Meth. Engng 2, pp. 5-32.

Jacobs, D.A.H. (1980) The exploitation of sparsity by iterative methods, these proceedings.

Jennings, A. (1966) A compact storage scheme for the solution of symmetric linear simultaneous equations, Comput. J. 9, pp. 281-285.

Kung, H.T. and Leiserson, C.E. (1979) Systolic arrays (for VSLI), in Duff and Stewart (1979), pp. 256-282.

Lewis, J.G. and Poole, W.G. Jr. (1980) Ordering algorithms applied to sparse matrices in electrical power problems, in Electric Power Problems: the Mathematical Challenge, (Erisman, Neves and Dwarakanath eds.) SIAM Press.

Lin, T.D. and Mah, R.S.H. (1977) Hierarchical partition - A new optimal pivoting algorithm, Math. Prog. 12, pp. 260-278.

Markowitz, H.M. (1957) The elimination form of the inverse and its application to linear programming, Management Sci. 3, pp. 255-269.

Markowitz, H.M. and Perold, A.F. (1980) Sparsity and piecewise linearity in large portfolio optimization problems, these proceedings.

Meijerink, J.A. and van der Vorst, H.A. (1977) An iterative solution method for linear systems of which the coefficient matrix is a symmetric M-matrix, Math. Comp. 31, pp. 148-162.

Munksgaard, N. (1980) Solving sparse symmetric sets of linear equations by preconditioned conjugate gradients, TOMS 6, pp. 206-219.

Nepomiastchy, P., Ravelli, A. and Rechenmann, F. (1978) An automatic method to get an econometric model in a quasi triangular form, IRIA Report No. 313.

Parter, S.V. (1961) The use of linear graphs in Gauss elimination, SIAM Rev. 3, pp. 119-130.

Pottle, C. (1979) Solution of sparse linear equations arising from power system simulation on vector and parallel processors, ISA Transactions 18, pp. 81-88.

Reid, J.K. (1977) Sparse Matrices in The State of the Art in Numerical Analysis, (D. Jacobs ed), Academic Press, pp. 85-146.

Reid, J.K. (1980) Frontal methods for solving finite-element systems of linear equations, these proceedings.

Rose, D.J. (1970) Symmetric elimination on sparse positive definite systems and the potential flow network problem, PhD Thesis, Applied Math. Harvard.

Scott, D.S. (1980) The Lanczos algorithm, these proceedings.

Toint, Ph.L. (1980) Towards an efficient sparsity exploiting Newton method for minimization, these proceedings.

Vinsome, P.K.W. (1976) Orthomin, an iterative method for solving sparse sets of simultaneous linear equations. Paper No. SPE 5729, 4th SPE Symposium on Numerical Simulation of Reservoir Performance.

DISCUSSION

PROFESSOR M.J.D. POWELL (Cambridge). Am I right in thinking that you do not consider the values of the non-zeros during the preordering phase? If so, how do you guarantee stability?

DUFF. Yes. In general the preordering is performed without reference to the non-zero values. However, for some preorderings, for example the permutation to block triangular form, stability will be ensured if you perform a stable decomposition within the diagonal blocks. Very often, of course, the preordering is based on some physically realizable tearing of the structure where the physical stability of the separate units leads to subsequent numerical stability in the solution process. We would certainly advocate monitoring the stability should there be any doubt.

PROFESSOR A.F. PEROLD (Harvard). Many attempts have been made to utilize the underlying structure of problems in linear programming but, no matter how ingenious the strategies, such attempts have not been as successful as the Markowitz criterion (with threshold pivoting) applied to the structure as a whole. Indeed, we do encounter stability problems with these structure-based schemes.

DUFF. Of course, these stability problems would not occur if our standard block triangularization algorithms were applied to the LP basis. In such a case, we have practical experience of considerable gains in efficiency over using Markowitz on the entire basis.

MR. J.J. DU CROZ (NAG). I strongly endorse your comments on interactive computing and feel that the full potential of these new sparse matrix techniques will only be realized when an engineer can interact with the package at the terminal.

MR. M.H. GOLDWATER (British Gas). I think we have to be careful not to be carried away by oversophisticated interative systems. There is a grave danger that the interface will confuse rather than assist the practising engineer.

MR. D.B. SZYLD (NYU). In the solution of general linear equations how do you ensure instability does not cause an incorrect solution?

DUFF. We monitor the growth in matrix elements which gives some indication of the accuracy of the decomposition. Furthermore, it is easy for the user to employ iterative refinement to improve his solution or to obtain an accurate solution to an ill-conditioned system. We recommend running with an increased value of the threshold parameter u should trouble occur.

DR. A.M. ERISMAN (Boeing). We have written a sparse version of the LINPACK condition estimator and found it very successful in determining the conditioning of our equations.

DR. P.G. THOMSEN (Lyngby). Clearly, you must do a lot of complicated programming with complex data structures. Have you considered using other languages than Fortran for this?

DUFF. Yes. While I was at Newcastle, I used the record and reference structures of ALGOLW which are ideal for handling linked lists, for example. Although I found the code easy to write, the execution time of the resulting compiled code was considerably slower than that of equivalent compiled Fortran code. Also, I am very concerned about the portability of my codes and am still of the opinion that Fortran is the only universal language.

SPARSE MATRIX PROBLEMS IN ELECTRIC POWER SYSTEM ANALYSIS

A.M. Erisman

(Energy Technology Applications Division, Boeing Computer Services Company, Seattle, Washington 98124)

1. INTRODUCTION

The electric power industry does a significant amount of computer analysis to assure its ability to meet customer demands in a reliable, stable, economic manner. Because of the interconnected nature of today's power systems, the computer models tend to be very large, and the proper accounting of sparsity is essential for realistic analyses. In the transient stability model, for example, there is one very large sparse matrix connecting models of the power plants and the loads through the large network. The problem naturally decomposes, however, and the power plant models may be treated separately from the network. A significant amount of work has been directed towards the network problem exploiting sparsity directly, and indirectly through tearing. Much less work has been directed towards the equations of the power plants, though they often represent a significant portion of the computation.

In this paper we briefly review transient stability and some other electric power systems models for the purpose of establishing essential characteristics of the sparse matrix problems. We then analyze separately the solution of the two types of sparse matrix problem from this application. We indicate the effect of the characteristics on the solution process and show why the commonly used methods for sparse matrix problems in structural analysis, for example, are unsuccessful here.

Sparse matrix research has grown from its roots in particular disciplines such as electric power systems analysis. The integration of these ideas towards a unified sparse matrix field, ignoring the characteristics of the applications, has not been as successful as was anticipated. One reason for this is that a matrix is more than sparse. Other characteristics, which tend to vary with the matrix origin, also have a large influence on the effectiveness of computational algorithms applied to the matrix. Some classification of the sparse matrix properties, relating distinctions in structure to computational performance, is needed to further the usefulness of general purpose work.

2. SPARSE MATRICES IN POWER SYSTEMS

The major components of what is usually referred to as the "power system" are the power plants, the transmission network, the ties to neighbouring systems, the electrical loads, and the system control. The power plants are the large central stations where fuel (nuclear, fossil, hydro) is converted to electrical energy through the generators. The electrical loads for this discussion refer to distribution centres (for example, a small city or industrial complex) from which individual customers are served. The tie lines allow interchange of energy between neighbouring systems. This system is run by a control centre (or centres) which schedules generation to meet the load economically, subject to physical constraints. A detailed diagram of this system is shown in Figure 1. We observe that the loads (located near the user) are usually physically separated from the generators (located near the fuel source). The transmission network is often large both geographically and in terms of numbers of lines and junctions. A simplified schematic in Figure 3 will be referred to later when we discuss our model structure.

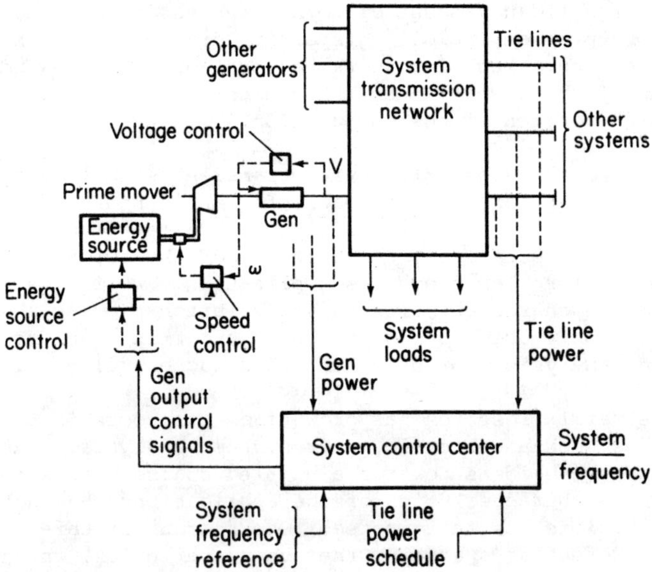

Figure 1 A conceptual model of a power system, Anderson and Dembart (1980).

The power system is modelled for many different purposes, from the development of growth plans for generation and transmission 30 years in the future to the determination of generation requirements only a few hours away. Both planning and operations strategies involve economic considerations, reliability considerations and contingency planning, for example, what to do when a major line goes out, a generator is taken off line, or the load grows unexpectedly in a particular area. We discuss the modelling problems physically and conceptually first. Later in this section we show the structure of the defining equations.

Models of interest are both static and dynamic. Power flow is an example of a static model: for a given load on the network, determine a generation plan to meet this load. This mathematical model is a large system of nonlinear algebraic equations made up of the nonlinear load models and the large linear network model. The sparsity of these equations is determined by the topology of the network. It has many solutions, and one is sought with realistic constraints. The optimal power flow seeks the "best" of all feasible power flow solutions within economic and reliability constraints. The derivation of the equations and solution methods commonly used in the power industry for this problem are discussed in detail by Stott et al (1980).

Dynamic models, for example transient stability and long term dynamics, are concerned with power system stability questions. Here the dynamic models of the generation system are coupled with the nonlinear load models and the linear constraints of the network. Stability problems deal with the ability of the power system to withstand disturbances which cause it to move from one operating point (power flow solution) to another. The disturbances may be induced by control centre action and/or faults (short circuits, line outages, etc.) on the system. Since generators are synchronous machines, they must operate at the same average speed to deliver power to the network in an acceptable way. Sudden changes in load due to disturbances may tend to accelerate or decelerate one or more of the turbine shafts; if this acceleration is not arrested quickly, the generator may be tripped off by relays. Transient stability is concerned with maintaining generation through controlling action. When instability results, other slower responding interactions can take place; failure of corrective action can lead to blackouts. This is called the long term dynamics problem. Anderson and Dembart (1980) give a detailed discussion of the development of the equations and solution methods used by the power industry for transient stability analysis. The models for long term dynamics are somewhat different, since they must include effects which can be ignored in the transient stability model, but the general structure of the equations, which is our main concern in this paper, is basically the same.

For the transient stability model, then, we are concerned with coupled nonlinear algebraic/differential equations for each of the power plants

$$\begin{pmatrix} \underline{\dot{\hat{x}}} \\ \underline{0} \end{pmatrix} = \underline{f}(\underline{x}, v_k) \tag{1}$$

$$i_k = g(\underline{x}, v_k). \tag{2}$$

Only a subset of the vector $\underline{x}$, denoted $\underline{\hat{x}}$, contains dynamic variables. Also included in the model are nonlinear algebraic load constraints of the form

$$i_m = h(v_m) \tag{3}$$

and linear network constraints relating voltage $\underline{v}$ and current $\underline{i}$ with the admittance matrix Y,

$$Y\underline{v} = \underline{i}. \tag{4}$$

Here $\underline{v}$ and $\underline{i}$ are vectors of length corresponding to the number of equations in the network. The quantities i_m, v_m, i_k and v_k are the contributions of current and voltage at the point in the network where the load and power plant, respectively, are connected and are components of $\underline{v}$ and $\underline{i}$ in (4). The vector $\underline{x}$ corresponds to the internal variables of the power plant.

The trapezoidal rule is the preferred integration method for solving the differential equations, for reasons discussed by Anderson and Dembart (1980). Applying this to the differential equations produces a large nonlinear system to be solved for $\underline{x}$ and $\underline{v}$ at each time step. Applying Newton's method to this nonlinear system yields a sparse, linear system where the Jacobian matrix has the structure shown in Figure 2.

The sparsity structure has a number of special features which can most easily be understood when it is compared with Figure 3, a simplified schematic version of Figure 1.

The sparsity pattern of each A_i matrix, corresponding to the structure of the individual power plants connected to the network, is discussed in the next section. We show there that these matrices are unsymmetric and very different from the network in their sparsity structure.

Y is a symmetric, complex matrix where the sparsity pattern is the one corresponding to the graph of the network. $\hat{Y}$ differs

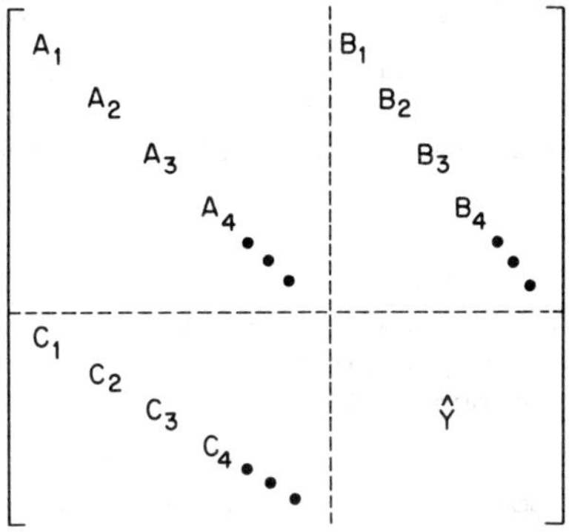

Figure 2 Jacobian sparsity structure

from Y in that it adds to Y the Jacobian terms from the load, on the diagonal at those nodes in the network where the loads are attached. The border of the Jacobian is special in that it has only one column and one row for each power plant, that is B_i has only one non-zero column. Thus, when Gaussian elimination is applied to the Jacobian, in the order shown, the resulting submatrix after eliminating all of the A_i submatrices has the same sparsity pattern as Y. Changes to $\hat{Y}$ are on the diagonals at those nodes where the power plants are attached.

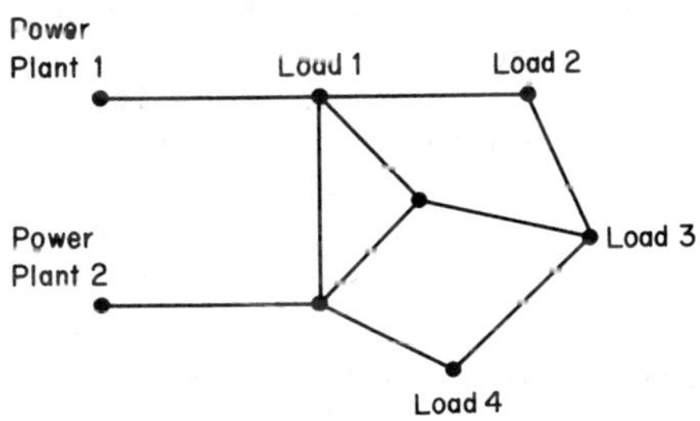

Figure 3 Schematic of the power system

From a sparse matrix point of view, each P_i matrix defined below may be factorized independently:

$$P_i \equiv \begin{bmatrix} A_i & B_i \\ C_i & * \end{bmatrix} = L_i U_i$$

where the sparsity pattern is the unsymmetric one corresponding to the individual power plants.

The lower right entry in U_i may then be added to $\hat{Y}$ at the appropriate diagonal, resulting in Y'. The resulting Y' differs from $\hat{Y}$ only in some diagonal entries. Thus this part of the sparse matrix problem is structurally the same as the network sparse matrix problem.

In summary, we have motivated the structure of two sparse matrix problems in power systems analysis. We have shown that the two interconnected problems decompose into independent subproblems; that corresponding to the P_i matrices and that corresponding to the Y matrix. The next two sections deal with these sparsity matrix problems independently because they have very different characteristics. Typically for U.S. stability studies, there are 100-1000 power plants each of order 20-200 variables, and a network of order 1000-6000. Usually the power plant models are reduced to about order 20, however, and this reduction problem is also discussed in the next section. The 1000-6000 order for the network includes a number of interconnected networks with the tie lines, see Figure 1.

3. SPARSITY IN THE POWER PLANT MODELS

A detailed block diagram representing a machine, a portion of a power plant, is shown in Figure 4. For those not familiar with block diagrams and their relationship to equations, we briefly discuss this. We then use these equations to define a sparse matrix representation. From this we discuss the computation of initial conditions for variables internal to the power plant, and briefly introduce the use of the sparse matrix structure in model reduction and the solution of the nonlinear equations. We conclude this section by summarizing some of the characteristics of sparse matrices representing power plants.

In the block diagram, each block is numbered and represents an equation relating the variables entering and leaving the block. For example, equation 9 is

$$x_4 = R_A x_{17}.$$

Equation 2 is more complex:

$$x_{16} = f_2(x_1, x_2, x_4, x_5, x_{28}, x_{29})$$

where f_2 is a nonlinear function denoted G(N7), an internal designation. Some of the equations involve dynamics, but for this discussion we assume that some integration rule (typically the trapezoidal rule) has been applied. Hence the equation

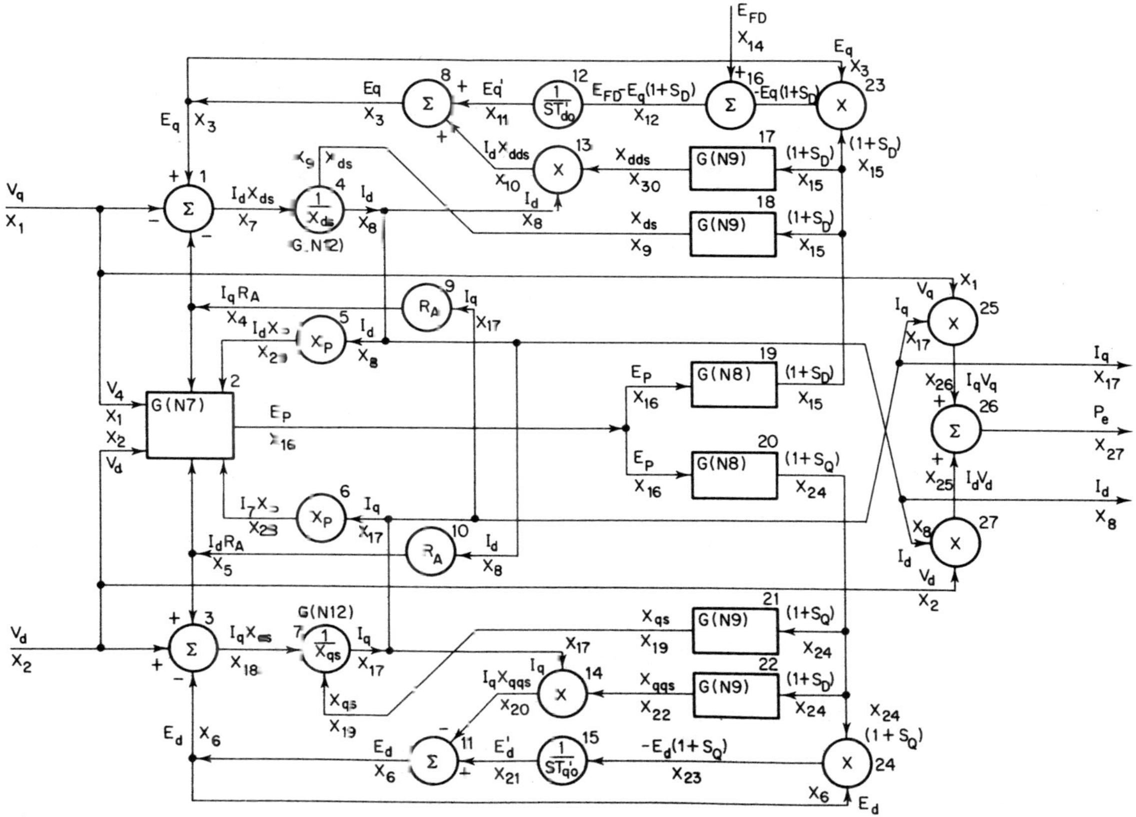

Figure 4 A block diagram for a machine model

becomes algebraic involving the variable at a previous time point, which may be treated as a known parameter.

Figure 5 shows a sparse matrix representation of the block diagram in Figure 4, where an x indicates that that variable is involved in the equation. The output variable for each equation is indicated by ⓧ.

VARIABLES

EQUATIONS	1	2	3	4	5	6	7	8	9	10	11	12	13	14	15	16	17	18	19	20	21	22	23	24	25	26	27	28	29	30
1	X		X	X			(X)																							
2	X	X		X	X											(X)												X	X	
3		X			X	X												(X)												
4							X	(X)	X																					
5								X																					(X)	
6																	X											(X)		
7																	(X)	X	X											
8			(X)							X	X																			
9				(X)													X													
10					(X)			X																						
11						(X)														X	X									
12											(X)	X																		
13								X		(X)																				X
14																	X			(X)		X								
15																					(X)	X								
16												(X)	X	X																
17															X															(X)
18									(X)						X															
19															(X)	X														
20																X								(X)						
21																			(X)					X						
22																						(X)		X						
23			X										(X)		X															
24						X																	(X)	X						
25	X																X									(X)				
26																									X	X	(X)			
27		X						X																	(X)					

Figure 5 The sparsity pattern for the model in Figure 4

In solving the transient stability problem, the starting point is to determine initial conditions for the system. Initial voltage and current distribution for the network can be determined from a power flow computation, so it remains to determine initial conditions for the internal variables $\underline{x}$ for each power plant. We will show how the sparse matrix representation in Figure 5 may be used to solve this problem.

From the power flow computation we have initial values for the interface variables. In this case that includes $x_1, x_2, x_8, x_{14}, x_{17}$, and x_{27}. We should like to define a sequence of steps to compute the remaining x_i values. To do this, we seek to put the equations in "precedence order", using a chemical engineering term discussed by Westerberg et al (1979). That is, we will attempt to order the equations so that when

$$x_i = f_i(x_{i_1}, \ldots, x_{i_k}),$$

is encountered, all of the x_{i_j}, $j=1,..k$, are known and x_i may be simply evaluated.

If this could be accomplished, it would be a very simple algorithm. First, the equations and variables would be reordered so that the output variables for the equations appear on the diagonal, as in Figure 6. Then, since the final six variables are known and the final three equations are redundant (because the output variables are known), we seek a symmetric permutation of the upper left 24×24 submatrix which transforms it to triangular form.

Now, if such a transformation exists, it is easy to compute. Select an equation with no unknown input variables. Order that equation and its output variable first, add that variable to the known list, and pick the next such equation. It is obvious that if the transformation can be performed, this algorithm will determine it, and if the algorithm breaks down, the transformation does not exist.

Unfortunately, the algorithm breaks down for this particular case, and in general. The algorithm reduces our sample problem to the one shown in Figure 7, and there is no remaining equation where the output variable is unknown and the input variables are all known.

Observe that even though the algorithm breaks down, it is still helpful. We could now solve the remaining nonlinear equations simultaneously (equations 1, 8, 12, 16, and 23 for variables 7, 3, 11, 12, and 13) and determine the initial conditions more easily than solving the original system simultaneously. Alternatively, however, we could relax the restriction of requiring the output variable to be on the diagonal. For some functions f it may be easy to solve for x given y in

$$y = f(x).$$

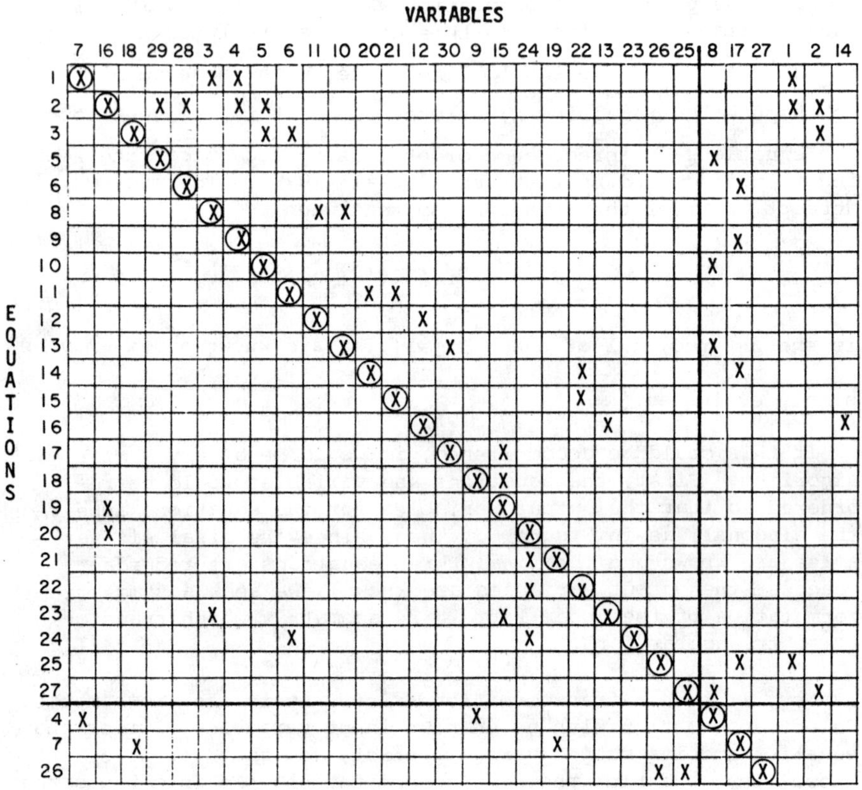

Figure 6 Reordering of Figure 5 placing outputs on the diagonal

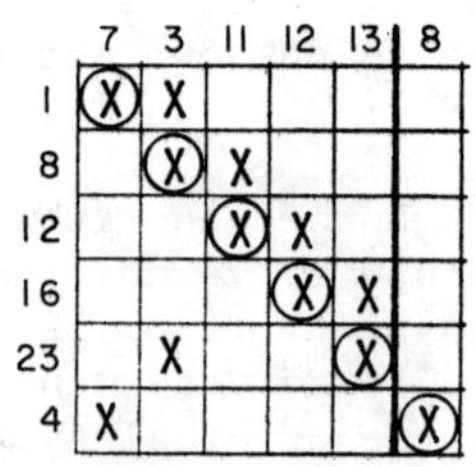

Figure 7 Remaining equations with unknown outputs also have unknown input

By doing this whenever possible, for example whenever f^{-1} is single valued, we may avoid completely the problem of solving nonlinear equations simultaneously.

Allowing an input variable to be determined for certain functions has been enough to allow the precedence ordering of all the equations in the models we have worked with. In the above example, equation 4 is used to determine x_7, then equation 1 is used to determine x_3. From this point we can go back to having output variables on the diagonal. The final result is shown in Figure 8.

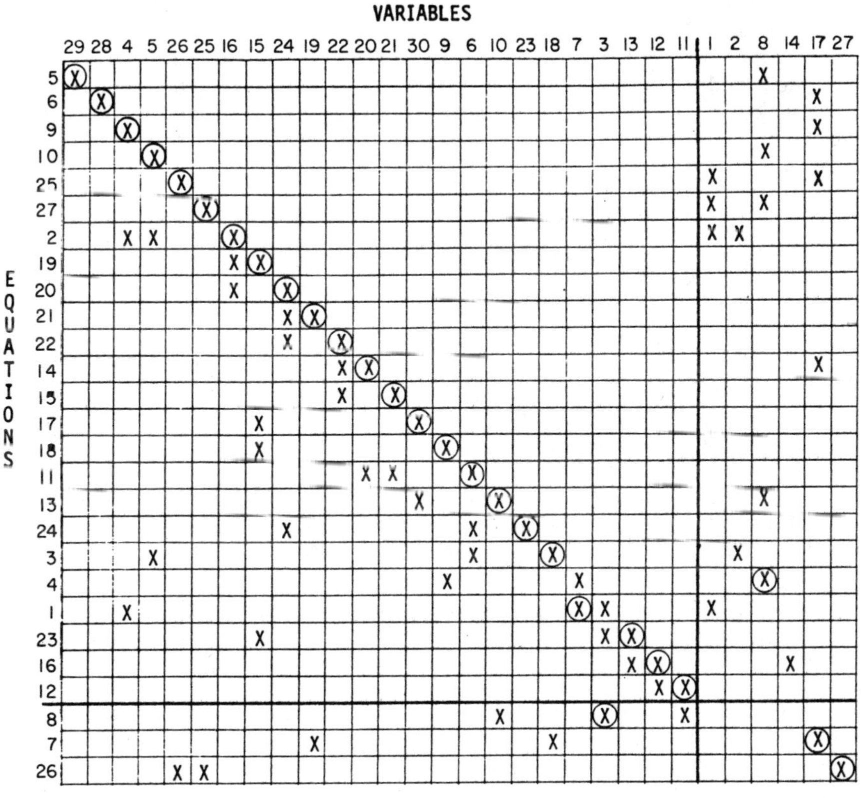

Figure 8 The reordered matrix from Figure 5

The resulting algorithm is rather simple, seeking to triangularize by permutations a square submatrix of a rectangular matrix, giving priority to symmetric permutations. The implications for power plant models is important, however, since the underlying sparse structure of these models has traditionally been ignored. The initial conditions problem has traditionally been treated with specialized code for each model, thus each model modification required new code for computing initial conditions.

Recognizing and using the underlying sparse structure for power plant models can also aid the model reduction process. Model reduction generally takes place for two distinct reasons: special values of parameters and the elimination of some internal variables which are not of interest. Both of these reductions are made with the objective or reducing computations involving the model. Again the typical treatment of this problem in electric power codes has been to perform these reductions by hand and end up with separate models for each special case. It may be possible to do some or all of the useful reduction using the sparse structure. Again the work in chemical engineering on flow sheet models, see Westerburg et al (1979), is similar and may provide insight on this problem.

To see how special parameter values lead to simpler models, suppose for example $R_A = 0$ in the model shown in Figure 4. Then equations 9 and 10 are eliminated, variables x_4 and x_5 are eliminated, and equations 1, 2 and 3 are simplified. This reduction could be handled in a general way by performing a computational step as a part of preprocessing, using specific model parameters with the model topology given by Figure 5.

Some model "reductions" which reduce the number of equations and variables in a model actually add to the computational complexity. For example, if x_4 was not of interest, both x_4 and equation 9 could be eliminated by including the computation $R_A x_{17}$ in both equations 1 and 2. Note that this actually increases the work.

It is not likely that the sparse structure could be used to replace model reduction now accomplished through "engineering judgment". But computing or avoiding the mechanical reduction steps should allow for all the advantages of efficiency with reduced models and the advantage of maintaining fewer models.

There are other ways in which the sparse structure could be used to computational and modelling advantage for power plants. These illustrations should be sufficient to establish the usefulness of the structure. Summarizing some of the characteristics of these sparse matrices, we observe

(1) they are very sparse,

(2) they are of moderate (20-200) order,

(3) they require many solutions (for example 300 models each with 300 time points), and

(4) they are very unsymmetric.

In addition, the nonlinear nature of the equations is important in how the sparsity is exploited. Finally, the sparse representation of the equations is useful for dealing with the structure of the equations, not just for factoring the matrix.

4. SPARSE MATRICES FROM THE NETWORK

While the sparse matrix problem from the power plants has not been investigated to a great extent for power systems, power systems analysts have provided leadership to other fields in exploiting sparsity in the network. Early work on general direct methods for symmetrically structured problems was done by Sato and Tinney (1963) and Tinney and Walker (1967) while tearing/diakoptics strategies were used even earlier. A review of the tearing approach, including some discussion of Kron's work, is done by Happ (1980).

In this section we review sparse matrix methods, particularly the effective sparse ordering strategies, for solving the network equations. Tinney's second strategy introduced in the 1967 paper has been an industry standard. Lewis and Poole (1980) have found that it still appears to be the best approach for these networks, in spite of all of the work on the sparse ordering problem since that time. Here we review some of the results of Lewis and Poole, examining some of the characteristics of the sparsity of the network. Also in this section we include a brief discussion of tearing and compare this with the direct solution.

A network schematic, typical in structure though smaller than in practice, is shown in Figure 9. The matrix corresponding to this graph has a symmetric pattern with nonzero diagonal as shown in Figure 10. This connectivity would be typical of a submatrix of a much larger network.

The second algorithm from Tinney and Walker (1967) is also known as minimum degree. It chooses as first pivot the diagonal in the row with the fewest number of nonzero off-diagonals. Gaussian elimination is performed with this pivot, the sparsity pattern is updated, and the selection process is repeated.

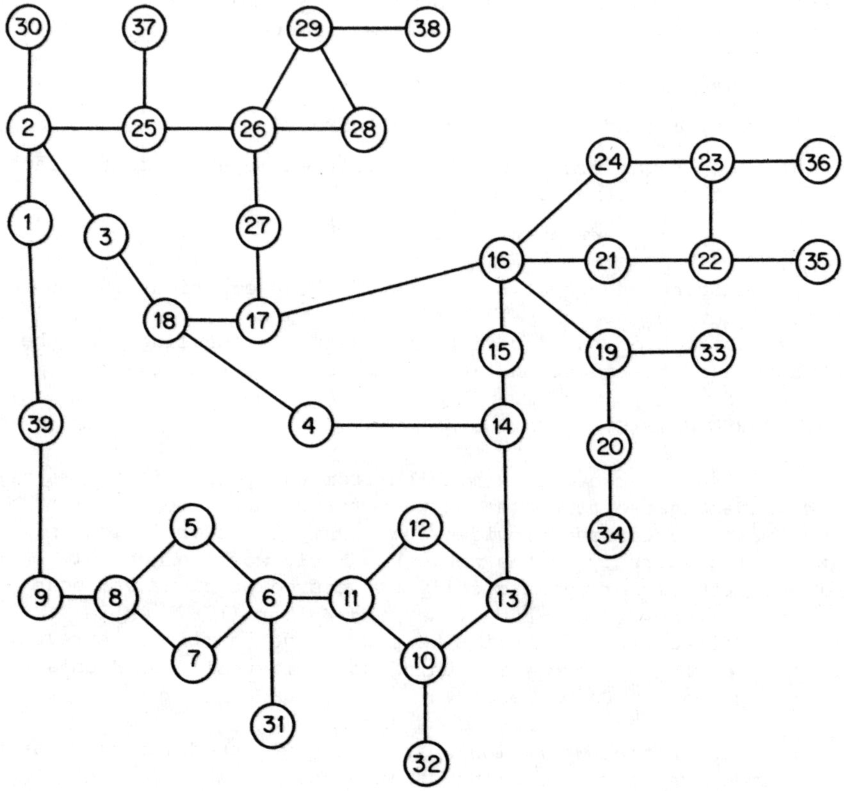

Figure 9 A 39-bus network pattern

Ties are broken arbitrarily. This description of the algorithm does not mean it is implemented this way, a discussion which we delay until later. In comparing this with other proposed orderings, there are four critical elements: storage required, factor time, solve time, and the time required to compute the ordering.

The best algorithm will generally optimize some weighted sum of these four elements, but the choice of weights is problem and computer dependent. Storage requirements limit the size of problems that can be solved in core, and may also affect solution cost if core usage is a part of the accounting algorithm. Factor time often is the dominant element, but when a modified Newton method is being used and there are many more solves than factors, solve time may become more important. When only one matrix of a particular sparsity pattern is to be solved, the time to determine the order may destroy any savings achieved by a reduced factor time. Jennings (1977, pp. 164-166) showed that this is often true in structures problems. Hence, for comparison purposes, we present all four numbers.

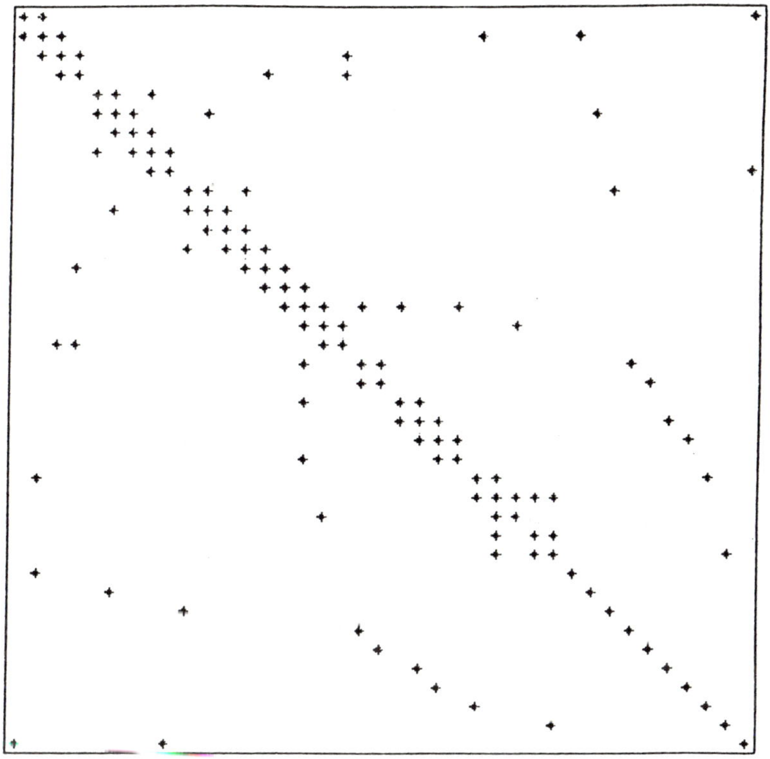

Figure 10 Sparse matrix pattern for Figure 9

To compare different orderings on a set of power systems network matrices, Lewis and Poole (1980) made use of SPARSPAK, developed by George and Liu (1978b). This allows a comparison on the same computer (IBM 3033 in this case) with a common implementation.

A detailed discussion of the various orderings in SPARSPAK is outside the scope of the paper, and the interested reader is referred to George and Liu (1978b). The MIN ordering is the minimum degree, or second ordering of Tinney. PRO is the Reverse Cuthill-McKee algorithm with the Gibbs, Poole and Stockmeyer (1976) starting point, which has been very successful for problems in structural analysis. NSD is a generalized nested dissection ordering. Nested dissection for regular grids produces an optimal ordering, see George (1973) and can do better than minimum degree by a substantial margin on these problems, see Duff et al (1976). The heuristic selection of dissection has been successful for structures problems. The other two, OWD (one way dissection) and RQT (refined quotient tree) are variations of NSD which have also been successfully applied to structures problems.

The two test cases given here present realistic power networks from the United States. The 1723 order problem is from the Bonneville Power Administration representation of the western U.S. electric power network while the 5300 order problem from American Electric Power Company is a representation of the east central U.S. network. Results of the tests are summarized in Tables 1 and 2. Total storage for the methods is the sum of that required to store the matrix and all overhead storage used by the method.

We make several observations about these test results. While it is true that the ordering time for minimum degree was greater than the others, that overhead time for these problems paid significant dividends, even for the solution of only one matrix problem with that sparsity structure. The time for determining the minimum degree ordering is based on the fast method of Goerge and Liu (1978a). As Duff (1980) shows, the fast ordering makes a significant difference compared with a standard implementation of minimum degree requiring updating of the sparsity pattern for each elimination. The impact of this on the extremely sparse network matrices would probably be less dramatic than Duff's figures. In the transient stability calculation, however, order time is less important because many factorizations are done over the time integration interval.

The relative sizes of factor time and solve time for minimum degree are rather typical of our experience with a large number of network problems. On problems like this with extreme sparsity, we need to be careful with the standard assumption that forward and backward substitution are basically free given the factorization (based on the n^3 vs. n^2 comparison in the dense case).

Further discussion of these problems is found in Lewis and Poole (1980). Here we consider two additional aspects of the network sparsity problem: some comments on the extreme sparsity and some comments on the performance of NSD.

The extreme sparsity of the 5300 order case is seen more clearly from examining the partial results of the minimum degree algorithm. Initially, there were a number of degree one eliminations, as we would expect comparing the 39 order example pictured in Figure 9, where no fill can result. Then a large number of degree two nodes were identified, etc. These figures are summarized in Table 3. No node of degree four was used as pivot until the order of the reduced matrix was 1776. Finally, at this point the remaining sparsity pattern has a complexity

TABLE 1

Matrix of order 1723 with 4788 off-diagonal non-zeros

	MIN	PRO	NSD	RQT	OWD
Overhead Storage (words)	17,416	5,170	19,458	13,900	11,003
Matrix Storage (words)	12,626	161,966	34,538	23,110	49,440
Factor time (sec)	.079	7.964	.815	2.405	3.168
Solve time (sec)	.031	.207	.062	.126	.135
Order time (sec)	.475	.079	.292	.194	.144

TABLE 2

Matrix of order 5300 with 16,542 off-diagonal non-zeros

	MIN	PRO	NSD	RQT	OWD
Overhead Storage (words)	53,140	15,901	63,880	40,852	32,463
Matrix Storage (words)	33,363	677,845	108,873	118,751	340,672
Factor time (sec)	.480	*	*	*	*
Solve time (sec)	.128	*	*	*	*
Order time (sec)	1.997	.313	1.196	.635	.526

* data not available

comparable to a relatively sparse structures problem. It would be interesting to switch at this point to one of the other orderings to see if a more efficient over-all solution can be achieved, but this has not been tried.

TABLE 3

Degree selection for 5300 bus problem

Order of submatrix	First selected node of degree
5300	1
5028	2
2986	3
1776	4

The performance of the NSD ordering was somewhat disappointing. NSD seeks to identify a partitioning of the problem so that the matrix A has a block-bordered diagonal form (BBDF) such as shown in Figure 11, for example.

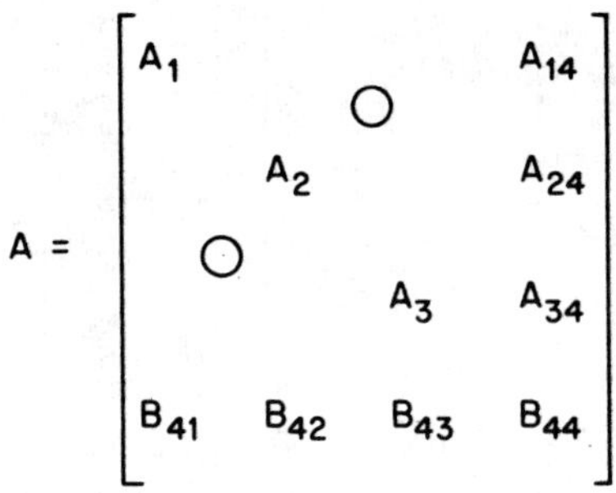

Figure 11 The BBDF determined by dissection

By nesting the dissections, the A_i diagonal blocks also have this structure. There are several reasons why more competitive (with minimum degree) results were expected. One is that the results of nested dissection on regular grids were very favourable where the border (the dimension of B_{44}) is relatively large ($2n-1$ compared with a matrix size of n^2). It was anticipated that the irregular network pattern may allow smaller dissection sets, making this ordering even more effective.

A second reason may be more a wish than an expectation. Minimum degree introduces a significant amount of scattering in the matrix, as illustrated in Figures 12 and 13. (This matrix is a minor modification to the 1723 order test case of Table 1. The modification reduced the order to 1624. These pictures were produced by John Lewis using Iain Duff's program at Harwell.) Hence, this ordering is not particularly appropriate for either out-of-core solutions for very large problems or for vectorized solutions on a pipeline computer. The BBDF ordering is more appropriate for both of these. It is unknown whether a revised heuristic for NSD would improve the performance. Barry et al (1978) record similarly disappointing results using the Sangiovanni-Vincentelli et al (1977) ordering to achieve the BBDF form for the same networks.

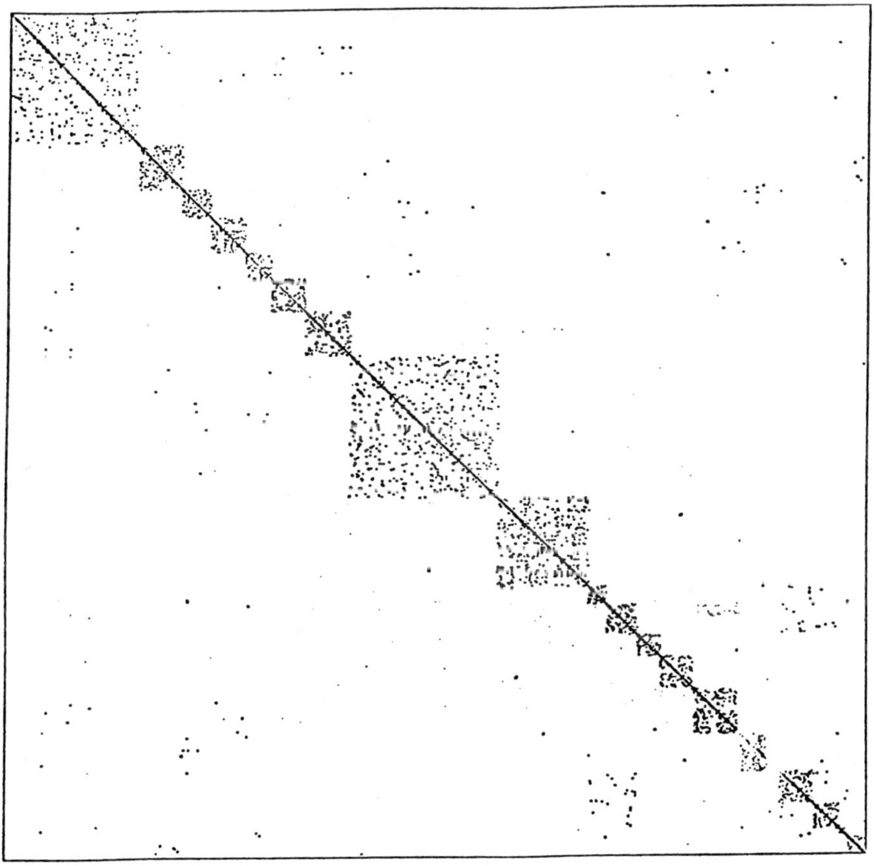

Figure 12 The sparsity pattern for the 1624-bus case

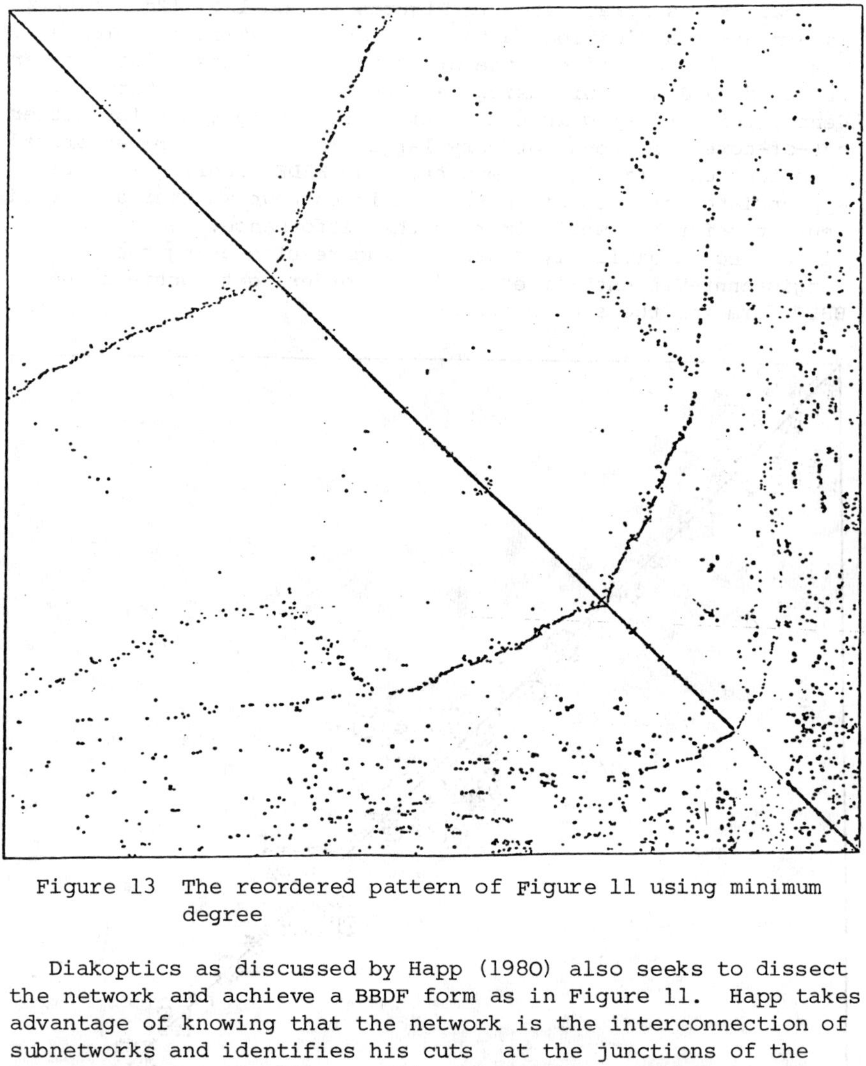

Figure 13 The reordered pattern of Figure 11 using minimum degree

Diakoptics as discussed by Happ (1980) also seeks to dissect the network and achieve a BBDF form as in Figure 11. Happ takes advantage of knowing that the network is the interconnection of subnetworks and identifies his cuts at the junctions of the interconnecting tie lines. This seems like a very reasonable strategy for several reasons, but unfortunately we do not know of any computational comparison between the BBDF form obtained in this way and minimum degree applied to the entire matrix.

One of the attractive features of this dissection choice is that it would seem to be right computationally, since we would expect more interconnections within a particular system and fewer to a neighbouring system. We could also represent each system in condensed form:

$$G_i = \begin{bmatrix} A_i & B_{i4} \\ B_{4i} & D_i \end{bmatrix}$$

where D_i is the appropriate contribution to B_{44}. That is,

$$D_1 \oplus D_2 \oplus D_3 = B_{44}$$

where the D_i are of perhaps different dimensions, but are added by placing them in the appropriate position in B_{44}. In this case, each G_i may be factored independently as

$$\begin{bmatrix} A_i & B_{i4} \\ B_{4i} & D_i \end{bmatrix} = \begin{bmatrix} L_{11}^{(i)} & O \\ L_{12}^{(i)} & L_{22}^{(i)} \end{bmatrix} \begin{bmatrix} U_{11}^{(i)} & U_{12}^{(i)} \\ O & U_{22}^{(i)} \end{bmatrix}$$

using one of the alternate partitioned factorizations, see George (1974). The solution of the entire problem then requires factoring

$$B_{44} = D_1 \oplus D_2 \oplus D_3$$

where

$$D_i = L_{22}^{(i)} U_{22}^{(i)}$$

Note that the G_i matrix here is a partition of the $\hat{Y}$ network matrix, discussed in this section, and is not to be confused with the F_i matrix on page 35.

With this approach we are doing more than solving the entire problem, but we make use of area information and solve sub-problems independently. Unfortunately, the effectiveness of this kind of strategy is only speculation at this point because there has not been a careful study of its relative performance. The discussions of Happ (1980) are more concerned with the physical interpretation of partial results and the elaborate interconnection strategy rather than simply evaluating the computational performance treating the BBDF as a partitioned matrix problem.

We conclude the network discussion by summarizing some of the sparsity characteristics of the matrices representing power system networks:

(1) They are extremely sparse (3-4 nonzeros per row independent of order).

(2) They are large (1000-5000 is typical, but larger problems are of interest).

(3) The entries are complex (though sometimes treated as 2x2 real).

(4) They are symmetrical in structure, though not necessarily symmetric in value.

At the present time we conclude that minimum degree is the most suitable ordering strategy, though physical dissection or a revised automated dissection strategy looks promising.

5. DIRECTIONS AND NEED FOR FUTURE RESEARCH

Some areas of additional research have been indicated in the previous two sections. The sparse matrix structure for power plant models offers promise for the improved representation and reduction of models as well as the efficient solution of equations resulting from them. Some work on hybrid algorithms (for example, partial reduction using minimum degree, then another strategy) and in tearing or partitioning strategies for the network problem is also indicated.

Additionally, the power industry is looking carefully at the potential for parallel computing (distributed computing, pipeline computers, array processors) to help them solve bigger problems faster. A meeting for the purpose of investigating the potential application of parallel computing to the power industry (see Anderson (1977)), was largely devoted to the sparse matrix problem, since the fundamental computational problem involves solving the sparse equations. A second meeting on the same subject, sponsored by the Electric Power Research Institute, was held in Dallas, Texas in October 1980.

There are several reasons why this industry continues to need more capability to solve problems faster and to solve bigger problems. Tighter operating margins are required because of the very high fuel costs combined with rate control, requiring better understanding of the system. In some places, more detailed models (for example, three phase power flow) are indicated which lead to larger, more complicated computations. To allow for more efficient operations through the exchange of power between neighbouring systems, more interconnections are being introduced leading to bigger problems. In the U.S., for example, the entire western system is interconnected, as is the eastern system, and some connections between these are being added.

For the operations purposes there is a need for training simulators requiring the capability to perform realistic transient stability in real time, for example. Much faster computation will have to be achieved to realize this goal. Better tools are also required on-line for operations purposes.

Of course sparse matrix computation is not the only bottleneck to achieving these goals. Improved data acquisition and state estimation, for example, are also required to allow the operator to simulate events in his system in its present state and to evaluate proposed actions. But the sparse matrix problem is one of the barriers between the capabilities and needs of the industry.

6. GENERAL CONCLUSIONS

The discussion of the sparse matrix problems in electric power systems models has two purposes. It illustrates an important class of problems where improved methods are really needed. The discussion also indicates some important distinction between the characteristics of these sparse matrices and those from some other applications.

Very significant advances have been made through recognizing sparsity at all. General purpose software discussed, for example, by Duff (1980) allows any user with a sparse matrix to be able to exploit sparsity readily without painful coding. This very important step has been achieved only in the last decade.

We now recognize that a matrix is more than sparse, however. And while sparsity produced gigantic savings compared to dense methods (factors of 100 and greater) and allowed solving problems otherwise intractable, use of additional characteristics may offer these savings again. Table 1 shows a factor of 100 savings between two sparse orderings, for example, and Table 2 shows problems which were intractable under one ordering and easy under another. Clearly the offering of a choice is a step in the right direction.

Perhaps this can be taken a step further. Rather than a choice of orderings, and classification of matrices as "structures" matrices, "power plant" matrices, "network" matrices, "linear programming" matrices, etc., we can extract essential features of both orderings and matrix types. If reasonable classification of matrix types can be made, and a general understanding of best methods for each type can be achieved, then general purpose software may be able to provide large gains over today's software. Classifying sparse matrices according to solution characteristics has already been done to some extent for iterative methods, so this is not really a new approach.

Unfortunately, looming in the background is the collection of different computer architectures which are becoming more and more important. Since these affect the choice of "best" sparse algorithms, it will be interesting to see if order can be achieved in this multiparameter (characteristics, computer architecture, algorithms) optimization problem.

ACKNOWLEDGEMENTS

I would like to thank Benjamin Dembart and W.F. Tinney for helpful discussions in the preparation of this paper. I also acknowledge the suggestions and comments of the editor, Iain Duff.

REFERENCES

Anderson, P.M. (1977) ed. Exploring Applications of Parallel Processing to Power System Analysis Problems, EPRI EL-566-SR.

Anderson, P.M. and Dembart, B. (1980) Computational aspects of transient stability analysis, In Erisman et al (1980).

Barry, D.E., Pottle, C. and Wirgau, K.A. (1978) Technology assessment study of near term computer capabilities and their impact on power flow and stability simulation programs, EPRI EL-946.

Duff, I.S. (1980) A Sparse Future, These proceedings.

Duff, I.S. and Stewart, G.W. (1978) Sparse Matrix Proceedings 1978, SIAM Press.

Duff, I.S., Erisman, A.M. and Reid, J.K. (1976) On George's nested dissection method, SIAM J. Numer. Anal., 13, pp. 686-695.

Erisman, A.M., Neves, K.W. and Dwarakamath, M.H. (1980) eds, Electric Power Problems: The Mathematical Challenge, SIAM Press.

George, A. (1973) Nested dissection of a regular finite element mesh, SIAM J. Numer. Anal., 10, pp. 345-363.

George, A. (1974) On block elimination for sparse linear systems, SIAM J. Numer. Anal., 11 pp. 585-603.

George, A. and Liu, J.W.H. (1978a) A quotient graph model for symmetric factorization", In Duff and Stewart (1978).

George, A. and Liu, J.W.H. (1978b) User Guide for Sparspak: Waterloo Sparse Linear Equations Package, Report CS-78-30, Computer Science Dept., Waterloo University.

Gibbs, N.E., Poole, W.G. and Stockmeyer, P.K. (1976) An algorithm for reducing bandwidth and profile of a sparse matrix, SIAM J. Numer. Anal., 13, pp. 236-250.

Happ, H.H. (1980) The application of diakoptics to the solution of power systems problems, In Erisman et al (1980).

Jennings, A. (1977) Matrix Computation for Engineers and Scientists, Wiley.

Lewis, J.G. and Poole, W.G. (1980) Ordering algorithms applied to sparse matrices in electric power problems, In Erisman et al (1980).

Sangiovanni-Vincentelli, A., Chen, L.K. and Chua, L.C. (1977) An efficient heuristic cluster algorithm for tearing large-scale networks, IEEE Trans., CAS-24, pp. 709-717.

Sato, N., and Tinney, W.F. (1963) Techniques for exploiting the sparsity of the network admittance matrix, IEEE Trans. PAS-82, pp. 944-949.

Stott, B., Alsai, O. and Marinho, J.L. (1980) The optimal power flow problem, In Erisman et al (1980).

Tinney, W.F. and Walker, J.W. (1967) Direct solution of sparse network equations by optimally ordered triangular factorization, Proc. IEEE., 55, pp. 1801-1809.

Westerberg, A.W., Hutchinson, H.P., Motard, R.L. and Winter, P. (1979) Process Flowsheeting, Cambridge University Press.

DISCUSSION

PROFESSOR A. JENNINGS (Belfast). In some of your comparisons you used the variable band technique. Could you tell me what ordering was employed since that is very critical to the efficiency of this scheme? Also, have you tried eliminating all nodes of degree up to 2 before using the variable band scheme?

ERISMAN. We used the reverse Cuthill-McKee ordering with the starting node a pseudo-peripheral node chosen by the Gibbs, Poole, Stockmeyer algorithm. This is, I believe, the best general ordering for profile or variable band schemes. I only briefly mentioned hybrid techniques in my talk. They are certainly worth investigating further but we have not done this.

DR. I.S. DUFF (Harwell). Typically how many of your equations are non-linear and how strong are these non-linearities?

ERISMAN. Many of the power plant equations are non-linear in only a few variables and these non-linearities can be very strong. Network non-linearities are generally not as strong. It may be very advantageous to use a variable type scheme which essentially pivots on a combination of sparsity and level of dependency and non-linearity.

DR. J. LEWIS (Boeing). I would like to comment on the remark of Jennings. For the problems we have been considering and about which Al Erisman spoke, straight minimum degree will be hard to beat. Typically, the factors are less than twice as dense as the original matrix. Hybrid methods are certainly possible but I doubt if minimum degree followed by the variable band scheme would be competitive.

ERISMAN. One has only to see the vast difference in the number of non-zeros stored in the various schemes and consequent work in factorization and solution to realize the truth of John Lewis' remark. The total storage belies much of the superiority of minimum degree.

DR. P.G. THOMSEN (Lyngby). How many solves do you typically do for each factor?

ERISMAN. We generally take about five time steps before recomputing the Jacobian, but notice how close the factorization and solution times are for our very sparse problems. In many applications it may be advantageous to factorize more often.

MR. D.B. SZYLD (NYU). You talked of difficulty in data transmission. I feel you may have over-exaggerated the problem of transferring a lot of data to a large computer rather than using interface nodes.

ERISMAN. We consider the use of interface nodes for two reasons other than data transmission. In a network made up of different utilities, some may be unwilling to release the details of their internal system. Computationally the use of interface nodes is closely related to nested dissection and may produce a better overall ordering anyway.

TOWARDS AN EFFICIENT SPARSITY EXPLOITING NEWTON METHOD FOR MINIMIZATION

Ph. L. Toint

(Department of Mathematics, Facultes N.D. de la Paix (Namur))

ABSTRACT

The paper surveys some recently proposed algorithms for unconstrained minimization when the second derivative of the objective function is sparse. Updating and estimation procedures are considered from the point of view of efficiency. Special attention is given to the case where the Hessian has a band structure. A new strategy for the choice of the step is also discussed and some numerical results on a specially designed test function are presented.

1. INTRODUCTION

In the last year, a number of proposals were made to allow unconstrained minimization problems of many variables to be solved efficiently. Fortunately a large proportion of these high dimensional problems have some sparsity structure that can be taken into account in order to reduce the storage needed for the algorithms used as well as to speed up convergence. The sparsity that is of interest here is the sparsity that is present in the second derivative matrix of the objective function.

Assume indeed that we consider the problem of minimizing the function $f(\underline{x})$ $(f:R^n \to R)$ over the whole space. In many cases one can observe that the gradient components, $D_i f(.)$ or $g_i(.)$ say, do not usually depend on all the n variables. But if $g_i(.)$ does not depend on the variable x_j, this implies that the (i,j) (and the (j,i) by symmetry) entry of the Hessian matrix of f(.), $[D^2 f(.)]_{i,j}$ is identically zero for all $\underline{x}$ in R^n. One obviously wishes to exploit that supplementary information about the objective function and a natural way to do so is to compute (or to update) only the non-zero entries of the Hessian (or approximate Hessian) matrix. If we keep in mind an iteration of the form

$$B^k(\underline{x}^{k+1}-\underline{x}^k) = -\underline{g}(\underline{x}^k), \tag{1}$$

that is

$$\underline{x}^{k+1} = \underline{x}^k-(B^k)^{-1}\underline{g}(\underline{x}^k), \tag{2}$$

we therefore want to apply this strategy to the matrices $\{B^k;\ k=1,2,\ldots\}$ that are supposed to approximate the matrices $\{D^2f(\underline{x}^k);k=1,2,\ldots\}$

Up to now, two directions have been followed to take the sparsity of $D^2f(.)$ into account.

The first natural idea is to use the second derivative matrix in a Newton procedure while calculating only the entries that might be non-zero. If the true second derivatives are not available, or are too expensive to be computed, one may use estimations based upon a difference scheme in the gradient values. Here again there are two possibilities. If the gradient components can be computed separately, then the (i,j) and (j,i) entries of the Hessian are obtained from

$$[D^2f(\underline{x})]_{i,j} = [D^2f(\underline{x})]_{j,i} = (g_i(\underline{x}+h\underline{e}_j)-g_i(\underline{x}))/h \tag{3}$$

where h is a suitably chosen steplength and where i is less than or equal to j. In this case, it is clear that one cannot hope to reduce the work for obtaining one estimate of the Hessian.

However, it is often the case that the gradient components are not available separately. It may be because the user has not supplied them or because many components involve a single quantity that should be computed once only to avoid excessive cost. In this situation, one has to design a strategy that will compute as few gradient differences as necessary, by taking sparsity into account. Such a strategy has been proposed for Jacobian matrices by Schubert (1970) and more recently for Hessian (i.e. symmetric) matrices by Powell and Toint (1979).

The second way of taking advantage of any sparsity present in the second derivatives is to extend the classical Quasi-Newton approach. This type of method builds an approximation of the Hessian matrix at the current point by updating the approximation at the previous point of the iteration. To exploit sparsity in these procedures, we therefore have to design symmetric updates that will preserve sparsity in the current approximating matrix while forcing the Quasi-Newton equation

$$B^{k+1}(\underline{x}^{k+1}-\underline{x}^{k}) = \underline{g}(\underline{x}^{k+1})-\underline{g}(\underline{x}^{k}) \tag{4}$$

to be satisfied. In (4), k is the iteration index and $\underline{x}^k$ is the approximation to the solution at iteration k. Such sparse and symmetric updates have been recently proposed by Toint (1977, 1980a, 1980b) and Shanno (1978).

Another important feature for an efficient algorithm whose general iteration is defined by (1) is the way the step selection is performed. In some cases, we indeed want to solve (1) completely, but in other cases we might consider that this method involves too much computing and is not worth the effort, especially in the early iterations of the algorithm, when the matrix B^k is still quite different from the Hessian matrix at the optimum.

The purpose of this paper is to report the author's views and experiments in the design of a composite algorithm, incorporating estimations and updating methods to revise the Hessian approximation, as well as a new step selection strategy and its possible numerical improvement.

The second section briefly surveys the available sparse and symmetric updates that can be used in this context, and sparsity exploiting estimation procedures are considered in section 3. Some remarks on the practical implementation of one of these methods are presented in section 4, together with a special discussion of the band structured case. Section 6 presents a class of test functions that are used to illustrate the different points that are made in Section 5 concerning the combination of updates and estimations as well as the proposed step selection procedure. Finally, section 7 presents some numerical experiences while section 8 offers some concluding remarks.

2. SPARSE AND SYMMETRIC UPDATING PROCEDURES

We now turn to the general problem of finding a numerically realistic procedure that forces equation (4) to be satisfied while preserving symmetry and any sparsity that is present in the true Hessian matrix. For the sake of clarity, we will temporarily drop the iteration indices and consider that we want to find a matrix B* defined by

$$B^* = B + E \tag{5}$$

In this equation, B is the actual approximation to the Hessian and E is a correction matrix that satisfies the relations

$$E\underline{s} = \underline{r}, \tag{6}$$

$$E = E^T, \tag{7}$$

and

$$E_{i,j} = 0 \qquad (i,j) \in I, \tag{8}$$

with I and r defined by

$$I = \{(i,j) \text{ such that } \forall \underline{x} \in R^n, [D^2 f(\underline{x})]_{i,j} = 0\}, \tag{9}$$

and

$$\underline{r} = \underline{y} - B\underline{s} \tag{10}$$

where $\underline{s}$ and $\underline{y}$ are two non-zero vectors of R^n that respectively stand for the change in $\underline{x}$ variables and the corresponding change in gradients. Observe then that (6) plays the role of (4). We assume that conditions (8) are consistent with the symmetry requirement (7) and also that the sparsity constraints (8) do not apply to the diagonal, that is

$$\forall i=1,\ldots,n \qquad (i,i) \notin I. \tag{11}$$

This assumption is made because it is natural in the setting of unconstrained optimization and because it simplifies some existence statements in the following discussion.

It is clear that relations (6)-(8) do not define the correction E completely. The remaining degrees of freedom are then taken up by imposing that the new matrix B* is as close as possible to the given approximation B in some norm. This is expressed by the requirement that

$$\|E\| \text{ is minimum subject to conditions (6)-(8)} \tag{12}$$

where $\|.\|$ is a suitably chosen matrix norm that makes the solution to the mathematical program (12) possible in a reasonable amount of computation. Up to now, only one class of norms has been used in this setting, the weighted Frobenius norms that are defined by

$$\|E\|_W^2 = \mathrm{Tr}(WEWE^T) \tag{13}$$

for a symmetric positive definite weighting matrix W. This variational approach has been proposed by Greenstadt (1970) and is extensively discussed by Dennis and Moré (1977).

The first practical solution to problem (12) has been proposed by Toint (1977) and uses the usual Frobenius norm that is defined by letting W equal the identity matrix in (13). In order to describe it, we define the vectors $\underline{s}(i)$ $(i=1,\ldots,n)$ componentwise by the relations

$$s(i)_j = \begin{cases} s_j & \text{if } (i,j) \notin I, \\ 0 & \text{if } (i,j) \in I. \end{cases} \tag{14}$$

With these vectors that reflect the sparsity pattern of the rows (and columns) of the Hessian, the (i,j)th element of the correction E can now be written as

$$E_{i,j} = u_i s(i)_j + u_j s(j)_i \qquad (i,j=1,\ldots,n) \tag{15}$$

where the vector $\underline{u}$ is in fact the Lagrange parameter associated with constraint (6) in problem (12) and can be computed as the solution of the linear system

$$Q\underline{u} = \underline{r} \tag{16}$$

with the matrix Q being defined elementwise by

$$Q_{i,j} = s(i)_j s(j)_i + \|\underline{s}(i)\|^2 \delta_{i,j} \tag{17}$$

where $\delta_{i,j}$ is the Kronecker delta.

Some observations are now appropriate:

1. The matrix Q defined by (17) is symmetric and has the same sparsity pattern as the original Hessian.
2. This matrix is positive definite if and only if all the vectors $\underline{s}(i)$ $(i=1,\ldots,n)$ are non-zero.
3. If some vector $\underline{s}(k)$, say, is zero, then the updating problem may be reduced in size and this causes no trouble at all in practice.

It is possible to express the above relations in terms of projections with respect to the particular matrix norm defined by (13). Indeed, if we define, for an arbitrary matrix X, the operator $P_W^L(X)$ as the projection of X onto the affine subspace L with respect to the norm induced by W, then the matrix B* of (5) satisfies the following property

$$B^* = P_W^V(B) \tag{18}$$

where

$$S = \{X \in R^{n \times n} \text{ such that } X_{i,j} = 0 \text{ for } (i,j) \in I\}, \tag{19}$$

$$A = \{X \in R^{n \times n} \text{ such that } X\underline{s}=\underline{y} \text{ and } X=X^T\}, \tag{20}$$

$$V = \text{the intersection of } A \text{ and } S. \tag{21}$$

In this setting, the above solution is nothing but the matrix given by (18) with the choice W=I. The correction (15) then satisfies

$$B^* = P_I^V(B) \tag{22}$$

and can be written explicitly as

$$E = P_I^S(\underline{u}\underline{s}^T + \underline{s}\underline{u}^T) \tag{23}$$

where $\underline{u}$ is still the solution of (16)-(17).

But, if $P_I^V(.)$ is quite easy to obtain computationally, it is not the case with $P_W^V(.)$ when W is allowed to be a general symmetric positive definite matrix. This is the reason why another proposal has been made by Shanno (1978) and by Toint (1980a). The proposal is to replace (18) by

$$B^* = P_I^V(P_I^S(B+C)) \tag{24}$$

where C is any dense correction matrix, such as the numerically efficient BFGS update defined by

$$C = \underline{y}\underline{y}^T/\underline{y}^T\underline{s} - B\underline{s}\underline{s}^T B/\underline{s}^T B\underline{s}. \tag{25}$$

The particular update (24)-(25) has proven to be computationally competitive with the simpler formula, which uses (18) with the choice W=I, on medium size problems. For larger dimensions, some difficulties may nevertheless appear.

Finally, another sparse update has been proposed by Toint (1980b) that remains in the variational framework of (18) while using a non-diagonal weighting matrix W. More precisely, W is allowed to be any symmetric positive definite rank two modification of the identity whose inverse can be written in the form

$$W^{-1} = I + a\underline{s}\underline{s}^T + b(\underline{s}\underline{w}^T + \underline{w}\underline{s}^T) \tag{26}$$

where a and b are two arbitrary constants and $\underline{w}$ is an arbitrary vector. This procedure is hoped to be more scaling independent than the previous ones, but this is still a conjecture since the necessary numerical experiments are not yet completed. Therefore, we will, in the sequel, restrict our attention to the first two updating methods described in this section.

3. SPARSE ESTIMATION PROCEDURES

After this quick survey of the available sparse and symmetric updates, we now examine some procedures that can be used for taking sparsity into account in the difference estimation of the Hessian matrix $D^2f(x)$. These procedures generalize the well-known strategy proposed by Curtis, Powell and Reid (1974) to the symmetric case.

Specifically, we consider the problem of choosing steps in R^n, $\underline{d}_1, \underline{d}_2, \ldots, \underline{d}_p$ say, so that all the non-zero elements of $D^2f(x)$ can be estimated from the differences

$$\underline{y}_k = \underline{g}(\underline{x}+\underline{d}_k) - \underline{g}(\underline{x}), \qquad (k=1,2,\ldots,p). \tag{27}$$

If we denote by H our estimate of $D^2f(\underline{x})$, then its elements are obtained from the equations

$$H\underline{d}_k = \underline{y}_k \qquad (k=1,2,\ldots,p), \tag{28}$$

but usually not all parts of those equations are used. Our purpose is to reduce the number of gradient evaluations by making p as small as possible. This reduction is based on the following simple observation. If we let $[D^2f(\underline{x})]_k$ be the k-th column of the matrix $D^2f(\underline{x})$, and if we assume that the row numbers of the unknown elements of $[D^2f(\underline{x})]_j$ are all different from the row numbers of the unknown elements of $[D^2f(\underline{x})]_k$, then the k-th and j-th columns of H can be computed from (28) using a single displacement vector that is some combination of the basis vectors $\underline{e}_k$ and $\underline{e}_j$.

The first method that is considered by Powell and Toint (1979) is called a direct method because no algebra is needed in the process of solving (28). As in the Curtis-Powell-Reid (CPR for short) strategy, the procedure divides the columns of $D^2f(\underline{x})$ into p disjoint groups according to the following rule:

CPR rule: Two columns are in the same group only if they have their non-zero components in different rows.

To each of these groups is associated a displacement vector $\underline{d}_k$ $(k=1,\dots,p)$ that has a non-zero component in the i-th position if and only if $[D^2f(\underline{x})]_j$ is in the k-th group. Symmetry is then taken into account as follows. Let $k(i)$ denote the number of the group that contains $[D^2f(\underline{x})]_i$ and let $H_{q,r}$ be a general off-diagonal element of H. Then the (q,r) and (r,q) entries of H are obtained from

$$H_{q,r} = H_{r,q} = [y_t]_v/h \tag{29}$$

where

$$t = \min\{k(q),k(r)\} \tag{30}$$

$$v = \begin{cases} r & \text{if } t=k(q) \\ q & \text{if } t=k(r), \end{cases} \tag{31}$$

and where h is the corresponding stepsize.

The practical performance of the method (in terms of the size of p) can be substantially improved by applying an a priori symmetric permutation that orders the rows and columns by decreasing number of unknown non-zeros.

The efficiency of this procedure compared with the CPR strategy when applied to a symmetric structure is best illustrated by the arrow-head case. Assume indeed that the Hessian under consideration has an arrow-head sparsity pattern, that is only the diagonal and the last row and column contain non-zero entries. Then it is easy to see that the CPR procedure will require n gradient differences (because the last row is full) while the direct method described above will only require two differences.

Despite the interest of this estimation algorithm, we now proceed by describing the only method that will be discussed in the minimization context, namely the substitution method. We only consider this procedure for possible incorporation in a general minimization method because it is more efficient with respect to the number of gradients that have to be computed. It is occasionally less exact than the direct method (Fivet (1980)) but we found out that its performance was completely satisfactory for our purpose.

In introducing the substitution method, consider the case when the Hessian matrix is tridiagonal. We build two displacement vectors of the form

$$\underline{d}_1^T = (h,0,h,0,h,0,\ldots) \tag{32}$$

and

$$\underline{d}_2^T = (0,h,0,h,0,h,\ldots). \tag{33}$$

Equation (28) for k=1 now yields

$$\begin{aligned} &H_{1,1}h = [y_1]_1 \\ &(H_{2,1}+H_{2,3})h = [y_1]_2 \\ &H_{3,3}h = [y_1]_3 \\ &\text{etc.} \end{aligned} \tag{34}$$

while for k=2 it gives

$$\begin{aligned} &H_{1,2}h = [y_2]_1 \\ &H_{2,2}h = [y_2]_2 \\ &(H_{3,2}+H_{3,4})h = [y_2]_3 \\ &\text{etc.} \end{aligned} \tag{35}$$

It is easy to observe that (n+2) unknowns, namely $H_{1,1}, H_{2,2}, \ldots, H_{n,n}, H_{1,2}$ and $H_{n,n-1}$, are defined directly by the above relations. We also clearly have a very simple substitution process that links $H_{j,j-1}$ and $H_{j+1,j}$ and one of our equations is redundant. A possible substitution method will then ignore the direct definition of $H_{1,2}$ and will compute all the non-zero entries of H using only two gradient differences. Hence p=2 is sufficient for the tridiagonal case, while it may be easily checked that p=3 for the direct method described above.

The process that is discussed in the last paragraph for tridiagonal structures can be extended to general symmetric sparsity patterns, as expressed in the following proposition.

Proposition 1: Let L be a lower triangular matrix whose sparsity pattern in the lower triangle and on the diagonal is the same as that of the Hessian matrix $D^2f(\underline{x})$. Assume that the columns of L are divided into p groups according to the CPR rule. Assume also that p gradient differences are computed along displacement steps corresponding to these groups. Then it is possible to obtain all the unknown elements of H by an algebraic substitution that uses these p differences.

It is now clear why we talked about a "substitution" method. Specifically, we will refer to substitution methods when the computational process that is used for obtaining the entries of H only involve algebraic substitutions. The particular procedure considered in proposition 1 will be called "lower triangular substitution".

We only outline the proof of proposition 1, to give the main idea of lower triangular substitution methods. A detailed proof can be found in Powell and Toint (1979), together with a complete discussion of the other results mentioned in this section.

This proof is constructive and shows that we can proceed as follows:

1. Estimate the last row and column of H using only direct relations. This is possible because the last row is entirely in the pattern of L.
2. By using the values of the entries of the last column, compute the elements of the penultimate row and column of H. This is possible because the equations that define one unknown of this row from the gradient differences only involve this unknown and elements of H that are on the same row but above the diagonal, and hence have already been determined.
3. Proceed inductively with rows (and columns) (n-2) to 1.

In order to improve the procedure, consider again the arrow-head structure given by

$$\begin{matrix} * & O & O & O & * \\ O & * & O & O & * \\ O & O & * & O & * \\ O & O & O & * & * \\ * & * & * & * & * \end{matrix}$$

then one observes that the lower triangular substitution method will require n gradient differences, because the last row of H and L is full. But, although permutations do not change dramatically the number of groups for the CPR strategy applied to a square structure, this is not true at all when it is applied to a lower triangular part of another structure. Indeed, if we simply interchange the first and last rows and columns in our example, giving

$$\begin{matrix} * & * & * & * & * \\ * & * & O & O & O \\ * & O & * & O & O \\ * & O & O & * & O \\ * & O & O & O & * \end{matrix}$$

we see that the number of groups is reduced to 2. In general, we have

$$p \geqq r(L) \tag{30}$$

where r(L) is the maximum number of non-zero elements in a row of L. We therefore wish to permute H symmetrically in order to reduce r(L) as much as possible. This goal can be achieved with the following permutation procedure

1. Let j=n.
2. Consider the leading jxj subpattern of H. Find i, the row number of one row of this subpattern that has the minimum number of unknown elements.
3. If i<j, exchange the i-th and j-th rows and columns of H.
4. If j>2, set j to j-1 and return to 2.
5. Otherwise, stop.

To be specific, it is possible to prove that this permutation actually minimizes r(L) over all the symmetric permutations of H. Furthermore, it can be implemented in a cheap computational way, as discussed in the next section.

A major property of this lower triangular substitution method is that the number of interdependence relations between the different unknowns is quite small. That is a consequence of the following result:

Proposition 2: In a lower triangular substitution method, the matrix element $H_{i,j}$ $(i \geq j)$ depends on the non-zero entry $H_{q,r}$ $(q \geq r)$ only if

$$q > i \tag{37}$$

and

$$\{k(q),k(r)\} = \{k(i),k(j)\} \tag{38}$$

This important property has two immediate corollaries:

Proposition 3: If there is no zero element on the diagonal of H, then there are no substitutions in the equations that define the diagonal entries of H.

Proposition 4: In a lower triangular method, the general non-zero matrix element $H_{i,j}$ $(i \geq j)$ depends on at most (n-i) other matrix elements.

This last property leads to some interesting bounds for the error $|[D^2 f(\underline{x})]_{i,j} - H_{i,j}|$. These are developed in Powell and Toint (1979).

4. IMPLEMENTING THE LOWER TRIANGULAR SUBSTITUTION METHOD

Let us now restrict our attention to the lower triangular substitution method and consider how it can be implemented on the computer.

The first thing we have to decide is the way our sparse matrix will be represented in the machine's memory. We will assume that the size and structure of the matrix is such that all the non-zero elements of its lower triangular part (including the diagonal) can be kept in core, so that our algorithm does not have to bother about transferring data between core and backing store. This restriction is somewhat alleviated by the fact that we choose a column representation of this lower triangular part, since that form of storage is suitable for easy out-of-core manipulations. Specifically, we let

m=number of non-zero elements in the lower triangle (L), (39)

H(i)=actual numerical values of the elements of L (i=1,...,m), (40)

IROW(i)=row number of the non-zero element H(i) (i=1,...,m), (41)

IDIAG(i) = position of the i-th diagonal element in H(.) (42)
(i=1,...,n),

where we assume that the m non-zero elements of the lower triangle L are ordered column after column in the vector H(.). This column representation is by no means original but it is quite natural in our setting. One can also point out that the algorithm can be transformed to accept a similar row oriented representation.

The second important design consideration is the following: as we are likely to estimate the Hessian matrix several times during the minimization process, we naturally wish to reduce as far as possible the amount of work that has to be done for each estimation. Indeed we can perform the pattern analysis once and for all at the beginning of the optimization procedure and then set up information sufficient to make the subsequent estimation processes computationally cheap.

Finally, we have to make sure that all the bookkeeping operations involved in the pattern analysis and in the actual estimations do not become too important compared with the total work of the minimization algorithm. In this respect, we consider as acceptable for the a priori analysis and estimation process computing work which grows as m and n, but we want to avoid $O(nm)$, $O(m^2)$ or $O(n^2)$ procedures.

The second and last considerations have decided us to build, during the analysis phase, new structural information (namely supplementary IROW and IDIAG vectors) corresponding to the structure that results from the permutations discussed in the preceding paragraph, and "tracking" information that will allow us to pass from the permuted structure to the original one and back. The final storage requirements for the analysis process will then be

- 3 integer m dimensional vectors, namely IROW and its permuted counterpart plus a "tracking" vector,
- 5 integer n dimensional vectors, namely IDIAG and its permuted counterpart, a vector to keep the group numbers (k(i)) of the permuted structure, another "tracking" vector and a working space vector,

while the actual estimation will need

- 1 real m dimensional vector, namely H(.),
- 3 integer m dimensional vectors as above,

- 5 real n dimensional vectors, namely the vector of variables at the current point, the gradient at the current point, the stepsizes for the different variables and two working space vectors,
- 4 integer n dimensional vectors, namely those mentioned above minus the working space vector.

We now briefly outline the successive steps for both the analysis phase and the estimation phase.

The pattern analysis algorithm may be summarized as follows:

1. Symmetric permutation:
   (a) initialize the process [O(n)],
   (b) compute the number of unknowns in each row and its maximum [O(m)],
   (c) test whether permutations are useful. If not, skip them [O(1)],
   (d) order the rows by decreasing number of unknowns [O(n)],
   (e) select the row with minimum number of unknowns [O(1)],
   (f) in this row, find the unknowns that leave L [O(1)],
   (g) revise the counts of unknowns and various pointers [O(1)],
   (h) if all rows have not been considered, go to step e.
2. Set up the information about the permuted structure:
   (a) compute the new column numbers and the new counts of unknowns by columns [O(m)],
   (b) compute the new information about the diagonal and the new maximum number of unknowns in a column [O(n)],
   (c) reorder the newly built information to collect together the elements of the same column [O(m)],
3. Apply the CPR strategy on the lower triangle of the permuted structure [O(m)][1].

The estimation process itself then has the following structure

1. For each group (1,...,p),
   (a) build the displacement vector [O(n)],
   (b) evaluate the perturbed gradient,

---

[1]If we consider the number of groups to be independent of n.

(c) store the corresponding differences in gradient in H [O(n)],

2. Working from the last row of H to the first, for each non-zero element in L,

(a) determine the group of the column under consideration [O(1)],

(b) find the correction term in the substitution for the element under consideration [O(1)][2],

3. Divide the corrected differences by the appropriate stepsizes [O(m)].

Observe that we did not give any order of complexity for the evaluation of the perturbed gradients : this only depends on the complexity of the gradient computation and is independent of the estimation process itself.

Finally, we wish to comment on the particular behaviour of these algorithms when they are applied to a band sparsity pattern (bandwidth = $2\ell+1$). In this case, the first important observation is that no permutations are needed in the pattern analysis phase, so that the information about the "permuted" structure is identical to that related to the original structure. Then, clearly, the space that is needed to store this information and the "tracking" information is no longer required. Furthermore, the estimation process also becomes simpler (because we know a priori the group structure of the columns) and we can omit the information about the diagonal provided that we know the bandwidth. As a consequence, we make, besides the non negligible economy in computing, another economy in storage, namely 2m+4n integer locations in both phases of the method.

As mentioned above, band structured matrices may exceptionally cause some precision problems, due to the fact that the substitutions involved always have their maximum number of terms, i.e. (n-1). When the dimension becomes large this may be troublesome. We see two possible ways of alleviating that difficulty:

1. When the number of gradient evaluations is of paramount importance, the best we can do is to reduce the length of the substitutions by a factor of 2. This is achieved by starting two substitutions, one from the last row and one from the first (see relations (34)-(35)).

---

[2]Because the number of non-zero elements in one row is independent of n

2. When we can afford one more gradient evaluation in the estimation process, we can use it to "slice" the band into bits that can be estimated separately. The maximum length of a substitution in these bits is then of order ℓ^2 instead of order n. The computation is somewhat more involved and has not been thoroughly tested yet. We nevertheless feel that it might be useful for particularly difficult examples of band structure.

5. DESIGN CONSIDERATIONS FOR A GENERAL SPARSE MINIMIZATION ALGORITHM

The general minimization algorithms we are interested in all have the same structure for their main iteration. Indeed, given $\underline{x}^k$ an approximation to the solution, $f(\underline{x}^k)$, $\underline{g}(\underline{x}^k)$ and B^k an approximation to the Hessian matrix $D^2f(\underline{x}^k)$, they proceed as follows:

1. search for a direction $\underline{s}^k$ in some "confidence region" defined by

$$\|\underline{s}^k\| \leq v^k, \tag{43}$$

where the radius v^k defines a ball in which we expect the quadratic

$$F(\underline{x}) = f(\underline{x}^k)+\underline{g}(\underline{x}^k)^T(\underline{x}-\underline{x}^k)+\frac{1}{2}(\underline{x}-\underline{x}^k)^T B^k(\underline{x}-\underline{x}^k) \tag{44}$$

to approximate $f(\underline{x})$ sufficiently accurately. The point $\underline{x}^k+\underline{s}^k$ should be such that $F(\underline{x}^k+\underline{s}^k)$ is less than $F(\underline{x}^k)$.

2. define a new trial point by

$$\underline{x}^{k+1}=\underline{x}^k+t\underline{s}^k \tag{45}$$

for some $t>0$.

3. compute the function and gradient values at this point.

4. revise the B^k to produce a new approximation B^{k+1}.

5. define the new confidence region by computing a suitable radius v^{k+1} and go back to 1.

When talking of "step strategy", we will, in this framework, refer to the detailed implementation of steps 1 and 2 while the words "Hessian revision" will cover that of step 4. Step 5 will not be discussed in detail below because we found the method proposed by Powell (1970) and already used in Toint (1978) quite satisfactory. All the procedures that we consider follow that implementation.

We now comment on the main problems that arise when considering an algorithm of the form just described.

The first question that we investigate is the choice of the update formula that we want to use in an efficient procedure for sparse minimization problems, and the way this sparse updating is mixed with the estimations of the true Hessian matrix. Clearly, if the cost of estimating the Hessian at the current point by a sparsity exploiting procedure is comparable with the cost of computing a sparse update, then a pure Newton procedure can be used quite efficiently. This issue depends on the relative costs of updating and estimation, and is therefore very much dependent on the way the linear system (16) is actually solved: this will be discussed further in the sequel. We now assume that solving (16) is cheap with respect to the estimation of $D^2f(\underline{x})$, so we do not want to estimate the Hessian when an updating is satisfactory.

As far as the updating procedure is concerned, we wish to choose the most efficient one between (22) and (24)-(25). One can argue that the performance of a particular update is not independent of the step strategy that is used, and consequently that one has to test both updates in all possible configurations. However, the numerical tests given below show that (24)-(25) is more efficient than (22) for one particular strategy and supplementary tests not presented in this paper show that this behaviour does not depend much on the step strategy that is used, so that we think our conclusion is quite general. However, it might be interesting to mention that the advantage of (24)-(25) over (22) is often marginal, especially when a good step strategy is considered.

If the updating procedure is chosen, we still have to determine a practical method to decide when the true Hessian $D^2f(\underline{x}^k)$ must be estimated. As this estimation requires several additional gradient evaluations, we certainly do not wish to perform it too often. In practice, we impose a prescribed number of updating iterations between two estimations of $D^2f(\underline{x}^k)$. This number is chosen as

$$n_{upd} = \max[4, 3p/2] \tag{46}$$

where p is, as in section 3, the number of gradient values that are needed for one estimation. Although this particular choice is empirical, we found that it provides a reasonable balance between estimation and updating.

Another interesting quantity to monitor is the normalized residual

$$\tilde{r}^k = \|\underline{g}(\underline{x}^{k+1})-\underline{g}(\underline{x}^k)-B^k(\underline{x}^{k+1}-\underline{x}^k)\|/\|\underline{x}^{k+1}-\underline{x}^k\|. \tag{47}$$

Indeed we know from the theory (see Dennis and Moré (1974)) that this quantity must decrease to zero if we expect superlinear convergence of our algorithm. It actually describes the accuracy of our quadratic prediction of $f(\underline{x})$ along the step just made. Since we already observed in Toint (1978) that estimation helps especially when we approach the quadratic neighbourhood of the solution, we may therefore avoid the estimation if this quantity is large. On the other hand, if it is really very large, this may well mean that the quadratic F(.) is very unlike f(.) and, in this case, we may decide to restart the method with a more correctly estimated Hessian approximation. If $\tilde{r}^k$ is very small, we may expect the modification to B^k to be of little importance and proceeding with that approximation may be quite satisfactory, even if doing so theoretically prevents superlinear convergence. Finally, when the normalized residual is decreasing sufficiently, that is when

$$\tilde{r}^k \leqq \frac{1}{2}\tilde{r}^{k-1}, \tag{48}$$

we may assume that the updating procedure is doing a good job and avoid estimation.

We also learned (Toint (1978)) that estimation is not desirable in regions where the function f(.) is not convex. To take this remark into account, we may easily check for non-convex behaviour of f(.) along the last step. This can be done by testing the inequalities

$$f(\underline{x}^{k+1})-f(\underline{x}^k)-\underline{g}(\underline{x}^k)^T(\underline{x}^{k+1}-\underline{x}^k) > 0 \tag{49}$$

and

$$f(\underline{x}^{k+1})-f(\underline{x}^k)-\underline{g}(\underline{x}^{k+1})^T(\underline{x}^{k+1}-\underline{x}^k) > 0. \tag{50}$$

However, we certainly do wish to start the algorithm with a correctly scaled Hessian approximation, so an estimation of $D^2f(\underline{x}^0)$ seems appropriate.

We now turn to the motivation of the proposed new step strategy.

In step 1, we mentioned that the new point $\underline{x}^k+\underline{s}^k$ should actually decrease the value of F(.). Clearly, if B^k is positive definite and if the minimum of (44) is in the confidence region (43), the new point should be identical to this minimum, as is usual in most Quasi-Newton methods. However, this requires the complete solution of the linear system (1). Furthermore, we have no guarantee, when we start the linear system solver, that the above two conditions are satisfied. We may indeed find that B^k is indefinite or that the minimum is not in our confidence region (43). In those cases, we wish to stop the solution of (1) as soon as these events are detected and nevertheless obtain a suitable step from the calculation that has been performed. One has also to consider that the linear system involved is symmetric and sparse and that its right hand side is our only definite information about the behaviour of the objective function.

We think that a suitable way of taking these remarks into account is provided by the following procedure: we apply a conjugate gradient algorithm to (44) starting from $\underline{x}^k$, and stop it when its inner iteration points leave the ball (43). This has the advantage of first minimizing F(.) along the gradient and then defining a path that will possibly lead us to the minimum of our quadratic approximation, if this minimum exists and is in (43). Hence the full linear system solution is only performed when useful and the method otherwise stops at an acceptable point, that lies somewhere in (43) below the minimum of F(.) along the gradient. Moreover, conjugate gradient procedures are of special interest when the dimension is large, since we then expect to perform much less than the theoretical n iterations.

One also has to decide what to do if a negative curvature search direction is encountered in the conjugate gradient iterations. This may happen because the objective function truly has negative curvature, but can also happen when the updates to the matrix B^k have produced an indefinite matrix, despite the actual positive definiteness of $D^2f(\underline{x}^k)$. In this last case, taking such a direction into account may produce a very poor step $\underline{s}^k$. As a consequence, we decide to avoid those problems and to set $\underline{s}^k$ to the minimum of F(.) in the ball (43) along $-\underline{g}(\underline{x}^k)$ when a direction of non-positive curvature is encountered. If f(.) has the same curvature, we do not degrade our step strategy much by doing so. This safeguarded procedure will be called "help with gradient" in the tests that are presented below.

Another important decision about our algorithm is to fix whether or not a line search procedure should be incorporated in step 2. As shown by the tests below, this procedure seems desirable, because it provides a safeguard against bad approximations to $D^2f(\underline{x})$ that may occur if the behaviour of the function is changing very quickly as the iterations proceed.

When combining these ideas on Hessian revision and step strategy, we do, in practice, take some supplementary precautions. If the linear search process has been non trivial (that is $\underline{x}^k+\underline{s}^k$ was not acceptable) and if it has produced another trial point in less than 4 function evaluations, we suspect that the behaviour of the objective function differs from that of our quadratic approximation, but that the updating procedure is still satisfactory. We then do not force estimation of the Hessian. On the other hand, if the line search procedure requires more function evaluations, then it is probably because the quadratic approximation is extremely poor, and in this case we do estimate the second derivatives. Following the same idea, we may also suspect the updating procedure of producing a poor approximation when no estimation has occurred in the last few iterations and no Quasi-Newton step has been taken. In our implementation, estimation is therefore imposed when the last estimation is older than $3n_{upd}$ iterations and no Quasi-Newton step has been computed in the last $2n_{upd}$ iterations.

We end these general considerations about the algorithm's structure by discussing a substantial numerical improvement that can be used in connection with the conjugate gradient procedure of step 1, viz. the use of a preconditioning technique. It is now well known that such techniques do shorten the computing time of general conjugate gradient routines considerably by clustering the eigenvalues of the problem.

There are two possibilities of incorporating a preconditioning technique into our general algorithm. The first one is obviously the preconditioning of $F(\underline{x})$ in step 1. We may replace (44) by the new quadratic

$$\hat{F}(\underline{w}) = f(\underline{x}^k)+(K\underline{g}(\underline{x}^k))^T(\underline{w}-\underline{x}^k)+\frac{1}{2}(\underline{w}-\underline{x}^k)^T(KB^kK^T)(\underline{w}-\underline{x}^k) \qquad (51)$$

and we are now interested in steps of the form

$$\underline{s}^k = (K^T)^{-1}(\underline{w}-\underline{x}^k), \qquad (52)$$

where we choose K non-singular and such that KB^kK^T has a smaller condition number than B^k itself. However, the quadratic (51) has no longer its gradient identical to the gradient of the objective function, as is the case with (44). This is troublesome because we do want to take this gradient information specifically into account. We therefore prefer to consider $\hat{F}(.)$ instead of F(.) only after the minimization along $-\underline{g}(\underline{x}^k)$ has been performed. This procedure is theoretically supported by some requirements on the step strategy that are made in Powell (1975).

The second possibility of applying a preconditioning technique in our general algorithm is for the solution of (16) when the updating procedure is used for the Hessian revision. Indeed, this system may be efficiently solved by a conjugate gradient procedure because we know it is symmetric and positive definite, and also because this allows us to avoid the explicit storage of the matrix Q where the products involved can be computed directly from (17). This is an important question, since we have already observed that the cost of solving (16) is critical in the decision between estimation and updating.

In the numerical tests presented below, K was chosen at iteration k as

$$K = \text{diag}[|B^k_{i,i}|^{-0.5}]. \tag{53}$$

It is a very simple preconditioning indeed, setting all the diagonal elements of KB^kK^T to +1 or -1. It has the further advantage of being symmetric, and of being cheap to apply and to store. However, the tests below show that, despite its simplicity, it can be very efficient. The preconditioning that is used in the solution of (16) is similar, except that we first normalize the right hand side.

6. A SPARSE TEST FUNCTION FOR MINIMIZATION

In order to evaluate the performance of a minimization algorithm that uses sparsity, we need some example functions on which the behaviour of the procedures can be illustrated and compared. Some suitable test functions have already appeared in the literature (see Toint (1978) or Shanno (1978) for example), but they may be difficult to use for a comparison because they sometimes have a number of attractive local minima, or because they are generalizations to higher dimensions of wildly non-linear functions and their minimization becomes, as the dimension grows, more and more time consuming. It is interesting to observe that, in Shanno (1978), the tests use such functions but the "largest" number of variables that is considered is 30!

We now introduce a class of test functions that can be used with arbitrary sparsity patterns, any dimension larger than 2, and whose conditioning at the solution can be chosen. This class is described by the relations

$$f(\underline{x}) = q(\underline{x}) + \sum_{\substack{(i,j)\notin I \\ i<j}} l_{i,j}(x_i, x_j) \tag{54}$$

where q(.) is the quadratic function defined by

$$[Q(\underline{z})]_{i,j} = z(i)_j z(j)_i + \|\underline{z}(i)\|^2 \delta_{i,j}, \qquad (i,j=1,\ldots,n) \tag{55}$$

$$q(\underline{x}) = \underline{x}^T Q(\underline{z}) \underline{x} \tag{56}$$

with the vectors $\underline{z}(i)$ (i=1,...,n) defined from a given non-zero vector $\underline{z}$ as in (14), and where $l_{i,j}(x_i, x_j)$ is a linking function between variables x_i and x_j. The sparsity pattern of $D^2f(.)$ is specified by the set I that contains, as before, the pairs of indices of zero elements in this matrix.

The general design of this class of test problems follows the idea that in many applications there is a quite simple overall objective that involves all the variables and a number of possibly more complicated links between them.

From the practical point of view, we now force (54) to have zero as a minimum, and also force its Hessian at this solution to be the Hessian of the quadratic q(.). The first of these constraints will make possible a detailed analysis of the asymptotic behaviour of the algorithm when it approaches the solution, and the second one will allow us to modify the conditioning of the problem at the optimum by simply modifying the matrix Q(z), that is the vector $\underline{z}$ itself. A way to obtain these properties is to force all the linking functions $l_{i,j}(.,.)$ to have a zero gradient and zero second derivative matrix at the origin.

In the tests that are presented below, we used the following linking functions for all (i,j) not in I

$$l(x,y) = 5[y^2-1+\cos(x)]^2 + x^4/(1+x^4) \tag{57}$$

and

$$l(x,y) = [y^3-3xy+x^3]^2 + x^4/(1+x^4). \tag{58}$$

More precisely, we will refer in the sequel to the functions B1, B2, B3, B4, R1, R2, R3 and R4 that are defined in the following way:

- the functions B1, B2, B3 and B4 use a sparsity pattern that is tridiagonal (this information is contained in the set I, given a priori).
- the functions R1, R2, R3 and R4 use a randomly generated sparsity pattern containing 5 non-zero elements per column on average.
- the functions B1, B3, R1 and R3 use in (55) a vector $\underline{z}$ whose i-th component is equal to unity.
- the functions B2, B4, R2 and R4 use instead a vector $\underline{z}$ whose i-th component is given by

$$z_i = 1+100i/n.$$

- the functions B1, B2, R1 and R2 use in (54) the linking function (57) while functions B3, B4, R3 and R4 use (58).

We may finally mention that, although we are not sure that zero is the only minimum of (54)-(56) with (57) or (58), we never found another one in the numerical tests.

7. SOME NUMERICAL TESTS AND COMMENTS

We now present some of the numerical tests that were performed with sparse unconstrained optimization which led the author to the considerations of section 5. All the results presented here were obtained on the DEC 2060 machine of the Facultes ND de la Paix in Namur (Belguim) (wordlength of 36 bits) using double precision arithmetic. The stopping criterion for the procedures considered was to ask for the gradient norm to be smaller than or equal to 10^{-6} and the starting point for the functions B1,....,R4 was chosen as

$$[\underline{x}^0]^T = (2,2,2,2,\ldots,2). \tag{59}$$

except if stated otherwise.

But, before presenting our results, we still need to introduce the practical algorithms that were tested. They will be referred to as A1 to A6 and feature the different methods discussed in section 5. They are written explicitly in Table 7.1.

TABLE 7.1

Features of the algorithms considered

	-- HESSIAN REVISION ---			-- STEP STRATEGY --		
	corr.	mix.	help with gradient	line	prec.	Eps
A1	none	no	no	no	no	no
A2	BFGS	no	no	no	no	no
A3	BFGS	no	no	yes	no	no
A4	BFGS	yes	no	yes	no	no
A5	BFGS	yes	yes	yes	yes	no
A6	BFGS	yes	yes	yes	yes	yes

In this table, the heading "corr." refers to equation (24): "none" means, in this context, that the update (22) is used while "BFGS" means that (24)-(25) are used instead. The heading "mix." concerns the possible estimation of the Hessian during the iterations, as discussed in Section 5. The heading "prec." refers to the use of the preconditioned quadratic (51), "line" refers to the presence of a line search procedure. Finally, the heading "Eps" indicates if the required accuracy in the solution of the system (1) is kept fixed (no) or if this accuracy increases as the norm of the gradient decreases (yes).

The first thing we want to test is the relative efficiency of the updates (22) and (24)-(25). As we said before, this should include comparisons of all the possible strategies but we simply compare the performance of algorithms A1 and A2 on the function B1. Further tests have been performed with essentially the same conclusion, so we only present this rather typical situation in Table 7.2[3].

This table presents the number of gradient evaluations needed for convergence. One can observe the superiority of A2 and A1, especially for dimensions in the range from 15 to 100. As a consequence, the updating procedure (24)-(25) is chosen for the subsequently tested algorithms.

---

[3]Starting point = (1,1,1,1,...,1)

TABLE 7.2

Gradient evaluations for A1, A2, A3 and A4 on B1

n	5	10	15	20	40	60	80	100	150
A1	22	29	58	57	58	96	111	>160	76
A2	39	32	38	32	38	44	37	39	48
A3	25	25	22	24	34	34	37	38	43
A4	16	16	17	23	16	15	16	19	15

The second interesting question we have to examine is the usefulness of a line search procedure. The related test is between A2 and A3, which is identical to A2 but features a cubic interpolation line search. The test is made with function B1 again and also appears in Table 7.2. Once more, this behaviour is typical and since a line search never deteriorated the performance and often enhanced it, we decided to incorporate this procedure in our general algorithm.

We now wish to admit Hessian estimations (by the more economical lower triangular substitution method) in order to reduce the overall number of gradients that have to be computed for the algorithm to reach the minimum within the prescribed accuracy. We expect from Toint (1978) that a reduction will occur, and this is confirmed by the results of A4 on function B1, appearing on the last line of Table 7.2. Algorithm A4 is the implementation discussed in section 5 mixing updates and estimations, and is therefore more complex than algorithm UNA of Toint (1978) in this respect. However, we observed that A4 did not converge in a reasonable number of gradient computations when applied to functions R1 and R3 for dimensions exceeding 10. Furthermore, the computing time that it needed to minimize the badly scaled functions B2 and B4 for large n was rather large, since the number of conjugate gradient iterations to compute a single Quasi-Newton step rose proportionally to the number of variables. These observations led the author to build a new method, namely A5, that incorporates a "help with gradient" procedure and preconditioned step strategy. Another method (A6) was also designed, that features the possibility of performing incomplete Quasi-Newton steps when far away from a solution by letting the accuracy that is prescribed for the solution of (1) depend on the norm of the gradient at the current point.

These two algorithms (A5 and A6) were then more thoroughly tested and compared with A4. As hoped, their behaviour was significantly better than the behaviour of A4, as is shown in Table 7.3. This table presents the average number of gradient evaluations that were needed by the different algorithms to converge to within the prescribed accuracy, these average numbers being computed for the eight functions B1 to R4. It can be seen in Table 7.3 that we solved with A5 and A6 the important problem of the convergence of A4. This is essentially due to the introduction in both algorithms of the "help with gradient" process that prevents the method from taking bad quadratic approximations too much into account. The difference of efficiency between A5 and A6 is clearly less considerable than that between A4 and A5. Nevertheless, the total of the average numbers for B1 to R4 shows a slight advantage for A6 over A5, probably due to more efficient use of the available information about the objective function. The behaviour of A5 and A6 with respect to the number of gradient evaluations is illustrated in Figure 7.1, where the advantage of A6 clearly appears. Triangles appear at the values corresponding to A5 and tilted squares appear at those corresponding to A6.

TABLE 7.3

Average evaluations for A4, A5 and A6

n	5	10	15	20	40	60	80	100	150	200
A4										
B1-B4	17	19	20	19	22	27	24	25	no conv. on B4	
R1-R4	17	no convergence on R1 and R3								
Total	17	....								
A5										
B1-B4	16	18	18	20	22	25	26	27	26	25
R1-R4	15	23	26	26	34	33	34	35	42	43
Total	16	21	22	23	28	29	30	31	34	34
A6										
B1-B4	17	18	20	20	22	24	25	27	27	27
R1-R4	17	22	25	25	28	21	28	27	33	32
Total	17	20	23	23	25	22	27	27	30	30

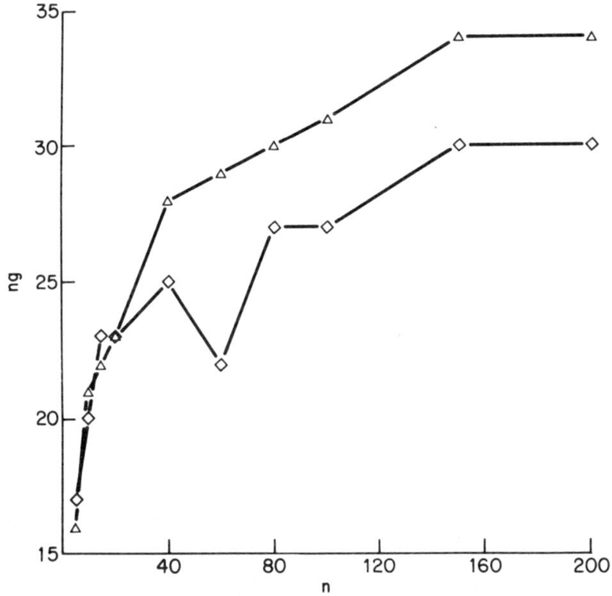

Figure 7.1 Average gradient evaluations for A5 and A6 (ng) versus the dimension (n).

Concerning the problem of computation time of A4 on large dimensional badly scaled functions (the condition number of B2 and B4 at the solution is of the order of 10^4), the improvement that is introduced by the preconditioning technique in the step strategy is even more spectacular. For example, the minimization of B2 with 200 variables using A4 required 123.55 iterations (of the conjugate gradient process that defines $\underline{s}^k$) per gradient evaluation, while A5 only required 7.25 iterations and A6 3.08 iterations. This radically different behaviour of A5 and A6 compared to A4, as far as the number of conjugate gradient iterations is concerned, is well illustrated in Figure 7.2 which shows the number of these internal iterations per gradient evaluation for function B2: triangles refer to A4 and tilted squares to A6. When we consider the comparison between A5 and A6, we observe in Table 7.4 and Figure 7.3 that A6 has a slightly better performance in terms of work in internal iterations. This table features the average number of internal conjugate gradient iterations taken on the eight functions B1 to R4 for n between 5 and 200. Figure 7.2 only shows the numbers corresponding to the rows of Table 7.4 that are labelled "Total": triangles mark the curve of A5 while tilted squares mark that of A6.

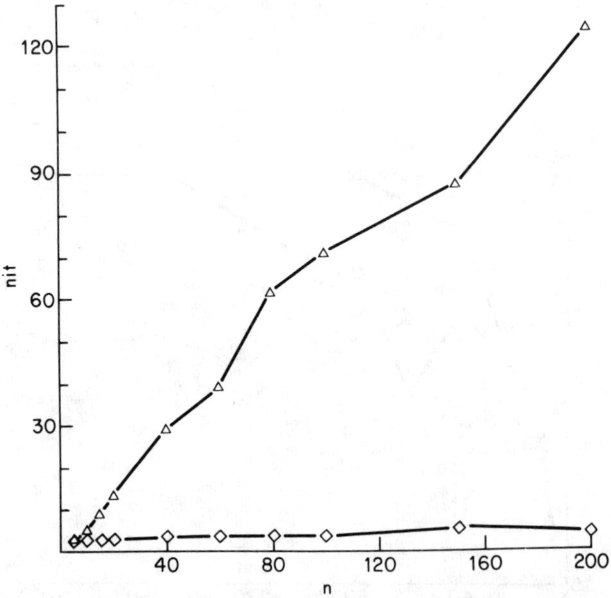

Figure 7.2 CG iterations per gradient on B2 for A4 and A6 (nit) versus the dimension (n).

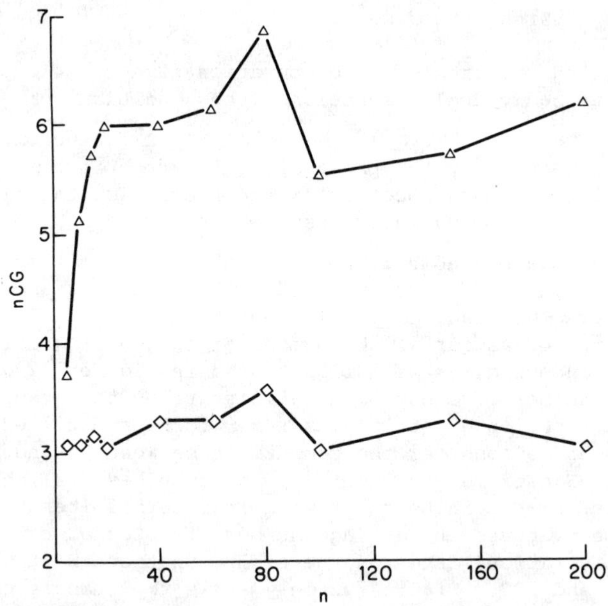

Figure 7.3 Average number of CG iterations per gradient (nCG) versus the dimension (n).

TABLE 7.4

Average number of CG iterations per gradient

n	5	10	15	20	40	60	80	100	150	200
A5										
B1-B4	3.8	5.5	6.8	6.9	7.4	7.6	7.6	6.5	6.8	7.8
R1-R4	3.7	4.8	4.7	5.0	4.6	4.7	6.2	4.6	4.8	4.6
Total	3.8	5.1	5.8	6.0	6.0	6.1	6.9	5.6	5.8	6.2
A6										
B1-B4	3.2	3.3	3.5	3.4	3.9	3.6	3.8	3.5	4.0	3.6
R1-R4	2.9	2.8	2.8	2.7	2.7	3.0	3.3	2.6	2.5	2.5
Total	3.0	3.0	3.1	3.0	3.3	3.3	3.6	3.0	3.3	3.0

We end this section by mentioning the reason why we did not compare the different algorithms with respect to the number of internal conjugate gradient iterations that are needed to solve (16) when an updating of B^k is performed. Because it was easier to take care of scaling at this point, a suitable diagonal preconditioning technique has been used for all the methods tested. It is remarkable that the number of iterations per update was extremely constant: for n greater or equal to 20, this number was nearly always between 15 and 22, regardless of the function, dimension or algorithm considered.

8. CONCLUSION

While we do not pretend to draw definitive conclusions about the design of an efficient sparsity exploiting Quasi-Newton algorithm, we think that we can learn something useful from the experiments and considerations made above.

The first important idea is that procedures that take sparsity and symmetry into account may be used with great success for estimating second derivative matrices in minimization algorithms. Whether or not they should be supplemented with updating methods depends on the relative costs of gradient evaluation and solution of a sparse symmetric and positive definite linear system of equations. Preconditioning techniques that reduce the condition number of that system may play an important role.

The second point that is worth noting is that general algorithms for the unconstrained minimization problem with a large number of variables require some safeguarded step strategy that will enforce convergence. Even if theoretical convergence is assured, it might indeed be very slow due to a poor Hessian approximation. This safeguard may be provided by returning to a simple gradient step when difficulties occur.

We also have to mention the important efficiency gain that can be made by using preconditioning not only in the solution of the linear system involved in the updates, but also in the step strategy itself. Other preconditionings for the approximate quadratic may prove to be quite valuable for the general behaviour of the algorithm. We have in mind, for example, preconditionings that are derived from the product forms of the well known dense Quasi-Newton updates applied to the identity matrix (see Davidon (1975) and Brodlie et al (1973) for such forms).

Finally, we wish to mention that new sparse updating procedures, like that of Toint (1980b), should be searched for and experimented on in the unconstrained optimization context. New sparsity exploiting estimation procedures could also contribute very much to the development of satisfactory general algorithms.

REFERENCES

Brodlie, K.W., Gourlay, A.R. and Greenstadt, J. (1973) Rank-one and rank-two corrections to positive definite matrices expressed in product form, J. Inst. Maths. Applics., 11, pp. 73-82.

Curtis, A.R., Powell, M.J.D. and Reid, J.K. (1974) On the estimation of sparse Jacobian matrices, J. Inst. Maths. Applics., 13, pp. 117-120.

Davidon, W.C. (1975) Optimally conditioned optimization algorithms without line searches, Math. Prog., 9, pp. 1-30.

Dennis, J.E. and Moré, J.J. (1974) A characterization of superlinear convergence and its application to quasi-Newton methods, Math. Comp., 28, pp. 549-560.

Dennis, J.E. and Moré, J.J. (1977) Quasi-Newton methods, motivation and theory, SIAM Rev., 19, pp. 46-89.

Fivet, E. (1980) Estimation de jacobien et hessien creux, Master's thesis, FUN Namur, Belgium.

Greenstadt, J. (1970) Variations upon variable metric methods, Math. Comp., 24, pp. 1-22.

Marwil, E. (1978) Exploiting sparsity in Newton-like methods, PhD thesis, Cornell University, Ithaca, New York.

Powell, M.J.D. and Toint, Ph. L. (1979) On the estimation of sparse Hessian matrices, SIAM J. Numer. Anal., 16, pp. 1060-1074.

Powell, M.J.D. (1970) A Fortran subroutine for unconstrained minimization requiring first derivatives of the objective func-function, Technical Report R-6469, AERE, Harwell, England.

Powell, M.J.D. (1975) Convergence properties of a class of minimization algorithms, In Nonlinear Programming 2, (O.L. Mangasarian, Meyer, R.R. and Robinson, S.M. eds), Academic Press, pp. 1-27.

Schubert, L.K. (1970) Modification of a quasi-Newton method for nonlinear equations with a sparse Jacobian, Math. Comp., 24, pp. 27-30.

Shanno, D.F. (1978) On variable metric methods for sparse Hessians, Technical Report 26, MIS, University of Arizona, Tucson.

Toint, Ph. L (1977) On sparse and symmetric matrix updating subject to a linear equation, Math. Comp., 31, pp. 954-961.

Toint, Ph. L. (1978) Some numerical results using a sparse matrix updating formula in unconstrained optimization, Math. Comp., 32, pp. 839-851.

Toint, Ph. L. (1980a) A note on sparsity exploiting quasi-Newton methods, Math. Prog., in press.

Toint Ph. L. (1980b) A sparse quasi-Newton update derived variationally with a non diagonally weighted Frobenius norm, Technical Report 80/1, Dept. of Maths., FUN Namur, Belgium.

DISCUSSION

DR. I.S. DUFF (Harwell). Some quasi-Newton methods force the approximation to the Hessian to be positive definite. Can you find updates to do this which at the same time maintain sparsity?

TOINT. It has been proved that such updates exist in all practically meaningful cases, but we know of no computational implementation of such a method that is competitive with the non positive definite sparse formulae.
However, what is important in the full case, when using a positive definite update such as BFGS, is that scaling invariance is obtained. With respect to scaling invariance, I now believe

a better class of algorithms can be obtained than those I described in my talk.

DR. I.S. DUFF (Harwell). Could you tell us how effective the substitution method for evaluating the Hessian is in practice in terms of gradient evaluations on typical structures?

TOINT. On band matrices with bandwidth $(2\ell + 1)$, we need only $\ell + 1$ gradient vectors while on random structures, with p non-zeros per column on average, the number of gradients that are needed is usually between p and p + 2, depending on the structure.

SPARSITY AND PIECEWISE LINEARITY IN LARGE PORTFOLIO OPTIMIZATION PROBLEMS

H.M. Markowitz

(IBM T.J. Watson Research Center, Yorktown Heights, N.Y.)

and

A.F. Perold

(Graduate School of Business Administration, Harvard University, Boston, MA.)

ABSTRACT

The general mean-variance portfolio selection problem is usually formulated as a parametric quadratic program that has a dense Hessian, viz. the covariance matrix of security returns. In applications with large numbers of securities, the covariances are estimated with the use of return generating models that require the estimation of far fewer coefficients. This paper discusses the exploitation of this structure using the technique of defining additional variables and constraints in order to transform the problem into one that is sparse. Then it is shown how piecewise linear transactions costs and upper and lower bounds can be implemented implicitly and in a unified way. This approach avoids the need to define separate variables for each piece of linearity, as is currently practised.

1. INTRODUCTION

The general portfolio optimization problem is the following: Given a universe of n securities (almost all of them risky) find the proportion of wealth to be invested in each so as to yield a portfolio with minimal risk for a given expected return. The problem was first posed as such by Markowitz (1959) and may be stated as the quadratic program

$$\begin{aligned} \text{minimize:}\quad & V = \underline{x}^T C \underline{x} \\ \text{subject to:}\quad & \underline{\mu}^T \underline{x} \geq E \\ & A\underline{x} = \underline{b} \\ & \underline{x} \geq \underline{0} \end{aligned} \tag{1}$$

where C is the n x n positive (semi) definite variance-covariance matrix of the security returns, $\underline{\mu}$ is the n-vector of expected returns, and the pair $A,\underline{b}$ constitutes a set of linear constraints. E and V are the mean and variance respectively of the portfolio's return.

The collection of all portfolios with minimal V for some E and maximal E for some V is called the efficient frontier (Figure 1). Each of these portfolios may also be characterized as being the solution of the quadratic program

$$\begin{aligned} \text{maximize:} \quad & \underline{\mu}^T\underline{x} - \lambda\, \underline{x}^T C\underline{x} \\ \text{subject to:} \quad & A\underline{x} = \underline{b};\ \underline{x} \ge \underline{0} \end{aligned} \tag{2}$$

for some $\lambda \ge 0$, where λ can be interpreted as the investor's tradeoff between risk and return. Rather than specifying E (or V) in advance, the investor will either specify λ and solve (2), or determine the whole frontier (solve (1) for each value of E) and then select a portfolio by directly comparing the mean-variance pairs. The latter is usually preferred in practice.

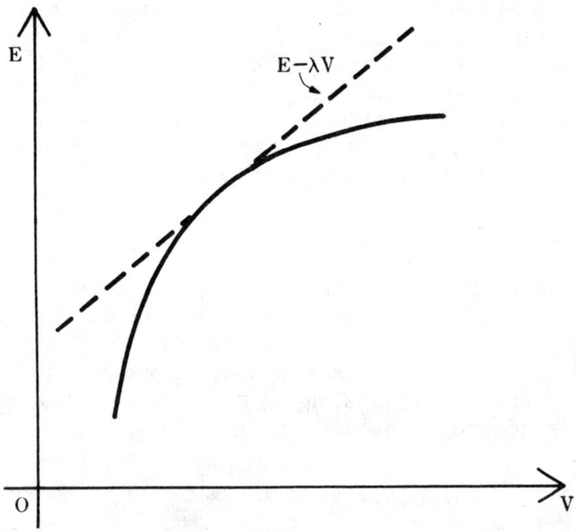

Figure 1

Often the investor already has a portfolio, say $\underline{z}$, and wishes to revise it because expectations of the future have changed. The optimization problem is the same as before except that due to transactions costs the expected return becomes

$$\underline{\mu}^T \underline{x} - \sum_{i=1}^{n} s_i (x_i - z_i)^- - \sum_{i=1}^{n} t_i (x_i - z_i)^+$$

where s_i and t_i respectively are the per unit costs of selling and purchasing stock in the ith security and where $(x)^+$ is the maximum of x and zero and $(x)^-$ is the maximum of -x and zero.

During the last decade, formal portfolio optimization of this kind has gained widespread acceptance by practising investment managers, primarily in the United States. This has created the need to develop techniques for large scale portfolio optimization, perhaps for as many as 1000 securities. Two problems are immediate: with 1000 securities, half a million covariances of returns have to be estimated. Even if this can be accomplished there is the question of economical computation and fast turn around.

To estimate the covariance matrix a number of simplifying models of security returns have been suggested that require the direct estimation of far fewer coefficients. While several optimization procedures exploiting this fact have been suggested, all either place further unnecessary restrictions on the class of problems that can be solved, or perform the calculations in a cumbersome manner.

In this paper we briefly review the models used to estimate covariances, and discuss the shortcomings of existing optimization methods. Then we show that probably the most effective method for solving these problems is an implementation of the original complementary pivot algorithm for quadratic programming due to Markowitz (1956) and Wolfe (1959), together with the idea of "sparsification", namely, reformulating the problem as one that is sparse by defining a few additional variables and constraints. This very simple approach has been used for single factor models (for example IBM (1969), Sharpe (1963)) and for multi-factor models (for example Sharpe (1970), but its greater generality and ease of implementation seem to have gone unnoticed. In the last part of the paper, we show how transactions costs can be easily implemented and extended to the more general convex piecewise linear case without the burdensome and currently practised procedure of splitting each variable into at least two others and so vastly increasing the size of the problem.

2. MODELS FOR THE ESTIMATION OF COVARIANCES

The most widely used model of security returns is the multi-factor model. This involves the selection of a few (say 10) common factors that impact the returns of all or a large group of securities simultaneously, and then making the assumption that any remaining variability in the return of a security is specific to that security only, and largely independent of events that impact other securities. The return on security i, R_i, may be written as

$$R_i = \mu_i + \beta_{i1}F_1 + \ . \ . \ . + \beta_{i\ell}F_\ell + \varepsilon_i$$

for some constants μ_i and β_{ik}, and random variables F_k and ε_i all with mean zero, the ε_i being uncorrelated with each other and the F_k. In this case the covariance matrix of the n securities may be written as

$$C = D + GPG^T \qquad (3)$$

where D is a diagonal matrix with i^{th} diagonal entry $\sigma_i^2 = \text{var}(\varepsilon_i)$, G is the n x ℓ matrix of coefficients β_{ik}, and P is the ℓ x ℓ covariance matrix of the factors (F_k). Often the factors are chosen to be uncorrelated, so that P is diagonal. The widely used model of Rosenberg (1974) has P of the same form as C.

A very different approach is a states of the world model as used for example by Hobman (1975) and recently generalized by Markowitz and Perold (1980). Here it is assumed that there are ℓ possible future states of the world, the sth occurring with probability p_s, s = 1,...,ℓ. Further, the return on security i in state s, R_{is}, is given by

$$R_{is} = \mu_{is} + \varepsilon_i$$

where the μ_{is} are constants and the ε_i are random variables satisfying $E(\varepsilon_i|s) = 0$, $E(\varepsilon_i^2|s) = \sigma_{is}^2$, and $E(\varepsilon_i\varepsilon_j|s) = 0$. The resulting covariance matrix has precisely the same form as in (3): D is a diagonal matrix with i^{th} diagonal entry $\sum_{s=1}^{\ell} p_s\sigma_{is}^2$,

G is the matrix of coefficients $(\mu_{is} - \sum_{t=1}^{\ell} p_t \mu_{it})$ and P is a diagonal matrix with s^{th} diagonal entry p_s.

These are perhaps the two most important models of security returns, and it is the form of the covariance matrix in (3) that will be exploited in the remainder of the paper. It is worth noting that this structure is also present in all other models that have been used to date in practice.

3. EXISTING OPTIMIZATION METHODS

One of the most widely used methods in current day practice is Von Hohenbalken's algorithm (Von Hohenbalken (1975), Rosenburg and Rudd (1976), and Wilshire Associates (1979)), an iterative method that makes successive approximations of the constraint set using simplices of increasing dimension. It very quickly gives a good <u>approximate</u> solution but is extremely slow in attaining the optimum exactly. Points on the efficient frontier are determined by solving (2) for several (for example, five) discrete values of λ. A similar method, due to Sharpe (1978), has the advantage of being extremely well suited for use out-of-core, even for an arbitrary covariance matrix, but can handle only upper and lower bounds on the variables, and a budget constraint ($\Sigma x_i = 1$). All other constraints have to be introduced with the use of penalty functions (fixed penalties) and, once again, only one point at a time can be determined on the efficient frontier.

There is essentially only one algorithm for computing the whole efficient frontier, namely the complementary pivot algorithm (Markowitz (1956), Wolfe (1959)) that solves (2) parametrically for all values of λ. (See section 4 for further elaboration on this point). Existing implementations of this method that take advantage of the covariance matrix structure in (3) can handle only very special cases: the codes of Williamson and Downs (1970) and IBM (1969) require G to be of rank one and can treat only a budget constraint. Pang (1980) requires C to be positive definite[1], and can also not treat general linear constraints. None of these can handle transactions costs without doubling the problem size.

Lastly, we mention the work of Elton et al (1976, 1977a, 1977b) who proposed simple ranking devices that can be used almost by hand. Like those in IBM (1969) or Williamson and

---

[1] With riskless assets and slack variables, C will always have rank less than n.

Downs (1970), however, these apply only to the case when G has rank 1.

4. REVIEW OF THE COMPLEMENTARY PIVOT ALGORITHM

Following Markowitz (1959, p. 338), the necessary and sufficient conditions for $\underline{x}$ to solve (1), for fixed E, are that there exist multipliers $\underline{\pi}$, $\underline{\eta}$, ρ which together with $\underline{x}$ satisfy

$$\begin{bmatrix} I & C & A^T \\ O & A & O \end{bmatrix} \begin{bmatrix} \underline{\eta} \\ \underline{x} \\ \underline{\pi} \end{bmatrix} = \begin{bmatrix} \underline{O} \\ \underline{b} \end{bmatrix} + \begin{bmatrix} \underline{\eta} \\ \underline{O} \end{bmatrix} \rho \qquad (4)$$

$$\eta_i x_i = 0, \; i = 1,\ldots,n, \quad \rho.(E - \underline{\mu}^T \underline{x}) = 0 \qquad (5)$$

$$\underline{x} \geq \underline{O}, \; \underline{\eta} \leq \underline{O}, \; \rho \geq 0. \qquad (6)$$

It is shown in Markowitz (1959) that the efficient frontier can be found by varying ρ over the interval $(0,\infty)$ with accompanying $\underline{x}(\rho)$, $\underline{\pi}(\rho)$, $\underline{\eta}(\rho)$ and $E(\rho)$ satisfying (4)-(6). Moreover, any solution to (4)-(6) for a given ρ is the solution of (2) with $\lambda = 1/\rho$; E and V are monotonically increasing as functions of ρ.

A typical iteration of the algorithm begins with a given value of ρ, say ρ^0, and a nonsingular basis

$$B = \begin{bmatrix} I & C_{\alpha\beta} & A^T_{\cdot\alpha} \\ O & C_{\beta\beta} & A^T_{\cdot\beta} \\ O & A_{\cdot\beta} & O \end{bmatrix},$$

a submatrix of the coefficient matrix in (4). The index sets α and β are disjoint and $\alpha \cup \beta = \{1,2,\ldots,n\}$. $\underline{x}$, $\underline{\pi}$ and ρ are determined by

$$B \begin{bmatrix} \underline{\eta}_\alpha \\ \underline{x}_\beta \\ \underline{\pi} \end{bmatrix} = \begin{bmatrix} \underline{O} \\ \underline{O} \\ \underline{b} \end{bmatrix} + \begin{bmatrix} \underline{\mu}_\alpha \\ \underline{\mu}_\beta \\ \underline{O} \end{bmatrix} \rho^0 \qquad (7)$$

$$\underline{x}_\alpha = \underline{O}, \; \underline{\eta}_\beta = \underline{O},$$

and it is assumed that

$$\underline{x}_\beta \geq \underline{O}, \ \underline{\eta}_\alpha \leq \underline{O},$$

i.e. relations (4)-(6) are satisfied.

The next step is to vary ρ, for example downwards if the procedure began with $\rho = +\infty$, while simultaneously adjusting $\underline{x}_\beta$, $\underline{\pi}$ and $\underline{\eta}_\alpha$ to satisfy (7). At some point, say $\rho = \rho^1 < \rho^0$, the further movement of ρ will cause a component of $\underline{x}_\beta$ or $\underline{\eta}_\alpha$ to violate its sign restriction. If this component is x_r, $r\varepsilon\beta$, then η_r is introduced into the basis in place of x_r, and the new basic solution is determined by the index sets

$$\beta_{new} = \beta_{old} \sim \{r\}, \ \alpha_{new} = \alpha_{old} \cup \{r\}.$$

If it is η_s, $s\varepsilon\alpha$, then x_s is introduced into the basis in place of η_s, the new basic solution then being determined by

$$\beta_{new} = \beta_{old} \cup \{s\}, \ \alpha_{new} = \alpha_{old} \sim \{s\}.$$

It is shown in Markowitz (1959) that this procedure yields a nonsingular basis at each step, that $\underline{x}$ and $\underline{\eta}$ remain feasible, and that ρ indeed varies over the full range $(0,\infty)$; the efficient frontier is given by the line segments between pairs of adjacent "corner" or basic portfolios. Note that this procedure may be used to define the efficient frontier, and therefore is the only algorithm to compute it, modulo getting started.

4.1 Getting started

There are essentially two methods for getting started. The most economical one is for the case occurring most often in practice, namely that $E=\underline{\mu}^T\underline{x}$ is bounded above on $\{\underline{x}: A\underline{x} = \underline{b}, \ \underline{x} \geq \underline{O}\}$. Here it is best to solve the linear program

$$\text{maximize} \quad \underline{\mu}^T\underline{x} : A\underline{x} = \underline{b}, \ \underline{x} \geq \underline{O}. \tag{8}$$

If (8) has a unique optimal solution (easily determined by inspection of the resulting vector of "reduced costs") this is also an efficient point, and a starting solution is immediately at hand for some sufficiently large ρ. Else, the minimum variance solution must be sought from among the optimal solutions to (8). This can be found using the same procedure as outlined above with the exception that in place of $\underline{\mu}$, an artificial $\bar{\underline{\mu}}$ is used for which the optimal solution obtained in (8) is efficient, and the variable corresponding to strictly positive reduced costs in (8) are deleted temporarily from the problem.

In the event that E is unbounded above, the more expensive procedure must be used of first determining the minimum variance solution with $\underline{\mu}^T\underline{x}$ unrestricted, using an artificial $\bar{\underline{\mu}}$ as indicated above, but now with all variables being considered. If this solution is unique (for example if C is positive definite) it is also efficient, and we have a starting solution with $\rho = 0$. If not, the starting solution must be sought from among those with minimum variance that yields the maximum E. This can be done by solving a linear program, again with certain variables temporarily omitted.

4.2 Basis changes

From the above description, it is clear that at each iteration, we require an explicit expression of the basic variables $(\underline{\eta}_\alpha, \underline{x}_\beta, \underline{\pi})$ in terms of ρ. By (7) we thus require two vectors $\underline{y}$ and $\underline{z}$ that satisfy

$$B\,\underline{y} = \begin{bmatrix} \underline{0} \\ \underline{0} \\ \underline{b} \end{bmatrix}, \quad B\,\underline{z} = \begin{bmatrix} \underline{\mu}_\alpha \\ \underline{\mu}_\beta \\ \underline{0} \end{bmatrix}, \tag{9}$$

for then

$$\begin{bmatrix} \underline{\eta}_\alpha \\ \underline{x}_\beta \\ \underline{\pi} \end{bmatrix} = \underline{y} + \underline{z}\rho. \tag{10}$$

From (10), the basis change can be determined by means of the usual minimum ratio test.

To perform the next iteration all that is required is the $\underline{y}$ and $\underline{z}$ satisfying (9) with respect to the new basis. Since the new basis differs from the old in a single column, these vectors can be updated by means of an elementary matrix as in the product form of the simplex method (Dantzig (1963)). This requires the solution of the set of equations

$$B\underline{w} = \underline{h} \tag{11}$$

where $\underline{h}$ is the column entering the basis (being the complement of the column leaving the basis). The new $\underline{y}$ and $\underline{z}$ are then given by

$$\underline{y}_{new} = W\underline{y}_{old}, \quad \underline{z}_{new} = W\underline{z}_{old}$$

where

$$W = I + \frac{1}{w_k}(\underline{w} - \underline{e}_k)\,\underline{e}_k^T,$$

k being such that the leaving column corresponds to the kth basic variable, and $\underline{e}_k$ being the kth unit vector.

We are left with the problem of finding a representation of B that allows (11) to be solved in an efficient and stable manner, and that can be easily updated from one iteration to the next.

5. EXPLOITING THE STRUCTURE OF C

We can now turn to the implementation of the above algorithm when the covariance matrix has the structure given in (3):

$$C = D + GPG^T.$$

To transform this into a sparse matrix problem notice that by defining new variables $\underline{u}$ and adding the constraints

$$G^T\underline{x} - \underline{u} = \underline{0},$$

the quadratic form in (1) becomes

$$\underline{x}^T C\underline{x} = \underline{x}^T D\underline{x} + \underline{u}^T P\underline{u}.$$

In the case of Rosenberg's model (Rosenberg (1974)), P, too, has the same form:

$$P = E + HFH^T$$

where E is diagonal and F is dense, symmetric positive definite. Proceeding as above, the quadratic form may now be written as

$$\underline{x}^T C\underline{x} = \underline{x}^T D\underline{x} + \underline{u}^T E\underline{u} + \underline{v}^T F\underline{v}$$

where in addition

$$H^T\underline{u} - \underline{v} = \underline{0}.$$

Problem (1) thus becomes

$$\begin{aligned} \text{minimize:}\quad & \underline{x}^T D \underline{x} + \underline{u}^T E \underline{u} + \underline{v}^T F \underline{v} \\ \text{subject to:}\quad & \underline{\mu}^T \underline{x} \geq E \\ & G^T \underline{x} - \underline{u} = \underline{0} \\ & H^T \underline{u} - \underline{v} = \underline{0} \\ & A\underline{x} \quad = \underline{b} \\ & \underline{x} \geq \underline{0}, \end{aligned} \tag{12}$$

where the first two terms in the minimand are weighted sums of squares. The resulting constraint matrix for the complementary pivot algorithm (cf (4)) is sparse, viz.

$$\left[\begin{array}{ccc|ccc} I & D & & & G & A^T \\ & E & & -I & H & \\ & & F & & -I & \\ \hline G^T & -I & & & & \\ & H^T & -I & & & \\ A & & & & & \end{array}\right] \tag{13}$$

In Rosenberg's model G is n x 47, H is 47 x 9, and F is 9 x 9. Furthermore, G is sparse, and H and F are dense. In a states of the world model, or a multi-factor model with uncorrelated factors, G is typically n x 10 and dense, and P (=E) is 10 x 10 and diagonal (that is,no need for F and H). A typical applications run might have n=500, and the constraint matrix A consisting of one or two dense rows (for example the budget constraint). Sometimes sparse sector constraints are also imposed, perhaps as many as 30. Upper and lower bounds are treated implicitly, as will be discussed in section 6.

In short, we have transformed a dense tableau with n + m rows, m being the number of explicit constraints, into one that is sparse with n + m + ℓ rows, where ℓ is the number of added variables. ℓ=56 in the case of the Rosenberg model, and will be the number of factors or states of the world when P is not further decomposable.

5.1 Use with a sparse matrix package

On a large computer, by far the easiest way to implement the reformulation (12) is to use the sparse matrix routines available as part of any large scale linear programming package. These commonly consist of an LU factorization routine, a routine to solve equations with respect to the LU factors, and a routine to update the LU factors after each change of basis.

Here, each basis has the form

$$B = \left[\begin{array}{cccc|ccc} I & & & & G_{\alpha.} & & A^T_{.\alpha} \\ & D_{\beta\beta} & & & G_{\beta.} & & A^T_{.\beta} \\ & & E & & -I & H & \\ & & & F & & -I & \\ \hline & G^T_{\beta.} & -I & & & & \\ & & H & -I & & & \\ & A_{.\beta} & & & & & \end{array}\right] \tag{14}$$

I II

Notice that the columns in section II of B always remain there. This is because both the dual variables on the equality constraints and the additional variables ($\underline{u},\underline{v}$) are unrestricted in sign, and can thus never "block". Any basis change is the addition of a column to β and the deletion of a unit column, or vice versa.

Whenever the basis change is the case of a unit column replacing a β column, the number of non-zeros in B decreases. This means that a factorization method placing as much weight as possible into U will be preferable, since during updating, the (denser) outgoing column in U will cause a net reduction in the total number of non-zeros in L and U. The LA$\emptyset$5 routines of Reid (1975), for example, have this property.

Besides ease of implementation, the use of general sparse matrix routines like those in Reid (1975) allows one to take full advantage of any sparsity present in G and A. As mentioned earlier, these can range from being completely dense to being rather sparse, depending on the particular constraints and model of covariance. The drawback of such an approach is that it fails to take into account the symmetry present in a large part of B and the fact that much of B (non-zero wise) never changes. We show next how exploiting this can save substantially on in-core storage, but makes implementation more complex.

5.2 Implementation out-of-core

The need to consider out-of-core implementations has arisen primarily because of the greatly increased use of in-house mini-computers. The following proposal requires a single read-only file in secondary storage, and in-core storage requirements of order at most $(m + \ell + r)^2$ where r (usually very small) is to be defined below.

We assume without loss of generality that F is diagonal[1]. Consider now the solution of (11), viz. $B\underline{w}=\underline{h}$. The most natural way to proceed is to perform Gaussian elimination along the diagonal of B in (14) up to the point where all the $\underline{\eta}_\alpha$, $\underline{x}_\beta$, $\underline{u}$ and $\underline{v}$ variables have been eliminated, or have been moved forward in a principal rearrangement because of zero or unstable diagonal pivots. Let the number of such skipped pivots be r. This leaves a "remaining matrix" in the elimination scheme that has the form

$$V = \begin{bmatrix} NN^T & M \\ M^T & Z \end{bmatrix}$$

and is of order $(m + \ell + r)$, where Z is the diagonal matrix of rejected pivots (usually Z will be zero) and (M,N) is a partition of

$$\begin{bmatrix} G^T_{\beta .} & -I & O \\ O & H & -I \\ A_{. \beta} & O & O \end{bmatrix}$$

with its columns suitably scaled. V will be treated in-core as a dense matrix.

Since, at each iteration, V may change in size up or down and is in addition modified by a matrix of rank one, we need to

[1] If F is not diagonal then diagonalize it beforehand and modify H appropriately.

choose a factorization for V that can be easily and stably updated, and yet is storage efficient. While the QR (Lawson and Hanson (1974)) and LU factorizations certainly qualify as far as updating is concerned, they require twice as much storage as a factorization exploiting the symmetry of V. A method that does accomplish the desired task is the symmetric indefinite factorization of Bunch and Parlett (1971):

$$V = L\tilde{D}L^T$$

where L is a permuted lower triangular matrix and $\tilde{D}$ has either 2 x 2 or 1 x 1 blocks on the diagonal, each 2 x 2 block having one positive and one negative eigenvalue. This can be updated from one iteration to the next, as indicated in Bunch and Kaufman (1977), with D requiring the work space of at most a five-diagonal matrix. In-core storage requirements for V would thus be limited to L and the five-diagonal matrix.

Note that the elimination prior to operating with V, and the back substitution involved thereafter can all be performed by having access to the diagonal entries of B and the columns of

$$\begin{bmatrix} G & -I & O \\ O & H & -I \\ A & O & O \end{bmatrix}$$

which can be stored on disk in packed form, say in a binary file, and then read repeatedly.

6 IMPLICIT TREATMENT OF UPPER AND LOWER BOUNDS AND TRANSACTIONS COSTS

As indicated in Section 1, the problem solved more commonly in practice is the portfolio revision problem where $\underline{x}$ is constrained to lie between upper and lower bounds, and a concave piecewise linear transactions cost is incurred for deviations of $\underline{x}$ from an existing portfolio. We show how both the bounds and transactions costs can be treated in a unified manner by working with the existing variables rather than by defining a separate variable for each piece of linearity.

In the following let the portfolio return be given by the separable function

$$f(\underline{x}) = \sum_{i=1}^{n} f_i(x_i)$$

where each $f_i(x_i)$ is concave and piecewise linear, and takes on the value $-\infty$ whenever x_i lies outside its bounds. For example, suppose that we currently have 10% of our portfolio in security number 3, and are willing to hold as little as 5% and no more than 20% in this security. If the current expected return on the security is 9% and brokerage commissions are incurred at a flat rate of 1% for both buying and selling, then $f_3(x_3)$ is given by

$$\begin{aligned} f_3(x_3) &= -\infty & &\text{if} \quad x_3 < .05 \\ &= .09x_3 - .01\,|x_3 - .1| & &\text{if} \quad .05 \le x_3 \le .20 \\ &= -\infty & &\text{if} \quad x_3 > .20 \end{aligned}$$

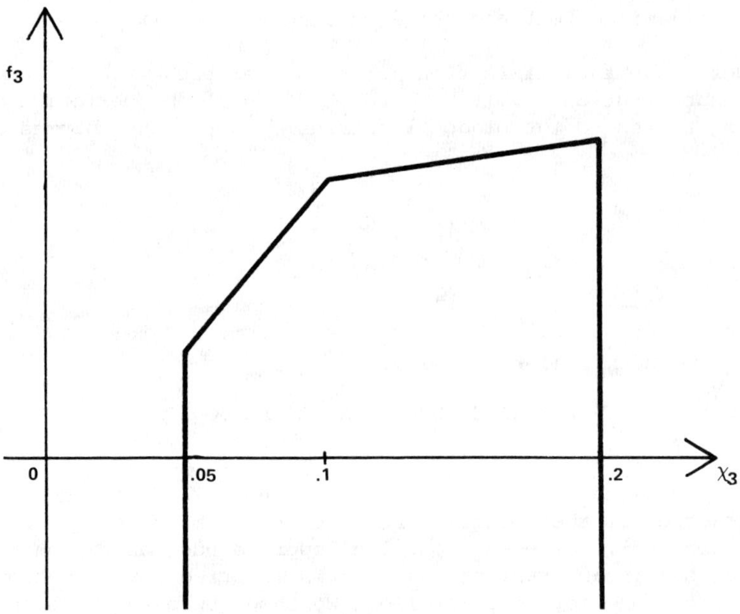

Figure 2

We allow the upper and lower bounds to take infinite values, or be equal.

The original problem (1) then becomes

$$\text{minimize}: \quad V = \underline{x}^T C \underline{x}$$
$$\text{subject to:} \quad f(\underline{x}) \geq E$$
$$A\underline{x} = \underline{b},$$

where now the inequality constraints on $\underline{x}$ have been taken into account by f. The optimality conditions for this problem are that there exist $\underline{x}$ and multipliers $\underline{\xi}$, $\underline{\pi}$, $\rho(\rho \geq 0)$ satisfying

$$C\underline{x} + A^T\underline{\pi} - \underline{\xi} = \underline{0}$$
$$A\underline{x} = \underline{b}$$

and

$$\underline{\xi} \in \rho.\underline{\partial f}(\underline{x}), \; \rho.(E - f(\underline{x})) = 0 \tag{15}$$

where the set $\underline{\partial f}(\underline{x})$ is the subgradient of f at $\underline{x}$. See for example, Rockafellar (1970).

It is easily checked that the earlier conditions (4)-(6) for the case of nonnegativity restrictions and no transactions costs are a special case of the above: For example, when $x_i > 0$, $\partial f_i(x_i) = \{\mu_i\}$ so that ξ_i becomes $\rho\mu_i$. This is equivalent to having, in (4), a right hand side value equal to $\rho\mu_i$, and $\eta_i = 0$. When x_i is at its lower bound, $\partial f(x_i) = [\mu_i, \infty)$ so that ξ_i lies in the interval $[\rho\mu_i, \infty)$. This is equivalent to having a right hand side value equal to $\rho\mu_i$, and $\eta_i \leq 0$. Sargent's extension (Sargent (1978)) to include upper bounds can also be immediately derived from (15).

More generally, it can be shown that the basic solutions to (15) have the following characterization: Each x_i is either out of the basis, $i\epsilon\alpha$, at one of the breakpoints defined by its piecewise linear return function f_i, or is in the basis, $i\epsilon\beta$, at some intermediate value. For each $i\epsilon\beta$, ξ_i is out of the basis, and assumes the value $\rho\nu_i$ where ν_i is the slope of $f_i(x_i)$ on the segment currently containing x_i. For each $i\epsilon\alpha$, ξ_i is in the basis and free to vary in the range $[\rho\nu'_i, \rho\nu''_i]$ where ν'_i and ν''_i are the slopes of f_i to either side of the breakpoint x_i.

Implementation of the general case is surprisingly straightforward if the slopes and breakpoints are stored consecutively in a packed array, and a basis status vector is maintained, indicating at which breakpoint or in which segment each x_i lies. The only other modification required is in the minimum ratio test (determining the leaving column) since the intervals of feasibility for ξ_i, namely $[\rho\nu'_i, \rho\nu''_i]$, are now functions of the parameter ρ. The change is simply to compute two ratios for ξ_i, one involving ν'_i and the other involving ν''_i.

7. CONCLUSIONS AND SUMMARY

This paper has proposed efficient solution techniques for a class of portfolio optimization problems that are being solved routinely by practising investment managers. The first recommendation was to sparsify the covariance matrix by defining a few additional variables and constraints. Any profundity in this suggestion lies perhaps in the fact that it is extremely simple to carry out, and that it appears to be the most effective alternative. Then we showed how to treat bounds and piecewise linear transactions costs implicitly and in a unified manner, without the need to more than double the size of the problem.

Both suggestions have been implemented on the DEC 10 at the Harvard Business School and found to work extremely well. However, direct and meaningful comparisons with other portfolio optimizers have yet to be carried out.

REFERENCES

Bunch, J.R. and Parlett, B.N. (1971) Direct methods for solving symmetric indefinite systems of linear equations, SIAM J. Numer. Anal., 8, pp. 639-655.

Bunch, J.R. and Kaufman, L. (1977) Indefinite quadratic programming, Technical Report No. 61, Dept. Computing Science, Bell Laboratories, Murray Hill, New Jersey.

Dantzig, G.B. (1963) Linear Programming and Extension, Princeton University Press.

Elton, E.J., Gruber, M.J. and Padberg, M.W. (1976) Simple criteria for optimal portfolio selection, J. Finance, 31, pp. 1341-1357.

Elton, E.J., Gruber, M.J. and Padberg, M.W. (1977a) Simple rules for optimal portfolio selection: the multi-group case, J. Financial and Quant. Anal., 12, pp. 329-344.

Elton, E.J., Gruber, M.J. and Padberg, M.W. (1977b) Simple criteria for optimal portfolio selection with upper bounds, Op. Res., 25, pp. 952-967.

Hobman, R.J. (1975) Setting investment policy in an ERISA environment, J. Portfolio Management, pp. 17-21.

IBM (1969) 1401 Portfolio Selection Program (1401-F1-04X) Program Reference Manual, IBM, New York.

Lawson, C.L. and Hanson, R.J. (1974) Solving Least Squares Problems, Prentice Hall.

Markowitz, H.M. (1956) The optimization of a quadratic function subject to linear constraints, Naval Res. Logist. Quart., 3, pp. 111-133.

Markowitz, H.M. (1959) Portfolio Selection, Efficient Diversification of Investments, Cowles Foundation Monograph 16, Yale University Press.

Markowitz, H.M. and Perold, A.F. (1980) Portfolio analysis with factors and scenarios, Working Paper HBS 80-47 Graduate School of Business Administration, Harvard University.

Pang, J.-S. (1980) A new and efficient algorithm for a class of portfolio selection problems, Ops. Res., 28, pp. 754-767.

Reid, J.K. (1975) Fortran subroutines for handling sparse linear programming bases, Report CSS 20, Computer Science and Systems Division, AERE Harwell, Oxon, England.

Rockafellar, R.T. (1970) Convex Analysis, Princeton University Press.

Rosenberg, B. (1974) Extra-market components of covariance in security returns, J. Financial and Quant. Anal., 9, pp. 263-274.

Rosenberg, B. and Rudd, A. (1976) Portfolio optimization algorithms: a progress report, Working Paper 42, Research Program in Finance, Graduate School of Business Administration, University of California, Berkeley.

Sargent, R.W.H. (1978) An efficient implementation of Lemke's algorithm and its extension to deal with upper and lower bounds, Math. Prog. Study, 7 pp. 36-54.

Sharpe, W.F. (1963) A simplified model for portfolio analysis, Management Sci., 9, pp. 277-293.

Sharpe, W.F. (1970) Portfolio Theory and Capital Markets, McGraw-Hill.

Sharpe, W.F. (1978) An algorithm for portfolio improvement, Research Paper No. 475, Graduate School of Business, Stanford University.

Von Hohenbalken, B. (1975) A finite algorithm to maximize certain pseudo-concave functions on polytopes, Math. Prog., 9, pp. 189-206.

Williamson, J.P. and Downs, D.H. (1970) Manuals for computer programs in finance and investments, Dartmouth College, Hanover, New Hampshire.

Wilshire Associates (1979) PRISM: portfolio review and investment system management, Wilshire Associates, Santa Monica, California.

Wolfe, P.S. (1959) The simplex method for quadratic programming, Econometrica, 27, pp. 382-298.

DISCUSSION

MR. R.E. BORLAND (NPL). Does not the actual use of your model by investment managers have an influence on the market itself?

PEROLD. Very little, if any, since there is so much subjectivity that goes into the estimation of the parameters. Besides, while larger and larger sums of money are being invested on the basis of mean-variance analysis, they do not yet constitute a fraction large enough to have any observable effects.

DR. A.M. ERISMAN (Boeing). One of your main considerations seemed to lie in developing techniques for use on small computers. With all the money that your investment managers have at their disposal why don't they buy or rent bigger machines?

PEROLD. The money is not their own, and it is very much in their interests to charge their clients as small a management fee as possible. Their clients always have the option of investing their money in a so-called "passive" portfolio, like a broadly diversified market index, at little or no cost. The investment managers thus have to outperform this passive portfolio by at least the management fee, in order to stay alive. Since large computers usually serve many different users, each optimization run incurs a substantial marginal cost. On a small computer the marginal cost is usually next to nothing, thus making it much easier to justify repeated sensitivity runs and the like.

DR. J.K. REID (Harwell). I would like to open up the discussion and ask what you currently consider to be the best techniques in linear programming?

PEROLD. Naturally such a broad question as that does not admit to a one-line answer. Many good techniques that will happily solve the "average" LP are part and parcel of a package like IBM's MPSX. However, there are problems like time-period (staircase) linear programs for which no truly satisfactory solution techniques have been found. Such problems are often numerically unstable and require disproportionately many simplex iterations. With respect to these, I would like to reiterate the comment I made after Duff's lecture regarding the benefits or otherwise of taking into account the structure of LP bases. It seems as if straight Markowitz does much better on these problems than do the bump and spike, and block factorization methods. On Dantzig's PILOT energy model, for example, Reid's LAØ5A code produced LU factors containing about 5000 nonzeros on a basis with initially 3500 nonzeros. The bump and spike codes in the MINOS and WHIZZARD packages, and certain block factorization techniques all produced factors with around 10,000 nonzeros. Moreover, these factorizations were all rather ill-conditioned, whereas that produced by LAØ5A was not.

DR. R.C. DANIEL (SCICON). We have found big gains in using supersparsity where we take advantage of similar valued elements in the tableau. Also, we find ourselves requiring to work out-of-core for most realistic problems.

PEROLD. The need to use supersparsity and/or work out-of-core has certainly diminished with the greater availability and decreased cost of large amounts of in-core storage. For small machines and for very large LP's, however, the need will remain. I am always a little sceptical, though, of the value of extremely large LP models (say with hundreds of thousands of variables) since one invariably ends up aggregating the answer. One might be just as well off aggregating the model. Very large LP models arising say by discretizing a continuous time model are probably best solved by a specialized procedure.

DR. H.M. LIDDELL (Queen Mary College). Have you any experience with array processors? We have a feasibility study under way at Queen Mary College although we have not yet found any encouraging results.

PEROLD. My computing environment has never included array processors although workers in other areas may have more common access to them.

(UNKNOWN). The NAG library has only routines for small linear programs. MPSX is not available to me so where else would I obtain routines for the large sparse case?

PEROLD. It all depends on what you want to pay. The Harwell routine LAØ5, written by Reid, is very good but only handles the basis matrices. For users of CDC machines, the APEX system is excellent.

EIGENVALUE METHODS AND THE ANALYSIS OF STRUCTURAL VIBRATION

A. Jennings

(Department of Civil Engineering, Queen's University of Belfast)

ABSTRACT

The well known tranformation methods (Givens, Householder, LR and QR) are not effective for solving large order eigenvalue problems occurring in the analysis of structural vibration because they cannot take advantage of sparsity. Mainly over the last ten years seven different classes of method have been developed for tackling this problem. These methods are reviewed and some properties of the Lanczos method are discussed. Convergence rates for the eigenvector predictions are given theoretical bounds, and the convergence rate is discussed in relation to alternative ways the method can be used to solve vibration equations.

INTRODUCTION

"Because vibration analysis has been one of the main sources of numerical eigenvalue problems there is a close historical link between them. However since eigenvalues now have more widespread application there is less liaison between the engineers interested in vibration and numerical analysts interested in eigensolutions. Furthermore, there are so many outlets for publication that very few are conversant with all the major developments. This paper may go some way in helping to rectify this disturbing situation".

This introduction to a recent review paper on vibration analysis (Jennings (1980)) is also appropriate for this paper which is aimed more specifically at sparse matrix eigenvalue techniques. As such it supplements the excellent review paper by G.W. Stewart (1976a).

I would like to set the scene about 1965. This date is a landmark for numerical analysts because of the publication of Wilkinson's treatise on the algebraic eigenvalue problem (Wilkinson (1965)). The Givens, Householder, LR and QR transformation methods were by that time well established. As with the Gaussian elimination method for the solution of simultaneous equations, the amount of computation required by all of these methods was proportional to n^3 where n is the order of the matrix. The methods were shown to have good numerical stability and the inevitable iterative phases in the computations were shown to have excellent convergence characteristics.

While the numerical analysts had been developing transformation methods for eigenvalue problems the structural engineers had been mainly devoting their efforts to static stress analysis problems. Finite element representations of large and complex structures were developed and the Gaussian elimination method was found to provide an adequate solution technique for quite large systems by making effective use of sparsity in the coefficient matrix. It was only natural for structural engineers to want to extend these methods to dynamic analyses of structures. After all, aircraft can be subject to the dynamic instability condition known as flutter (indeed many of the early accidents were due to this phenomenon (Collar (1978)), the possibility of wind induced dynamic instability of suspension bridges was highlighted by the collapse of the Tacoma Narrows bridge and wind induced vibration of other structures such as television masts and chimneys is possible (Rocard (1957)). Components of engines including large turbines can fracture due to resonance. Hydrodynamic forces in the region of propellers and engine out of balance can cause severe vibration problems in ships (Todd (1961)). Spacecraft are very light and flexible structures and have no surrounding air to help damp out any vibration caused by actuating motors. The most serious vibration risk of all comes from earthquakes. Not only could buildings collapse, dams could be breached and nuclear reactor vessels could be fractured if adequate steps were not taken to allow for the dynamic characteristics in design. However, the eigenvalue transformation techniques were not the answer to the structural engineers requirement, because when applied to sparse matrices they cause almost complete fill-in and hence become prohibitive when the number of equations is large. Since 1965 much effort of numerical analysts and structural engineers have been devoted to large sparse eigenvalue problems with the aim of producing algorithms which can exploit sparsity as much as possible.

2. UNDAMPED NATURAL VIBRATION

The basic variables in a structural analysis are normally the displacements at joints or nodes of the structure. Using these variables, the equations of unforced motion of an undamped structure can be reduced to the form

$$K\underline{x} = \omega^2 M\underline{x} \qquad (1)$$

where ω is the vibration frequency associated with the mode $\underline{x}$ (Przemieniecki (1968)). The mass and stiffness matrices M and K are determined by summing the contributions from all the component elements of the structure. Thus a triangular finite element connecting nodes 4, 7 and 9 of a structure would contribute to the elements of the stiffness matrix located in nine submatrices as shown in Figure 1. The selfweight of this element and of any non-structural material attached to the element will contribute to the corresponding elements of the mass matrix. Consequently the mass and stiffness matrices should have the same sparsity pattern although idealization of the mass or stiffness properties may affect this correspondence. For instance a pin-ended member will exhibit a mass coupling coefficient linking the lateral displacements of the two ends, whereas there will be no corresponding stiffness coupling coefficient.

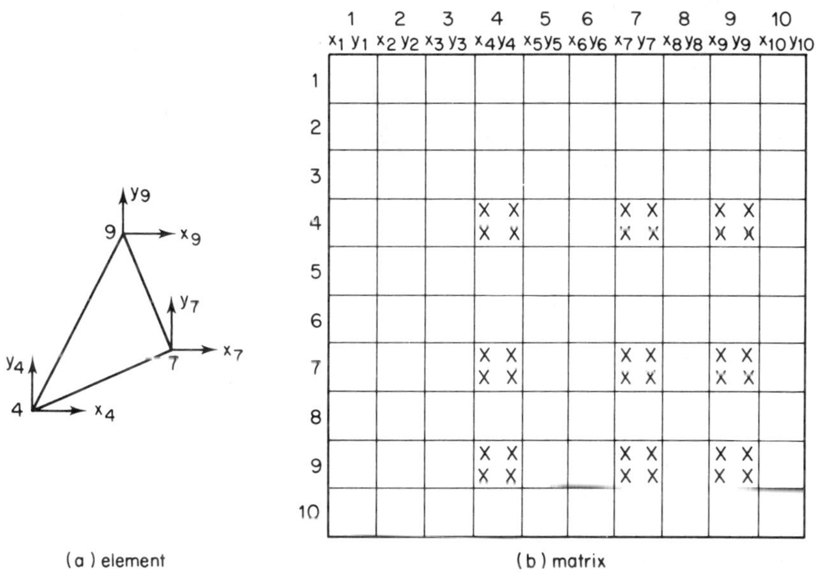

Figure 1 Contributions to the stiffness and mass matrices of a plane triangular finite element

The dynamic behaviour of a structure is less sensitive to the mass distribution than it is to the stiffness distribution and hence it is a feasible approximation to replace the distributed mass of the structure and the attached non-structural materials by a set of point masses located at the nodes (Figure 2). A diagonal mass matrix is obtained either by using the above approximation or a finite difference approximation. So that, although the treatment of distributed mass throughout the structure by the Rayleigh-Ritz principle leads logically to mass coupling coefficients in M (called the "consistent mass" formulation) it could be argued that it is a luxury which is worth indulging in only if the presence of the mass coupling coefficients do not substantially increase the amount of computation required for solution, for otherwise the same amount of computation devoted to solving the problem with a larger number of basic variables could be more profitable.

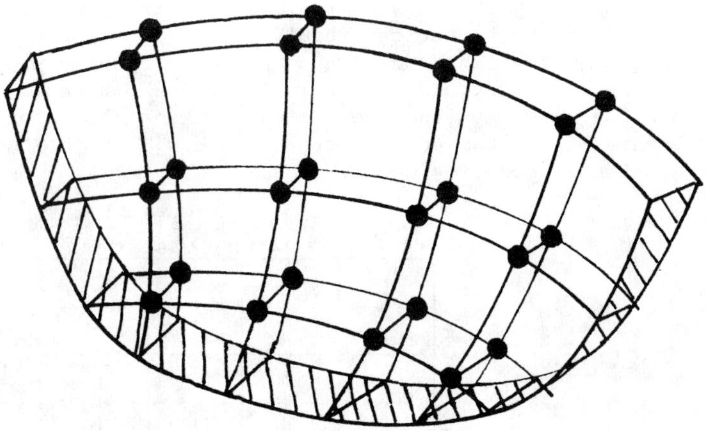

Figure 2 A simple lumped mass idealisation for an arch dam

The mass and stiffness matrices will normally be symmetric positive definite although the following exceptions can arise:

(a) if the structure is free to move bodily (for example an aircraft or a ship) then the stiffness matrix is positive semi-definite having rank n-r where r is the number of possible independent body movements,

(b) if no mass has been allocated to some node displacements then completely null rows and columns occur in the mass matrix.

Equations (1) therefore represent a symmetric linear generalised eigenvalue problem in which the values of ω are normally real and positive but could, under certain conditions be zero or infinite. The set of lower frequencies will be required, although occasionally interest is centred around the frequencies close to a specified value to discover whether or not an oscillatory disturbing force will cause resonance. Even in this latter case the specified frequency will be in the lower end of the spectrum since the approximations involved in the physical idealization cause the higher frequencies of the mathematical model to be inaccurate. Equation (1) may be converted to standard eigenvalue form by using either of the factorizations

$K = LL^T$ or $M = GG^T$ providing only that the matrix to be factorized is non-singular (Martin and Wilkinson (1968)). The resulting equations are

$$\left(L^{-1}ML^{-T}\right)\left(L^T\underline{x}\right) = \frac{1}{\omega^2}\left(L^T\underline{x}\right) \tag{2}$$

$$\left(G^{-1}KG^{-T}\right)\left(G^T\underline{x}\right) = \omega^2\left(G^T\underline{x}\right) \tag{3}$$

Because of the connectivity of our problems K and M are sparse and the variables can be ordered so that the factors L and G are in band form.

3. DIFFERENT SYMMETRIC EIGENSOLUTION PROCEDURES

Whenever a major problem arises it is not surprising to see different solutions being advocated. With the large sparse symmetric eigenvalue problem at least seven different lines of development can be recognised:

3.1 Condensation methods

The first approach to be used for vibration problems was mass condensation, otherwise known as the eigenvalue economiser method (Irons (1963), Guyan (1965), Ramsden and Stoker (1969), Zienkiewicz (1977)) in which the basic variables are restricted to be a linear combination of m auxiliary variables $\underline{z}$ ($m \ll n$). Thus with the relationship

$$\underline{x} = H\underline{z} \tag{4}$$

equation (1) may be reduced to the m-th order linear generalized eigenvalue problem

$$\left(H^TKH\right)\underline{z} = \omega^2\left(H^TMH\right)\underline{z} \tag{5}$$

Without discussing the method of obtaining a suitable transformation matrix, suffice it to say that sparsity in K and M can be utilized in the computation of H and in forming the reduced mass and stiffness matrices, but not in the solution of the reduced equations (5).

An alternative condensation technique (Vysloukh et al (1973), Hughes (1976)) can be applied when some variables have not been allocated any mass terms, with the result that equations (1) take the form

$$\begin{bmatrix} K_{aa} & K_{ab} \\ k_{ba} & K_{bb} \end{bmatrix} \begin{bmatrix} \underline{x}_a \\ \underline{x}_b \end{bmatrix} = \omega^2 \begin{bmatrix} M_{aa} & 0 \\ 0 & 0 \end{bmatrix} \begin{bmatrix} \underline{x}_a \\ \underline{x}_b \end{bmatrix} \tag{6}$$

and $\underline{x}_b$ can be eliminated to give

$$\left[K_{aa} - K_{ab} K_{bb}^{-1} K_{ba} \right] \underline{x}_a = \omega^2 M_{aa} \underline{x}_a \tag{7}$$

3.2 Simultaneous iteration

When applied to the standard eigenvalue problem

$$A\underline{q} = \lambda \underline{q} \tag{8}$$

the basic simultaneous iteration method may be specified as follows:

Starting with a set of m trial vectors compounded to form the n×m matrix U perform the multiplication

$$V = AU \tag{9}$$

Then determine the m×m matrix B such that

$$U^T UB = U^T V \tag{10}$$

and its full eigensolution

$$BP = P\Theta \tag{11}$$

The revised set of trial vectors for the next round of iteration may then be generated by normalising the reorientated vector set

$$W = VP \tag{12}$$

The eigenvalues specified in the diagonal matrix Θ converge to the dominant set of eigenvalues of A and the trial vectors converge to the corresponding eigenvectors.

Although simultaneous iteration is a development of the power method and Bauer's bi-iteration, the reorientation of the vectors through eigensolution of B is the step which gives the method good convergence characteristics for many problems (Jennings (1967), Clint and Jennings (1970) Rutishauser (1969) and G.W. Stewart (1969)). Applying the method to equation (2) it is possible to proceed without formally computing the matrix $L^{-1}ML^{-T}$ by performing the premultiplication step, equation (9), by means of forward substitution, multiplication and backsubstitution operations involving L, M and L^T respectively, (Jennings and Orr (1971), Jennings (1977a)). The matrix L may thus be stored in band form and M in a sparse packed form. If simultaneous iteration is applied directly to equation (3) convergence will be to the wrong set of eigenvectors. However if the eigenvalues are shifted by an amount $-\bar{\lambda}$ to give

$$\left(G^{-1}KG^{-T} - \bar{\lambda}I\right)\left(G^T\underline{x}\right) = \left(\omega^2 - \bar{\lambda}\right)\left(G^T x\right) \qquad (13)$$

the required eigenvectors will be obtained provided that $\bar{\lambda} \simeq \omega_n^2$, where ω_n is the highest frequency.

Subspace iteration (Dong et al (1972), Bathe and Wilson (1972),(1976)) is effectively the same as simultaneous iteration except that the use of transformed variables $L^T\underline{x}$ is avoided.

3.3 Lanczos' method

As with the power method, in Lanczos' method (Lanczos (1950)) a sequence of trial vectors are formed by premultiplication with the matrix A. However, each new vector is also orthogonalized with respect to the previous two vectors. The factors computed from the orthogonalization process then combine to form a tridiagonal matrix which after n iterations, theoretically, has the same eigenvalues as the original matrix. Although the method must be related to the stable Householder method (since the same tridiagonal matrix is obtained if the starting vector is chosen to be {1 0 ... 0}), it is known to go astray through breakdown of implicit orthogonality conditions between the vectors.

Taking account of the need for re-orthogonalisation Wilkinson (1965 p.395) said "It is difficult to think of any reason why we should use Lanczos's method in preference to Householder's". However with sparse matrices this method has the same advantage that simultaneous iteration has, namely that because the matrix A is only involved in premultiplication operations, sparsity is preserved. It will be discussed later in this paper (see also Scott (1980)).

3.4 Coordinate relaxation

Another method which should be considered is coordinate relaxation (Faddeev and Faddeeva (1963)). If $\underline{x}$ is a trial vector and $\underline{e}_i = (0 \ldots 0\ 1\ 0 \ldots 0)^T$ is the i-th column of the unit matrix, the effect of changing the ith element of $\underline{x}$ can be investigated by considering the vector $\bar{\underline{x}} = c_1\underline{x} + c_2\underline{e}_i$. If the linear generalized eigenvalue equation

$$A\underline{q} = \lambda B\underline{q} \tag{14}$$

is under consideration then the minimum and maximum values of the Rayleigh quotient for $\bar{\underline{x}}$ can be shown to occur when

$$\begin{bmatrix} \underline{x}^T A\underline{x} - \bar{\lambda}\underline{x}^T B\underline{x} & \underline{e}_i^T A\underline{x} - \bar{\lambda}\underline{e}_i^T B\underline{x} \\ \underline{e}_i^T A\underline{x} - \bar{\lambda}\underline{e}_i^T B\underline{x} & \underline{e}_i^T A\underline{e}_i - \bar{\lambda}\underline{e}_i^T B\underline{e}_i \end{bmatrix} \begin{bmatrix} c_1 \\ c_2 \end{bmatrix} = \begin{bmatrix} 0 \\ 0 \end{bmatrix} \tag{15}$$

If $A\underline{x} = \underline{u}$ and $B\underline{x} = \underline{v}$ then equation (15) simplifies to

$$\begin{bmatrix} \underline{x}^T\underline{u} - \bar{\lambda}\underline{x}^T\underline{v} & u_i - \bar{\lambda}v_i \\ u_i - \bar{\lambda}v_i & a_{ii} - \bar{\lambda}b_{ii} \end{bmatrix} \begin{bmatrix} c_1 \\ c_2 \end{bmatrix} = \begin{bmatrix} 0 \\ 0 \end{bmatrix} \tag{16}$$

The quadratic determinantal equation may be solved and the ratio c_1/c_2 corresponding to either the smaller or larger root computed, thus defining a modification to the trial vector. The process is repeated with different values of i and the smallest or largest root of the quadratic is chosen on each occasion in the hope of converging on to the eigenvector corresponding to the smallest or largest eigenvalue respectively. The particular advantage of this technique is that sparseness in both A and B is completely preserved. If applied to equation (1) with A=K and B=M the lowest vibration frequency may be obtained by choosing the lowest root at each step (see also Nesbet (1980)).

3.5 Determinant search

All of the above methods employ trial vectors to predict the eigenvector defining the mode shape corresponding to the lowest frequency and also perhaps some neighbouring eigenvectors and eigenvalues. It is however possible to start with an estimate for a frequency of the system and adjust this until the determinantal equation

$$|K - \omega^2 M| = 0 \tag{17}$$

is satisfied. The determinant is evaluated by factorization with the result that advantage can be taken of band structure in K and M (Bates (1973), Raju et al, (1974)). Two risks inherent in this process are as follows:

(a) since $K - \omega^2 M$ is not generally positive definite, there is a possibility that normal symmetric factorization will fail due to the presence of a zero pivot;
(b) if close or multiple frequencies occur it is possible for these to be missed in the search process.

3.6 Sturm sequence

The Sturm sequence principle applied to the dynamic equations (1) may be stated as follows: the number of changes of sign in the sequence of principal minors of $K - \bar{\omega}^2 M$ is equal to the number of frequencies which are less than $\bar{\omega}$. If $K - \bar{\omega}^2 M$ is tridiagonal or quindiagonal, recursive algorithms can be used to determine the signs of the principal minors (Evans (1975)) but with larger bandwidths it is necessary use either a symmetric factorization, which is possibly unstable or a special unsymmetric factorization which operates in an enlarged band store (Peters and Wilkinson (1969), Gupta (1970)). Sturm sequences are analysed for different trial frequencies and a bisection process is adopted to isolate the position of the required frequencies.

3.7 Band transformation

Crawford (1973) developed a method of reducing a symmetric band linear generalized eigenvalue problem to standard tridiagonal form by an extension of Rutishauser's band stripping technique (Rutishauser (1963), Schwarz (1968)). The amount of computation required for this method is proportional to $n^2 b$ (where n is the order of the system and b is the semi-bandwidth) and would be a competitive technique if all the eigenvalues were required. However it is not competitive for vibration problems where only partial eigensolutions are sought.

4. DEVELOPMENT OF STURM SEQUENCE AND SIMULTANEOUS ITERATION METHODS

Both Sturm sequence and simultaneous iteration methods (using stiffness matrix factorization) have been well developed and utilized for the solution of large order vibration problems. The computational requirement for Sturm sequence methods is proportional to nb^2h where h is the number of trial frequencies employed and n and b are as before. It is therefore suitable where the bandwidth can be kept particularly small. The most effective procedures make use of inverse iteration and/or determinantal search techniques to reduce the number of trial frequencies h (Gupta (1973), Schwarz (1977), Bathe and Wilson (1973a and 1976)).

The main term in the computational requirement of simultaneous iteration is nbmk where m is the number of simultaneous vectors, k is the number of iterations and n and b are as before. Both m and k will usually be larger when more frequencies are required. Thus, in comparison with Sturm sequence methods, simultaneous iteration methods tend to be advantageous when the bandwidth is large, particularly when only a few of the lower frequencies are required (Jennings and Orr (1971), Bathe and Wilson (1973b)). They also derive some benefit from a lumped rather than a consistent mass idealization. A comparison between mass condensation and simultaneous iteration (Jennings (1973)) reveals that numerical efficiencies are likely to be similar but simultaneous iteration gains advantage through

(i) permitting smaller matrices to be used
(ii) being able to start with random trial vectors and
(iii) allowing some degree of control over accuracy.

Chebyshev accelerations of simultaneous iteration have been carried out (Rutishauser (1969), Corr and Jennings (1973), Yamamoto and Ohtsubo (1976).

There is also the possibility of combining Sturm sequence and simultaneous iteration methods in hybrid form (Jennings and Agar (1977 and 1979)) and also Jensen's (1972) inverse iteration method of finding interior eigenvalues should be noted.

5. ON THE BASIC LANCZOS ALGORITHM

Although the Lanczos method started well back in the field of runners, it has more recently shown itself to have high potential, and must figure strongly in any assessment of available methods for the eigensolution of symmetric matrices. In some ways it is simple and elegant but in others it is complex and mysterious.

5.1 Polynomial equivalence

If the effects of round-off are ignored, after m iterations of the Lanczos algorithm, m mutually orthogonal vectors $Y = [\underline{y}_1 \underline{y}_2 \; ... \; \underline{y}_m]$ and an m × m tridiagonal matrix

$$T = \begin{bmatrix} \alpha_1 & \beta_1 & & & \\ \beta_1 & \alpha_2 & \beta_2 & & \\ & \beta_2 & \cdot\cdot & \cdot\cdot & \\ & & \cdot\cdot & \cdot\cdot & \beta_{m-1} \\ & & & \beta_{m-1} & \alpha_m \end{bmatrix}$$

will have been derived satisfying the equations

$$\begin{bmatrix} A \end{bmatrix} \begin{bmatrix} \underline{y}_1 \underline{y}_2 & \cdots\cdots & \underline{y}_m \end{bmatrix} = \begin{bmatrix} \underline{y}_1 \underline{y}_2 & \cdots & \underline{y}_m & \underline{y}_{m+1} \end{bmatrix} \begin{bmatrix} \alpha_1 & \beta_1 & & & \\ \beta_1 & \cdot\cdot & \cdot\cdot & & \\ & \cdot\cdot & \cdot\cdot & \cdot\cdot & \\ & & \cdot\cdot & \cdot\cdot & \beta_{m-1} \\ & & & \beta_{m-1} & \alpha_m \\ \hline & & & & \beta_m \end{bmatrix}$$

which can be specified in the form

$$AY = YT + \beta_m \underline{y}_{m+1} \underline{e}_m^T \tag{18}$$

where $\underline{e}_m$ is the mth column of the unit matrix (Householder (1964), Parlett (1980)). Also, because $\underline{y}_{m+1}$ is orthogonal to Y

$$Y^T AY = T \, . \tag{19}$$

Premultiplying equation (18) by A gives

$$A^2 Y = YT^2 + \beta_m \underline{y}_{m+1} \underline{e}_m^T T + \beta_m A \underline{y}_{m+1} \underline{e}_m^T . \tag{20}$$

Continuing with further premultiplications by A it can be shown that for any values of μ_i

$$\left[\prod_{i=1}^{m} (A - \mu_i \underline{I}) \right] Y = Y \prod_{i=1}^{m} (T - \mu_i I) + \beta_m \underline{y}_{m+1} \underline{e}_m^T T^{m-1} + W \tag{21}$$

where W is a matrix whose first column is null (by noting that $\underline{e}_m^T T^k$ has a null first element for $k < m-1$). If μ_i are the eigenvalues of T then, from the Cayley-Hamilton theorem

$$\prod_{i=1}^{m} (T-\mu_i I) = 0 \tag{22}$$

and since $\beta_m \underline{e}_m^T T^{m-1}$ has its first component equal to $\prod_{k=1}^{m} \beta_k$, the first column of the matrix equation (21) gives the following direct relationship between $\underline{y}_1$ and $\underline{y}_{m+1}$:

$$\left[\prod_{i=1}^{m} (A-\mu_i I)\right]\underline{y}_1 = \left[\prod_{k=1}^{m} \beta_k\right]\underline{y}_{m+1} \tag{23}$$

If the complete eigensolution of A is given by

$$AQ = Q\Lambda \tag{24}$$

where $\Lambda = \lceil \lambda_1 \lambda_2 \ldots \lambda_n \rfloor$ is the diagonal matrix of eigenvalues and Q is an orthogonal matrix of corresponding eigenvectors, such that $Q^TQ=I$, any of the Lanczos vectors may be expressed as a linear combination of the eigenvectors according to

$$\underline{y}_i = Q\underline{c}_i \tag{25}$$

Premultiplication of equation (23) by Q^T and substituting for $\underline{y}_i$ from equation (25) gives

$$\left[\prod_{i=1}^{m} (\Lambda-\mu_i I)\right]\underline{c}_1 = \left[\prod_{k=1}^{m} \beta_k\right]\underline{c}_{m+1} \tag{26}$$

The jth element of the diagonal matrix $\prod_{i=1}^{m} (\Lambda-\mu_i I)$ is the value of the characteristic polynomial ($f(\mu)$, say) of T at coordinate λ_j. Thus the coefficients $\underline{c}_{m+1}$ can be expressed in terms of $\underline{c}_1$ by multiplying the latter by the characteristic polynomial and normalizing (Figure 3a).

Consider the vector

$$\underline{y}_{im} = \left[\prod_{j\neq i}^{m} (A-\mu_j I)\right]\underline{y}_1 \tag{27}$$

On account of equation (23) and the orthogonality of Y to $\underline{y}_{m+1}$

$$Y^T(A-\mu_i I)\underline{y}_{im} = \underline{0} \tag{28}$$

and since, from equation (27), $\underline{y}_{im}$ must be a linear combination of the vectors $\underline{y}_1, \underline{y}_2, \ldots \underline{y}_m$ let

$$\underline{y}_{im} = Y\underline{p}_{im} \tag{29}$$

and from equations (19),(28) and (29)

$$(T-\mu_i I)\underline{p}_{im} = \underline{0} \tag{30}$$

Hence $\underline{p}_{im}$ is the eigenvector of T corresponding to μ_i and $\underline{y}_{im}$ is therefore the prediction for the corresponding eigenvector of A. With $\underline{y}_{im} = Q\underline{c}_{im}$ equation (27) gives

$$\underline{c}_{im} = \left[\prod_{j\neq i}^{m} (\Lambda-\mu_j I)\right]\underline{c}_1 \tag{31}$$

The use of this transformation to convert $\underline{c}_1$ to $\underline{c}_{1m}$ and $\underline{c}_{2m}$ is illustrated in Figures 3b and 3c.

5.2 *Minimization properties*

Consider the function

$$h = \underline{c}_1^T\left[\prod_{j=1}^{m} (\Lambda-\mu_j I)^2\right]\underline{c}_1 \tag{32}$$

Differentiating with respect to μ_i gives

$$\frac{\partial h}{\partial \mu_i} = -2\underline{c}_1^T\left[\prod_{j=1}^{m} (\Lambda-\mu_j I)\prod_{j\neq i}^{m} (\Lambda-\mu_j I)\right]\underline{c}_1 \tag{33}$$

However $\underline{y}_{m+1}^T\underline{y}_{im} = \underline{y}_{m+1}^T Y\underline{p}_{im} = 0$ and therefore

$$\underline{c}_{m+1}^T\underline{c}_{im} = 0 \tag{34}$$

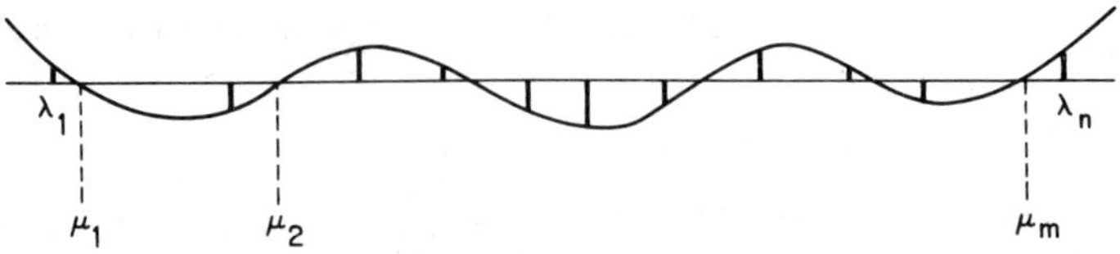

(a) Characteristic polynomial of T, f(μ)

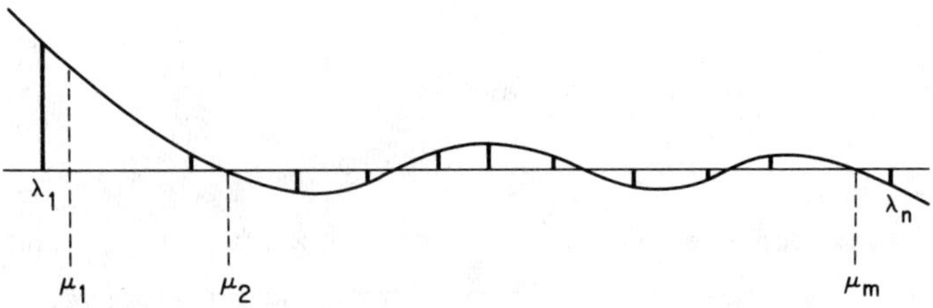

(b) Polynomial corresponding to lowest eigenvector

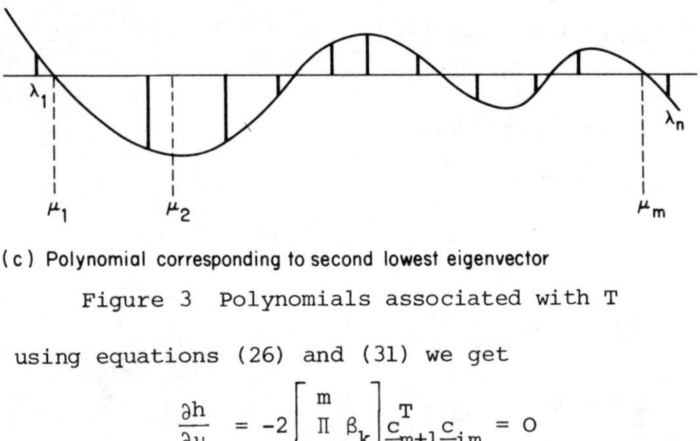

(c) Polynomial corresponding to second lowest eigenvector

Figure 3 Polynomials associated with T

and so using equations (26) and (31) we get

$$\frac{\partial h}{\partial \mu_i} = -2\left[\prod_{k=1}^{m} \beta_k\right] \underline{c}_{m+1}^T \underline{c}_{im} = 0 \tag{35}$$

Also

$$\frac{\partial^2 h}{\partial \mu_i^2} = 2\underline{c}_1^T \left[\prod_{j \neq i}^{m} (\Lambda - \mu_j I)^2\right] \underline{c}_1 > 0 \tag{36}$$

Hence the values μ_i are the ones which minimise the function h.

Since h may be written in the form

$$h = \sum_{j=1}^{n} c_{j1}^{2} f(\lambda_j)^2 \tag{37}$$

where c_{j1} is the jth element of $\underline{c}_1$, the polynomial $f(\mu)$ is the one which gives the least squares fit to the n points $(\lambda_j, 0)$ subject to weighting factors c_{j1}^2. (Note that f = 0 is not a permissible solution since the polynomial must have a unit value for the μ^m coefficient).

Furthermore it is possible to show that the eigenvector predictions also exhibit a minimal property. Consider the vector

$$\bar{\underline{y}}_{im} = \frac{1}{\prod\limits_{j\neq i}^{m} (\mu_i - \mu_j)} \underline{y}_{im} \tag{38}$$

From equation (27)

$$\left[\prod_{j\neq i}^{m} (\mu_i - \mu_j)\right] \bar{\underline{y}}_{im} = \left[\prod_{j\neq i}^{m} (\Lambda - \mu_j I)\right] \underline{y}_1 \tag{39}$$

Hence with $\bar{\underline{y}}_{im} = Q\bar{\underline{c}}_{im}$

$$\left[\prod_{j\neq i}^{m} (\mu_i \;\; \mu_j)\right] \bar{\underline{c}}_{im} = \left[\prod_{j\neq i}^{m} (\Lambda - \mu_j I)\right] c_1 \tag{40}$$

and squaring we have

$$\left[\prod_{j\neq i}^{m} (\mu_i - \mu_j)^2\right] \bar{\underline{c}}_{im}^T \bar{\underline{c}}_{im} = \underline{c}_1^T \left[\prod_{j\neq i}^{m} (\Lambda - \mu_j I)^2\right] \underline{c}_1 \tag{41}$$

Differentiating with respect to μ_k where $k \neq i$ and using equation (40) to eliminate $\underline{c}_1$ gives

$$-\frac{2}{\mu_i - \mu_k} \bar{\underline{c}}_{im}^T \bar{c}_{im} + \frac{\partial}{\partial \mu_k} (\bar{\underline{c}}_{im}^T \bar{\underline{c}}_{im}) = -2\bar{\underline{c}}_{im}^T (\Lambda - \mu_k I)^{-1} \bar{\underline{c}}_{im} \tag{42}$$

However $(A-\mu_i I)\bar{\underline{y}}_{im}$ is proportional to $\underline{y}_{m+1}$ and $(A-\mu_k I)^{-1}\bar{\underline{y}}_{im}$ is the product of Y and a coefficient vector (equations (23), (27) and (39)). Hence

$$\bar{\underline{y}}_{im}^T (A-\mu_i I)(A-\mu_k I)^{-1}\bar{\underline{y}}_{im} = 0 \tag{43}$$

and thus

$$\bar{\underline{c}}_{im}^T (\Lambda-\mu_i I)(\Lambda-\mu_k I)^{-1}\bar{\underline{c}}_{im} = 0 \tag{44}$$

By equating

$$(\Lambda-\mu_i I) = (\Lambda-\mu_k I) - (\mu_i-\mu_k)I \tag{45}$$

equation (44) gives

$$\bar{\underline{c}}_{im}^T \bar{\underline{c}}_{im} = (\mu_i-\mu_k)\bar{\underline{c}}_{im}^T (\Lambda-\mu_k I)^{-1}\bar{\underline{c}}_{im} \tag{46}$$

Substituting this result in equation (42) yields

$$\frac{\partial}{\partial \mu_k}(\bar{\underline{c}}_{im}^T \bar{\underline{c}}_{im}) = 0 \tag{47}$$

Differentiating equation (41) twice with respect to μ_k where $k \neq i$ and using equation (40) gives

$$\frac{2}{(\mu_i-\mu_k)^2}\bar{\underline{c}}_{im}^T \bar{\underline{c}}_{im} + \frac{\partial^2}{\partial \mu_k^2}(\bar{\underline{c}}_{im}^T \bar{\underline{c}}_{im}) = 2\bar{\underline{c}}_{im}^T (\Lambda-\mu_k I)^{-2}\bar{\underline{c}}_{im} \tag{48}$$

However

$$\bar{\underline{c}}_{im}^T (\Lambda-\mu_i I)^2 (\Lambda-\mu_k I)^{-2}\bar{\underline{c}}_{im} > 0 \tag{49}$$

from which it may be shown, using equations (45) and (46), that

$$(\mu_i-\mu_k)^2 \bar{\underline{c}}_{im}^T (\Lambda-\mu_k I)^{-2}\bar{\underline{c}}_{im} > \bar{\underline{c}}_{im}^T \bar{\underline{c}}_{im} \tag{50}$$

and hence equation (48) gives

$$\frac{\partial^2}{\partial \mu_k^2}(\bar{\underline{c}}_{im}^T \bar{\underline{c}}_{im}) > 0 \tag{51}$$

From equations (47) and (51) it is established that the values μ_k ($k \neq i$) are the ones which minimise the function $\underline{\bar{c}}_{im}^T \underline{\bar{c}}_{im}$. The transformation to convert the elements $\underline{c}_1$ to those of $\underline{\bar{c}}_{im}$ (equation (40)) is a polynomial in μ which passes through the point $(\mu_i, 1)$. Equations (47) and (51) show that it is the one which gives the least squares fit to the n points $(\lambda_j, 0)$ subject to weighting factors c_{j1}^2.

The above minimization properties can be used to show that theoretically the process will terminate with no error or no apparent error when m equals the number of distinct eigenvalues of A which have non-zero coefficients in the vector $\underline{c}_1$, and the interlacing property associated with the Sturm sequence of T can be used to show that $\lambda_i \leq \mu_i \leq \lambda_{n-m+i}$.

6. INFLUENCE OF THE EIGENVALUE SPECTRUM ON CONVERGENCE RATES

Kaniel (1966) has made use of the minimization characteristics to define minimum convergence rates for both the conjugate gradient and Lanczos methods, see also Parlett (1980). When applied to the vector converging to $\underline{q}_i$ the method adopted is to find a polynomial of order m which has unit value at λ_i and has absolute values less than or equal to ε, the machine precision, at all other eigenvalues λ_j. In this case the contaminating eigenvectors are scaled by at least ε over m iterations. The minimization principle is then invoked to claim that the equivalent Lanczos polynomial must perform at least as well as the one chosen and hence must satisfy a similar bound for the reduction in error.

Error bounds derived by Jennings (1977b) using double Chebyshev functions for the conjugate gradient method can also be applied to eigenvector prediction by the Lanczos method. If the eigenvalues in the interval λ_p to λ_q are closely spaced, and if one of the outlying smaller eigenvalues λ_j is to be determined, the number of iterations to achieve an error bound ε is given by

$$k = \sqrt{\frac{\lambda_q - \lambda_j}{\lambda_p - \lambda_j}} \left\{ \tfrac{1}{2} \log_e \left(\frac{2}{\varepsilon} \right) + \sum_{\substack{i=1 \\ i \neq j}}^{p-1} \left(1 + \tfrac{1}{2} \log_e \left[\frac{\Pi(\lambda_p - \lambda_j)}{4 |\lambda_i - \lambda_j|} \right] \right) \right\} + n - q \qquad (52)$$

TABLE 1

Number of Lanczos iterations required to achieve an error reduction $\varepsilon = 0.001$ (computed with $p = 4$). Regular spacing of lower eigenvalues.

Eigenvector	Contributions from terms in equation (52)				Total iterations required	
	λ_1	λ_2	λ_3	$\lambda_4 - \lambda_n$		
	0	1	2	3 - 200	Theoretical	Test result
q_1	-	12	9	31	52	53
q_2	12	-	12	38	62	66
q_3	8	13	-	54	75	80

TABLE 2

Number of Lanczos iterations required to achieve an error reduction $\varepsilon = 0.001$ (computed with $p = 4$). Irregular spacing of lower eigenvalues.

Eigenvector	Contributions from terms in equation (52)				Total iterations required	
	λ_1	λ_2	λ_3	$\lambda_4 - \lambda_n$		
	0	0.01	1.5	3 - 200	Theoretical	Test result
q_1	-	30	10	31	71	74
q_2	30	-	10	31	71	74
q_3	10	10	-	44	64	70

To obtain this formula from the conjugate gradient formulae it has been necessary to shift the origin to λ_j and take the modulus of $\lambda_i-\lambda_j$ since $\lambda_j > \lambda_i$ is now possible. If the lower eigenvalues are equally spaced, eigenvectors associated with the outer eigenvalues exhibit faster convergence rates, as shown in Table 1. However, if two eigenvalues are close, the convergence rate for each of their eigenvectors is impaired so that interior eigenvectors could be the first to converge (see Table 2). In both Tables 1 and 2 the contributions to k from the separate terms in equation (52) have been listed and test results have been included which have been obtained using a diagonal matrix for A, a starting vector with equal elements, and a tolerance criterion in which the infinity norm of the error in the eigenvector prediction is required to be less than 0.001.

The above formula can also be applied to convergence of the eigenvectors associated with the larger eigenvalues by reversing the spectrum. In any application of equation (52) where the spectrum is known, it is necessary to try different choices for p and q to find which give the best convergence bounds. The well known property that the outer eigenvalues converge first will not apply when the eigenvalues are more densely packed near the extremities of the spectrum. Indeed, a Chebyshev eigenvalue distribution is known to give a slow rate of convergence (Scott (1979)).

7. DEVELOPMENT OF THE LANCZOS METHOD

Just as the magician has to be watched carefully to see where the rabbits come from, so the Lanczos method has to be inspected carefully to see how so many orthogonal vectors can be generated from only one initial vector. In practice rounding errors may be magnified in the matrix multiplications and in the orthogonalisation processes so that the computed vectors must partially originate from this source. However "error" is really a misnomer in this context because Paige (1972) discovered that with a careful choice of algorithm accurate results could be obtained even without re-orthogonalisation. What goes wrong is that after an eigenvector is predicted accurately rounding errors start to create another copy of the same eigenvector (which, in the case of equal eigenvalues, may also contain a component of the hitherto undetected partner). The method is not wayward but over-zealous. The advantage of carrying out re-orthogonalisation with respect to earlier vectors is that multiple copies of the same eigenvector are avoided without endangering the late development of eigenvectors of multiple eigenvalues. However unless convergence is rapid re-orthogonalisation can be time consuming, and so Parlett and Scott (1979) (Scott (1980)) have developed a less cumbersome selective orthogonalisation process in which the current vector is intermittently orthogonalised with respect to only those eigenvector predictions which are close to convergence.

Just as the power method can be extended to simultaneous iteration, so the Lanczos method can be extended to the block Lanczos method (Underwood (1975), Golub and Underwood (1977), Cullum and Donath (1974), Cullum (1978)). It can be expected that block Lanczos will be more efficient in terms of matrix - vector multiplications than simultaneous iteration simply because the block Lanczos method retains and utilizes information about previous iterations which the simultaneous iteration method discards. However in assessing the relative merits of different methods it must be remembered that the use of block Lanczos requires eigensolutions of a matrix with a semi-bandwidth one greater than the number of simultaneous vectors. In comparison with the basic Lanczos method individual eigenvectors may not necessarily converge more rapidly when using the block method but the latter does have the advantages that convergence of close or equal eigenvalues is not normally delayed and that the matrix requiring eigensolution needs to be accessed less frequently. There may also be advantages for the block Lanczos method with computers capable of parallel processing.

Ruhe (1979) has developed a band Lanczos method which is theoretically slightly more efficient than block Lanczos. However it lacks the possible advantages to be gained from simultaneous processing of the vectors through the premultiplication. The standard Lanczos method may be applied in a form of inverse iteration in order to obtain eigenvectors associated with interior eigenvalues (Ericsson and Ruhe (1979)).

8. WHERE NOW WITH LARGE ORDER VIBRATION PROBLEMS?

At present large order vibration problems are normally solved by variants of equation (2) involving stiffness matrix factorization, because the required eigenvalues tend to be well separated from the main group. An example quoted by Ojalvo and Newman (1970) of a 44 variable problem is typical. The spectrum for values of $\lambda(=\frac{1}{\omega^2})$ is {0.200, 0.034, 0.020, 0.014, 0.010, 0.0073, ..., 0.000008} with the higher values required. Convergence bounds obtained from this eigenvalue spectrum using equation (52) suggest that all the first four eigenvectors should be predicted to at least three figure accuracy after 11 iterations, a result which has been confirmed by numerical tests. This is better than simultaneous iteration which would require about 10 iterations using five simultaneous vectors to obtain the same result, but part of this advantage is lost if the Lanczos iteration needs to be continued until it is established that none of the predicted eigenvalues is one of a multiple set.

On the other hand, if mass matrix factorization were adopted to solve the same problem using equation (3), the eigenvalue spectrum for values of $\lambda(=\omega^2)$ is {5.0, 30.9, 49.5, 71.7, 94.7, 37.5, ..., 47611, 53870, 63403, 125680} with the lower values required. Numerical tests and also application of the above convergence bound (equation (52)) reveal that approximations to the required eigenvectors will not achieve any significant accuracy before the number of iterations equals n. (For instance with $\lambda_p = 49.5$ and $\lambda_q = 47611$ the formula gives 164 iterations for convergence of the first eigenvector to three figure accuracy). In the case where more nodes are used in the analysis of a structure resulting in larger order equations, more higher frequency roots will be present so that the convergence rate will be even poorer and the terminal property will be less helpful. It is easily verified that the computation of the higher frequencies by simultaneous iteration using a shift technique would give a disastrously slow convergence rate.

It is therefore evident that when using the Lanczos method or simultaneous iteration with consistent mass formulations it is far better to factorize the stiffness matrix rather than the mass matrix (Ojalvo and Newman (1970), Jennings and Orr (1971)). In the case of lumped mass or finite difference idealizations where the mass matrix is diagonal, mass factorization has the attraction that sparsity is completely preserved. For very large systems this will allow storage space to be saved and also make the premultiplication operation considerably more rapid. I would expect these advantages always to be outweighed by the disadvantage of having an exceptionally poor convergence rate.

Coordinate relaxation has been extended in various ways (Ruhe (1974), Schwarz (1974a, 1974b and 1977), Shavitt et al (1973), Davidson (1975)) and Schwarz has shown that there is a close relationship between coordinate over-relaxation, when converging to the eigenvalue λ_1, and successive over-relaxation for simultaneous equations having a coefficient matrix $K-\lambda_1 M$. In cases where K is ill-conditioned the rate of convergence will be slow and it is possible for the wrong eigenvalues to be determined when the over-relaxation parameter is not sufficiently large. There have also been other methods proposed for finding vibration modes without factorizing the stiffness matrix (Downs, (1973), Muda (1973), Allman and Brotton (1968), McCormick and Noe (1977)) but as far as I know all will have poor convergence rates when K is ill-conditioned, and so it seems that for large order systems one should either choose a method which factorizes

K or else be prepared to face severe convergence difficulties.

9. OTHER EIGENVALUE PROBLEMS IN VIBRATION

In the vibration of marine and hydraulic structures such as ships and off-shore platforms the boundary integral method is very useful for determining fluid inertia effects. In these cases the mass matrix is more densely populated than the stiffness matrix and will be unsymmetric (Murray et al (1980)). It has been found satisfactory to replace the unsymmetric mass matrix by an equivalent symmetric matrix obtained by averaging it with its own transpose, the justification being based on the conservative nature of the system. However because the stiffness matrix has greater sparsity than the resulting mass matrix, vector iterative methods (either simultaneous iteration or Lanczos) which only factorize the stiffness matrix are to be preferred to Sturm sequence methods which require the factorization of $K-\mu M$.

If structural damping is assumed to be viscous its inclusion leads to a symmetric quadratic generalized eigenvalue problem

$$\lambda^2 M\underline{x} + \lambda C\underline{x} + Kx = \underline{0} \tag{53}$$

where C is the damping matrix, for which simultaneous iteration and Sturm sequence methods have been developed to take advantage of sparsity (Borri and Mantegazza (1977), Gupta (1974)). Also the inclusion of gyroscopic forces arising from the analysis of spinning structures yields equations which can be reduced to symmetric quadratic generalized eigenvalue form (Meirovitch (1974)). On the other hand, if fluid damping forces are present (as in aircraft flutter analysis) the C and K matrices will be unsymmetric and an unsymmetric eigensolution procedure must be used.

10. SPARSE UNSYMMETRIC EIGENSOLUTION PROCEDURES

The normal technique for solving flutter problems has been to find the lower vibration frequencies and corresponding modes for the undamped structure and then to use these modes as coordinates in the main investigation. Therefore the resulting unsymmetric eigenvalue problem is neither of large order nor sparse. However sparse unsymmetric eigensolution methods based on both Sturm sequence and simultaneous iteration methods have been developed for use in vibration problems (Gupta (1976), Dong (1977)). At present it seems that simultaneous iteration is the most well tried method available for large sparse unsymmetric matrices (Clint and Jennings (1971), Jennings and W.J. Stewart (1975), G.W. Stewart (1976b), W.J. Stewart and Jennings (1981)).

11. CONCLUSION

The analysis of structural vibration frequently leads to large sparse linear generalized eigenvalue problems. If factorization of the stiffness matrix or K-μM is accepted as part of the algorithm then Sturm sequence, simultaneous iteration or Lanczos methods can be used, which all have good convergence rates. Although it is possible to derive methods which avoid any matrix factorizations, convergence rates may be very poor. Large sparse unsymmetric eigenvalue problems occur less frequently but can be solved by using simultaneous iteration.

ACKNOWLEDGEMENTS

The author would like to thank Mr. D.C. Doran for numerical tests and Dr. M. Clint for comments on the script.

REFERENCES

Allman, K.J. and Brotton, D.M. (1968) A systematic relaxation method for the determination of the normal modes and the natural frequencies of vibrating systems, Int. J. Mech. Sci., 10, pp. 129-141.

Bates, C.J. (1973) A computational technique for the efficient handling of large matrices, Int. J. Num. Meth. Engng. 7, pp. 85-93.

Bathe, K.J. and Wilson, E.L. (1972) Large eigenvalue problems in dynamic analysis, ASCE J. EM. Div. 98, pp. 1471-1485.

Bathe, K.J. and Wilson, E.L. (1973a) Eigensolution of large structural systems with small bandwidth, ASCE J. EM Div. 99, pp. 467-479.

Bathe, K.J. and Wilson, E.L. (1973b) Solution methods for eigenproblems in structural mechanics, Int. J. Num. Meth. Engng. 6, pp. 213-226.

Bathe, K.J. and Wilson, E.L. (1976) Numerical Methods in Finite Element Analysis, Prentice Hall.

Borri, M. and Mantegazza, P. (1977) Efficient solution of quadratic eigenproblems arising in dynamic analysis of structures, Comp. Meth. Appl. Mech. and Engng. 12, pp. 19-31.

Clint, M. and Jennings A. (1970) Evaluation of eigenvalues and eigenvectors of real symmetric matrices by simultaneous iteration, Comput. J. 13, pp. 76-80.

Clint, M. and Jennings A. (1971) A simultaneous iteration method for the unsymmetric eigenvalue problem, J. Inst. Maths. Applics. 8, pp. 111-121.

Collar, A.R. (1978) The first fifty years of aeroelasticity, Aerospace 5(2), pp. 12-20.

Corr, B.R. and Jennings, A. (1973) Implementation of simultaneous iteration for vibration analysis, Comput. and Struc. 3, pp. 497-507.

Crawford, C.R. (1973) Reduction of a band-symmetric problem, CACM 16, pp. 41-44.

Cullum, Jane (1978) The simultaneous computation of a few algebraically largest and smallest eigenvalues of a large, sparse, symmetric matrix, BIT 18, pp. 265-275.

Cullum, Jane and Donath, W.E. (1974) A block Lanczos algorithm for computing the q algebraically largest eigenvalues and a corresponding eigenspace of large sparse real symmetric matrices, IEEE Conf. on Decision and Control, Phoenix, Ariz.

Davidson, E.R. (1975) The iterative calculation of a few of the lowest eigenvalues and eigenvectors of large real-symmetric matrices, J. Comput. Phys. 17, pp. 87-94.

Dong, S.B. (1977) A block-Stodola eigensolution technique for large algebraic systems with non-symmetrical matrices, Int. J. Num. Meth. Engng. 11, pp. 247-267.

Dong, S.B., Wolf, J.A., and Peterson, F.R. (1972) On a direct-iterative eigensolution technique, Int. J. Num. Meth. Engng. 4, pp. 155-161.

Downs, B. (1973) Direct iteration of ascending natural frequencies of vibration without inversion of the stiffness matrix, Int. J. Num. Meth. Engng. 7, pp. 554-556.

Ericsson, T. and Ruhe, A. (1979) The spectral transformation Lanczos method for numerical solution of large sparse generalized symmetric eigenvalue problems, Report UMINF-76.79, Univ. of Umeå, Sweden. (To appear in Math. Comp.)

Evans, D.J. (1975) Recursive algorithm for determining the eigenvalues of a quindiagonal matrix, Comput. J. 18, pp. 70-73.

Faddeev, D.K. and Faddeeva, V.N. (1963) Computational Methods of Linear Algebra, Freeman.

Golub, G.H. and Underwood, R. (1977) The block Lanczos method for computing eigenvalues ,in Mathematical Software III (J.R. Rice ed) Academic Press, pp. 361-377.

Gupta, K.K. (1970) Vibration of frames and other structures with banded stiffness matrix, Int. J. Num. Meth. Engng. 2, pp. 221-278.

Gupta, K.K. (1973) Eigenproblem solution by a combined Sturm sequence and inverse iteration technique, Int. J. Num. Meth. Engng. 7, pp. 17-42.

Gupta, K.K. (1974) Eigenproblem solution of damped structural systems, Int. J. Num. Meth. Engng. 8, 877-911.

Gupta, K.K. (1976) On a numerical solution of the supersonic panel flutter eigenproblem, Int. J. Num. Meth. Engng. 10, pp. 637-645.

Guyan, R.J. (1965) Reduction of stiffness and mass matrices, AIAA J. 3, pp. 380-380.

Householder, A.S. (1964) The Theory of Matrices in Numerical Analysis, Dover.

Hughes, T.J.R. (1976) Reduction scheme for some structural eigenvalue problems by a variational theorem, Int. J. Num. Meth. Engng. 10, pp. 845-852.

Irons, B.M. (1963) Eigenvalue economizers in vibration problems, J. RAeS. 67, pp. 526-528.

Jennings, A. (1967) A direct iteration method of obtaining latent roots and vectors of a symmetric matrix, Proc. Camb. Phil. Soc. 63, pp. 755-765.

Jennings, A. (1973) Mass condensation and simultaneous iteration for vibration problems, Int. J. Num. Meth. Engng. 6, pp. 543-552.

Jennings, A. (1977a) Matrix Computation for Engineers and Scientists, Wiley.

Jennings, A. (1977b) Influence of the eigenvalue spectrum on the convergence rate of the conjugate gradient method, J. Inst. Maths. Applics. 20, pp. 61-72.

Jennings, A. (1980) Eigenvalue methods for vibration analysis, Shock and Vib. Digest 12(2), pp. 3-16.

Jennings, A. and Agar, T.J.A. (1977) Hybrid Sturm sequence and simultaneous iteration methods, Symp. Appl. Comp. Meth. Engng., Univ. S. California, Los Angeles.

Jennings, A. and Agar, T.J.A. (1979) Progressive simultaneous inverse iteration for symmetric eigenvalue problems, Civ. Engng. Dept. Report, Queen's Univ. Belfast (To appear in J. Comput. Struc.)

Jennings, A. and Orr, D.R.L. (1971) Application of the simultaneous iteration method to undamped vibration problems, Int. J. Num. Meth. Engng. 3, pp. 13-24.

Jennings, A. and Stewart, W.J. (1975) Simultaneous iteration for partial eigensolution of real matrices, J. Inst. Maths. Applics. 15, pp. 351-361.

Jensen, P.S. (1972) The solution of large symmetric eigenproblems by sectioning, SIAM J. Numer. Anal. 9, pp. 534-545.

Kaniel, S. (1966) Estimates for some computational techniques in linear algebra, Math. Comp. 20, pp. 369-378.

Lanczos, C. (1950) An iteration method for the solution of the eigenvalue problem of linear differential operators, J. Res. Nat. Bur. Standards, 45, pp. 255-282.

Martin, R.S. and Wilkinson, J.H. (1968) Reduction of the symmetric eigenproblem $A\underline{x} = \lambda B\underline{x}$ and related problems to standard form, Numer. Math. 11, pp. 99-110.

McCormick, S.F. and Noe, T. (1977) Simultaneous iteration for matrix eigenvalue problems, Lin. Alg. and its Applics. 17, pp. 43-56.

Meirovitch, L. (1974) A new method of solution of the eigenvalue problem for gyroscopic systems, AIAA J. 12, pp. 1337-1342.

Muda, Y. (1973) A new relaxation method for obtaining the lowest eigenvalue and eigenvector of a matrix equation, Int. J. Num. Meth. Engng. 6, pp. 511-519.

Murray, M.A., Agar, T.J.A. and Jennings, A. (1980) Evaluation of virtual mass matrices for ship hulls using fluid singularities, Int. Conf. Recent Advances in Structural Dynamics, Southampton.

Nesbet, R.K. (1980) Large matrix techniques in quantum chemistry and atomic physics. These proceedings.

Ojalvo, I.U. and Newman, M. (1970) Vibration modes of large structures by an automatic matrix reduction method, AIAA J. 8, pp. 1234-1239.

Paige, C.C. (1972) Computational variants of the Lanczos method for the eigenproblem, J. Inst. Maths. Applics. 10, pp. 373-381.

Parlett, B.N. (1980) The Symmetric Eigenvalue Problem, Prentice-Hall.

Parlett, B.N. and Scott, D.S. (1979) The Lanczos algorithm with selective orthogonalization, Math. Comp. 33, pp. 217-238.

Peters, G. and Wilkinson, J.H. (1969) $A\underline{x} = \lambda B\underline{x}$ and the generalized matrix eigenvalue problem, SIAM J. Numer. Anal. 7, pp. 479-492.

Przemieniecki, J.S. (1968) Theory of Matrix Structural Analysis, McGraw-Hill.

Raju, I.S., Rao, G.V. and Murthy, T.V.G.K. (1974) Eigenvalues and eigenvectors of large order banded matrices, Comput. and Struc. 4, pp. 549-558.

Ramsden, J.N. and Stoker, J.R. (1969) Mass condensation: a semi-automatic method for reducing the size of vibration problems, Int. J. Num. Meth. Engng. 1, pp. 339-349.

Rocard, Y. (1957) Dynamic Instability, Crosby Lockwood.

Ruhe, A. (1974) SOR methods for the eigenvalue problem with large sparse matrices, Math. Comp. 28, pp. 695-710.

Ruhe, A. (1979) Implementation aspects of band Lanczos algorithms for computation of eigenvalues of large sparse symmetric matrices, Math. Comp. 33, pp. 680-687.

Rutishauser, H. (1963) On Jacobi rotation patterns, Proc. AMS Symp. in Appl. Math. 15, pp. 219-239.

Rutishauser, H. (1969) Computational aspects of F.L. Bauer's simultaneous iteration method, Numer. Math. 13, pp. 4-13.

Schwarz, H.R. (1968) Tridiagonalisation of a symmetric band matrix, Numer. Math. 12, pp. 231-241.

Schwarz, H.R. (1974a) The eigenvalue problem $(A-\lambda B)\underline{x}=\underline{0}$ for symmetric matrices of high order, Comput. Meth. Appl. Mech. Engng. 3, pp. 11-28.

Schwarz, H.R. (1974b) The method of coordinate over-relaxation for $(A-\lambda B)\underline{x}=\underline{0}$, Numer. Math. 23, pp. 135-151.

Schwarz, H.R. (1977) Two algorithms for treating $A\underline{x} = \lambda B\underline{x}$, Comput. Meth. Appl. Mech. Engng. 12, pp. 181-199.

Scott, D.S. (1979) How to make the Lanczos algorithm converge slowly, Math. Comp. 33, pp. 239-247.

Scott, D.S. (1980) The Lanczos algorithm. These proceedings.

Shavitt, I., Bender, C.F., Pipano, A. and Hosteny, R.P. (1973) The iterative calculation of several of the lowest eigenvalues and corresponding eigenvectors of very large symmetric matrices, J. Comput. Phys. 17, pp. 87-94.

Stewart, G.W. (1969) Accelerating the orthogonal iteration for the eigenvectors of a Hermitian matrix, Numer. Math. 13, pp. 362-376.

Stewart, G.W. (1976a) A bibliographical tour of the large, sparse generalized eigenvalue problem, in Sparse Matrix Computations, (J.R. Bunch and D.J. Rose eds.), Academic Press, pp. 113-130.

Stewart, G.W. (1976b) Simultaneous iteration for computing invariant sub-spaces of non-Hermitian matrices, Numer. Math. 25, pp. 123-136.

Stewart, W.J. and Jennings, A. (1980) LOPSI: A simultaneous iteration algorithm for real matrices, TOMS. To appear.

Todd, F.H. (1961) Ship Hull Vibrations, Edward Arnold.

Underwood, R. (1975) An iterative block Lanczos method for the solution of large sparse symmetric eigenproblems, Report CS-75-496, Stanford Univ.

Vysloukh, V.A., Kandidov, V.P. and Chesnokov, S.S. (1973) Reduction of the degrees of freedom in solving dynamic problems by the finite element method, Int. J. Num. Meth. Engng. 7, pp. 185-194.

Wilkinson, J.H. (1965) The Algebraic Eigenvalue Problem, Clarendon Press.

Yamamoto, Y. and Ohtsubo, H. (1976) Subspace iteration accelerated by using Chebyshev polynomials for eigenvalue problems with symmetric matrices, Int. J. Num. Meth. Engng. 10, pp. 935-944.

Zienkiewicz, O.C. (1977) The Finite Element Method, McGraw-Hill.

DISCUSSION

J.K. REID (AERE, Harwell). Should you not have stressed more the usefulness of condensation techniques to avoid having to work with a large problem at all? Often we only want information on the smooth eigenvalues which can be obtained quite nicely by using Rayleigh-Ritz on the smaller problem.

JENNINGS. Whereas condensation techniques have been used very effectively on many problems they have some disadvantages. One is that the effectiveness of the method depends on the choice of master displacements. If an operator makes a poor choice, which is a distinct possibility when the structure is complicated, some required modes may be inaccurately represented in the condensed system.

The simultaneous iteration method has similarities with condensation techniques, has a fairly similar numerical efficiency and has three advantages as listed in section 4 of my paper.

R.E. BORLAND (NPL). Have you ever encountered any problems where modes have been missed due to an inappropriate choice of starting vectors?

JENNINGS. I have only had one case which caused us a problem and even then we did not miss a mode. It was when we used a segment of the unit matrix as a starting vector for a cable vibration problem. It happened that six of the eight vectors involved in-plane displacements and only two out-of-plane displacements, whereas three of the required frequencies corresponded to out-of-plane vibrations. What happended was that the correct solution was obtained but after a lot more iterations than usual (the extra vector being generated from rounding errors). Now we use random trial vectors and we have never had any trouble. If one is concerned that a mode may be omitted, it is always possible to check this by one pass of the Sturm sequence method at the end of the calculation.

THE LANCZOS ALGORITHM

D.S. Scott

(Mathematics and Statistics Research Dept., Computer Sciences Division, Union Carbide Corporation, Nuclear Division, Oak Ridge Tennessee 37830, USA)

ABSTRACT

The Lanczos algorithm was originally used to tridiagonalize symmetric matrices but it was soon replaced by more effective methods based on explicit orthogonal similarity transformations. The algorithm was revived as an effective scheme for solving sparse symmetric eigenvalue problems. This paper surveys the historical development and the current state of the Lanczos algorithm.

1. THEORY

Given an n×n symmetric matrix A, a non-zero vector $\underline{r}_0$, $\beta_0 = \|\underline{r}_0\|$, and $\underline{q}_0 = \underline{0}$ then for j = 1,2,3,... the Lanczos algorithm computes†

1. $\underline{q}_j = \underline{r}_{j-1}/\beta_{j-1}$

2. $\underline{u}_j = A\underline{q}_j - \underline{q}_{j-1}\beta_{j-1}$

3. $\alpha_j = \underline{q}_j^*\underline{u}_j$

4. $\underline{r}_j = \underline{u}_j - \underline{q}_j\alpha_j$

5. $\beta_j = \|\underline{r}_j\|$ (stop if $\beta_j=0$).

Defining $Q_j = (\underline{q}_1,\underline{q}_2,\ldots,\underline{q}_j)$, an n×j matrix, and

†There are alternative implementations of the algorithm which will be discussed later but the one presented here is the best.

$$T_j = \begin{bmatrix} \alpha_1 & \beta_1 & & & 0 \\ \beta_1 & \alpha_2 & \beta_2 & & \\ & \beta_2 & & & \\ & & & & \beta_{j-1} \\ 0 & & & \beta_{j-1} & \alpha_j \end{bmatrix},$$

a j×j tridiagonal matrix, then it can be shown (see Parlett (1980b) for a good exposition) that

$$AQ_j - Q_jT_j = \underline{r}_j\underline{e}_j^*, \tag{1.1}$$

where $\underline{e}_j^* = (0,0,\ldots,0,1)$ is a j-vector, and

$$I - Q_j^*Q_j = 0. \tag{1.2}$$

The Lanczos vectors (the $\underline{q}$'s) form an orthonormal basis for the subspace span $(\underline{r}_0, A\underline{r}_0, \ldots, A^{j-1}\underline{r}_0)$ usually called a Krylov subspace and denoted $K_j(\underline{r}_0)$. Combining (1.1) and (1.2) we obtain

$$T_j = Q_j^*AQ_j,$$

which means that T_j is the projection of A onto the subspace $K_j(\underline{r}_0)$.

Using T_j it is possible (via the Rayleigh-Ritz procedure) to compute the optimal approximations to eigenpairs of A from the subspace $K_j(\underline{r}_0)$ as follows. Let $\Theta_j = \text{diag}\,(\theta_1^{(j)}, \theta_2^{(j)}, \ldots, \theta_j^{(j)})$ be the eigenvalues of T_j and let $S_j = (\underline{s}_1^{(j)}, \underline{s}_2^{(j)}, \ldots, \underline{s}_j^{(j)})$ be the corresponding matrix of eigenvectors, so that $T_j = S_j\Theta_jS_j^*$. Then the θ's are the optimal eigenvalue approximations (Ritz values) and the columns of $Y_j = (\underline{y}_1^{(j)}, \underline{y}_2^{(j)}, \ldots, \underline{y}_j^{(j)}) = Q_jS_j$ are the corresponding approximate eigenvectors (Ritz vectors).

Since the Lanczos vectors are orthogonal the process must terminate ($\beta_j = 0$) for some $j \leq n$. Originally, the algorithm was used as a way of tridiagonalizing A and early termination ($\beta_j = 0$ for $j < n$) was a nuisance requiring the selection of a vector orthogonal to all the current Lanczos vectors to continue the process. Now the algorithm is used primarily to find a few eigenpairs of a large matrix and early termination is a delight since the computed eigenpairs will be exact.

Much is known about the accuracy of Rayleigh-Ritz approximations obtained from Krylov subspaces. Explicit bounds were obtained by Kaniel (1966), corrected in Paige (1971), and improved in Saad (1980) (see Parlett (1980b) for a unified presentation) based on the characterization of $K_j(\underline{r}_0)$ in terms of polynomials

$$K_j(\underline{r}_0) = \{p(A)\underline{r}_0 | \deg p < j\}.$$

Three factors influence the accuracy of an approximation to an eigenpair (λ, $\underline{z}$):

1. $\underline{z}^*\underline{r}_0$, the component of $\underline{r}_0$ in the direction of $\underline{z}$. The larger the component the better the approximation.
2. The location of λ in the spectrum of A. The closer λ is to the edge of the spectrum the better the approximation.
3. The relative separation of λ from the other eigenvalues of A. The bigger the separation the better the approximation.

1 measures the situation at the beginning while 2 and 3 indicate the convergence rate as j increases. It is common for the Ritz approximations to well separated extreme eigenvalues to be correct to working accuracy for values of $j \ll n$ even when $\underline{r}_0$ is chosen randomly. On the other hand if $\underline{z}^*\underline{r}_0 = 0$ then the whole Krylov subspace $K_j(\underline{r}_0)$ will be orthogonal to $\underline{z}$ and $\underline{z}$ will never be detected. For simple eigenvalues this situation has probability zero but for multiple eigenvalues it is inevitable. In theory the Lanczos algorithm without modification is incapable of finding more than one eigenvector for any eigenvalue, namely the projection of the starting vector on the eigenspace, and so is unable to determine the multiplicity of any eigenvalue it finds.

It is of theoretical interest to note that for most matrices it is possible to choose a starting vector for which no eigenpair converges until j=n (Scott (1979a)). Its importance is that it gives insight into why such behaviour is never seen in practice.

The Lanczos algorithm is also an attractive method for solving generalized eigenvalue problems $A-\lambda M$ where M is positive definite. If $M = LL^*$ is the Cholesky factorization of M then the pencil $A-\lambda M$ can be reduced to an equivalent pencil $L^{-1}AL^{-*}-\lambda I$. The Lanczos algorithm can then be applied to the operator $L^{-1}AL^{-*}$ without forming the matrix explicitly. The Lanczos algorithm can even be applied directly to the pencil $A-\lambda M$ without factoring M provided that some method exists (for instance conjugate gradients) for solving equations $Mx = b$. The details are in section 15-11 of Parlett (1980b).

2. EARLY PRACTICE

When he introduced the algorithm (Lanczos (1950)), Lanczos was aware that the quantities computed by the algorithm in finite precision arithmetic would eventually diverge entirely from their theoretical counterparts. Equations (1.1) and (1.2) had to be modified to

$$AQ_j - Q_jT_j = \underline{r}_j\underline{e}_j^* + F_j \tag{2.1}$$

$$I - Q_j^*Q_j = H_j \tag{2.2}$$

where F_j and H_j account for the rounding errors. It was discovered that while (1.1) is always nearly satisfied ($\|F_j\|$ is tiny), the orthogonality among the Lanczos vectors was completely destroyed ($\|H_j\| \geq 1$).

So Lanczos suggested that the simple algorithm be modified so that the computed $\underline{r}_j$ is explicitly orthogonalized against all the earlier $\underline{q}$'s. This process is known as re-orthogonalization and with careful implementation it is sufficient to assure that the Lanczos vectors are orthogonal to working accuracy. This modification greatly increases the cost of the algorithm but allowed it to be successfully used to reduce symmetric matrices to tridiagonal form. With the advent of explicit orthogonal similarity transforms, more efficient methods of tridiagonalizing symmetric matrices were developed and interest in the Lanczos algorithm died.

The increased use of sparse matrices sparked new interest in the Lanczos algorithm. Of central importance is that the only way in which the matrix A enters the calculation is in the formation of the matrix-vector product $A\underline{q}_j$. Thus A may be stored in any compact way which allows such products to be computed efficiently. Also important is that only 3 n-vectors are needed in fast memory since the earlier $\underline{q}$'s can be put into secondary storage until they are needed to form the Ritz vectors. In fact, in Paige (1972), it is shown that with clever programming two n-vectors are sufficient. On the other hand full re-orthogonalization requires all the Lanczos vectors at each step and the use of this device nullifies the second advantage.

In practice the simple Lanczos algorithm (without re-orthogonalization) was found to behave quite oddly. After computing an accurate approximation to an eigenpair of the matrix A the algorithm "forgets" that it has done so and starts computing another. For larger values of j there will be two Ritz pairs which accurately approximate the <u>same</u> eigenpair of A. The Ritz vectors will be parallel which manifests the loss of linear independence among the Lanczos vectors. Still later a third copy will be found. The simple algorithm will grind away for ever computing more and more copies of extreme eigenvalues for each new interior one which converges. If A has multiple eigenvalues the repeated copies will contain components in the various orthogonal directions but determining the multiplicity requires computing all the Ritz vectors.

The loss of orthogonality in the algorithm left the users in a quandary. The use of full reorthogonalization is quite expensive and becomes more so as j increases. On the other hand allowing the orthogonality to dissipate required successful identification of the spurious copies as well as raising theoretical questions concerning the accuracy of the computed results. The most effective solution combined two ideas.

Iterative Lanczos periodically restarts the algorithm with a special starting vector chosen as some linear combination of Ritz vectors from the current subspace. If full re-orthogonalization is used then the start over point is determined by cost and storage considerations while some kind of monitoring of the loss orthogonality is required if no re-orthogonalization is performed (see Parlett and Scott (1979) for a discussion of monitoring the loss of orthogonality).

The block Lanczos algorithm replaces each Lanczos vector $\underline{q}_i$ by an $n \times p$ orthonormal matrix P_i. The resulting matrix T is block tridiagonal with $p \times p$ blocks. The off diagonal blocks can be chosen to be triangular so that T is a band matrix with $2p + 1$ diagonals. Block Lanczos has a theoretical advantage in that multiple eigenvalues can be found (up to the block size p).

In practice it allows multiplicities to be found simultaneously rather than sequentially. It has three further advantages. The loss of orthogonality tends to be directly dependent on the number of Lanczos steps taken. Thus with block Lanczos, when loss of orthogonality requires an iteration, the size of the current subspace will be larger so that more and better eigenvalue approximations will be available. More importantly, p vectors can be chosen for the new starting block which allows more information to be retained from one step to the next. Finally in many applications the matrix A will be in secondary storage and will have to be fetched (in pieces) to form the matrix-vector product $A\underline{q}$. In such a situation the cost of multiplying several vectors will be little more than the cost of multiplying one.

Iterative block Lanczos programs were successfully implemented by Cullum and Donath (1974) and by Underwood (1975). It should be noted that while block Lanczos has strong theoretical justification, iteration of the algorithm is employed only because of problems associated with loss of orthogonality and information is always lost when the current subspace is discarded.

3. THE RESULTS OF PAIGE

A deep understanding of the Lanczos algorithm was first obtained by Chris Paige in his PhD thesis (Paige (1971)). Unfortunately his thesis was never published and some of the most valuable results have become generally known only recently. We present here some of his most important discoveries.

He found that it is possible to monitor the convergence of the algorithm at step j by computing the residual norm $\|A\underline{y}_i - \underline{y}_i\theta_i\|$ of a Ritz pair $(\theta_i, \underline{y}_i)$ without computing the Ritz vector $\underline{y}_i$ at all! By definition $\underline{y}_i = Q_j\underline{s}_i$, where $(\theta_i, \underline{s}_i)$ is an eigenpair of T_j, and so multiplying equation (1.1) on the right by $\underline{s}_i$ we find

$$AQ_j\underline{s}_i - Q_jT_j\underline{s}_i = \underline{r}_j\underline{e}_j^*\underline{s}_i$$

$$A\underline{y}_i - Q_j\underline{s}_i\theta_i = \underline{r}_js_{ji}$$

$$A\underline{y}_i - \underline{y}_i\theta_i = \underline{r}_js_{ji}$$

where s_{ji} is the bottom element of $\underline{s}_i$. Taking norms we have

$$\|A\underline{y}_i - \underline{y}_i\theta_i\| = \beta_j|s_{ji}| \equiv \beta_{ji}. \tag{3.1}$$

In finite precision arithmetic it is equation (2.1) which the algorithm satisfies. Fortunately $\|F_j\|$ is always small in practice and specific bounds are given in Paige (1976). Thus β_{ji}, which can be computed by solving a small tridiagonal eigenvalue problem, is an inexpensive and accurate estimate of the residual norm of a Ritz vector. There remained the fear that the loss of linear independence among the Lanczos vectors might lead to a Ritz vector $\underline{y}$ with $\|y\| << 1$, which would invalidate the use of β_{ji} as an estimate of the accuracy of the eigenvalue. A complicated analysis in Paige (1971) which also appears in Scott (1978) shows that $\|\underline{y}\| << 1$ can happen only when there is a cluster of Ritz values all approximating the same eigenvalue of A. There will always be at least one Ritz vector in the cluster with $\|\underline{y}\| \simeq 1$, and, in any case, such a cluster of eigenvalues is always converged to working accuracy.

There is an alternative to the implementation of the Lanczos algorithm given in section 1. T_j is replaced by the nonsymmetric tridiagonal matrix

$$\begin{bmatrix} \alpha_1 & \eta_1 & & & \\ \beta_1 & \alpha_2 & \eta_2 & & \\ & \beta_2 & & & \eta_{j-1} \\ & & & \beta_{j-1} & \alpha_j \end{bmatrix}$$

where the η's are chosen as $\eta_i = \underline{q}_i^* A \underline{q}_{i+1}$. This ensures the orthogonality of $\underline{q}_i$ and $\underline{q}_{i+2}$ and since it seemed to preserve the orthogonality better, it was considered the better formula. Paige discovered (see Paige (1972)) that maintaining the symmetry of T was more important than maintaining local orthogonality between $\underline{q}_i$ and $\underline{q}_{i+2}$. The non-symmetric form can lose as much as half the available precision. (That is the Ritz values never converged to a relative accuracy of more than about $\sqrt{\epsilon}$, where ϵ is the relative machine precision). The symmetric implementation is also more efficient so there is no reason for further consideration of the nonsymmetric version.

Paige's most important discovery was the relationship between convergence and loss of orthogonality. In words:

loss of orthogonality $\Longleftrightarrow$ convergence.

This result became known soon after its discovery but the proof and some important intermediate results have only just appeared in the open literature (Parlett (1980b)).

The quantity of interest in monitoring loss of orthogonality is the j-vector $Q_j^* \underline{q}_{j+1}$ which would be zero in exact-arithmetic. Computational experience indicates that the elements of the vector grow in magnitude with increasing j but show no discernable pattern. The pattern can be seen only by making a change of basis by premultiplying $Q_j^* \underline{q}_{j+1}$ by S_j^* the eigenvector matrix of T_j.

$$S_j^* Q_j^* \underline{q}_{j+1} = Y_j^* \underline{q}_{j+1},$$

the inner products of the Ritz vectors with $\underline{q}_{j+1}$. Paige proved that

$$|\underline{y}_i^* \underline{r}_j| \; |s_{ji}| = \gamma_{ji} \tag{3.2}$$

where γ_{ji} is a messy expression involving n, j, A, and details of the arithmetic unit. Bounds for γ_{ji} exist and in practice

$$\gamma_{ji} \simeq \epsilon \, \|A\|$$

is always satisfied. Since $\underline{r}_j = \underline{q}_{j+1} \beta_j$ equation (3.2) can be rewritten as

$$\underline{y}_i^* \underline{q}_{j+1} = \gamma_{ji} / \beta_{ji}. \tag{3.3}$$

where β_{ji}, defined in equation (3.1), is the residual norm of $\underline{y}_i$ to working accuracy.

Thus orthogonality is lost only in the direction of converging Ritz vectors. Loss of linear independence occurs when a Ritz vector converges to working accuracy at which point a second Ritz vector appears essentially parallel to the first. The second Ritz vector perturbs the first away from the eigenvector (the β_{ji}'s are about $\sqrt{\epsilon}\|A\|$) and then the two vectors converge together until the $\beta_{ji} \simeq \epsilon \|A\|$ at which point a third copy appears.

Paige recommended that the simple Lanczos algorithm be run past the loss of linear independence until all the eigenvalues of interest had converged. This method was quite successful although it required the identification of the spurious copies. The identification is easy if the Ritz vectors are computed (since copies will be parallel instead of orthogonal) but it is not easy to make the identification without the Ritz vectors. The spurious copies also make it more difficult to determine when all the desired eigenvalues have converged.

4. SELECTIVE ORTHOGONALIZATION

Selective orthogonalization, the idea of orthogonalizing $\underline{q}_{j+1}$ against converged Ritz vectors only, was suggested to me in 1976 by Parlett, without knowledge of equations (3.3), as a way of suppressing repeated copies of Ritz vectors. The process seemed to be effective and the discovery of equation (3.3) in Paige's thesis gave a theoretical justification to the idea as well as the analytical tool needed for its analysis. The important ideas are in Parlett and Scott (1979) and a more detailed analysis in Scott (1978).

The basic idea is to maintain a level of independence among the Lanczos vectors measured by $\kappa_j = \|I - Q_j^* Q_j\|$. $\kappa_1 = \|1-\underline{q}_1^*\underline{q}_1\|$ will always be tiny ($\approx\epsilon$) since it just measures how well a vector can be normalized. κ_j is an increasing function of j and in the simple Lanczos algorithm κ_j will surpass 1 as the second copy of a Ritz vector appears and the Lanczos vectors become linearly dependent.

If $\kappa_j \approx \epsilon$ is to be maintained then full reorthogonalization is required but the important observation is that quite modest values of κ (say κ=0.01) are quite sufficient to prevent repeated copies from appearing. Thus selective orthogonalization allows the simple Lanczos algorithm to proceed as long as $\kappa_j < \kappa$ for some predetermined κ. It then steps in and performs the needed orthogonalization when κ_j threatens to grow bigger than κ.

Using equation (3.3) it can be shown that as long as

$$\beta_{ji} > \epsilon \|A\| / \kappa, \tag{4.1}$$

then

$$|\underline{y}_i^{(j)*} \underline{q}_{j+1}| < \kappa \text{ and}$$

$\|I - Q_{j+1}{}^* Q_{j+1}\| < (j+1)\kappa$. Except for the factor j + 1 this is the desired result. The factor j + 1 is never seen in practice and it will be omitted from further discussion.

Selective orthogonalization monitors the β_{ji} until some β_{ji} crosses the threshold, that is satisfies

$$\beta_{ji} < \epsilon \| A \| / \kappa . \quad (4.2)$$

The algorithm then pauses and computes the corresponding Ritz vector $\underline{y}_i^{(j)}$ by recalling the Lanczos vectors. $\underline{r}_j$ is then explicity orthogonalized against ("purged of") $\underline{y}_i^{(j)}$. It is also necessary to purge $\underline{r}_{j+1}$ of $y_i^{(j)}$. Then later $\underline{r}$'s will be essentially orthogonal to $\underline{y}_i^{(j)}$ due to the three term recurrence. It is only necessary to monitor (bound) $|\underline{y}_i^{(j)*}\underline{r}_k|$ for $k>j$ and to purge again (twice) when $|\underline{y}_i^{(j)*}\underline{r}_k|>\kappa$. The details are in Parlett and Scott (1979).

If suppressing repeated copies were the sole criterion for the choice of κ then a value like $\kappa = 0.01$ would be sufficient. However there are two compelling reasons to insist that equation (3.3) be satisfied by the unconverged Ritz vectors even after some Ritz vector has crossed the threshold. Firstly equation (3.3) provides the only inexpensive way of measuring the loss of orthogonality in the direction of a given Ritz vector and secondly it is only equation (3.3) which guarantees that $|\underline{y}^*\underline{q}_{j+1}|>\kappa$ will be satisfied by only a few Ritz vectors at most. This means that the orthogonalization of $\underline{r}_j$ against one Ritz vector must not significantly perturb the angles between $\underline{r}_j$ and the other Ritz vectors. This leads to a more stringent constraint on the value of κ as shown by the following analysis.

We need names for $\underline{r}_j$ and β_j before and after the orthogonalization and we will use $\underline{r}_j'$ and β_j' for the quantities before the orthogonalization. For convenience assume that only one Ritz vector $\underline{y}_m$ has crossed the threshold and needs to be purged from $\underline{r}_j'$. Then

$$\underline{r}_j \equiv r_j' - \underline{y}_m \xi_m,$$

where

$$\xi_m = \underline{y}_m^* \underline{r}_j', \quad \text{and so for } k \neq m,$$

$$\underline{y}_k^* \underline{r}_j = \underline{y}_k^* \underline{r}_j' - (\underline{y}_k^* \underline{y}_m) \xi_m.$$

In exact arithmetic both ξ_m and $\underline{y}_k^* \underline{y}_m$ would be zero but for our purposes only their product must be tiny. The factor $\underline{y}_k^* \underline{y}_m$ can be bounded directly by κ. The analysis of ξ_m is more complicated but the gist of it is that

$$|\xi_m| = |\underline{y}_m^* \underline{r}_j'| = |\cos \angle (\underline{y}_m, \underline{r}_j')| \beta_j'$$

and large values of $|\cos \angle (\underline{y}_m, \underline{r}_j')|$ can only occur in combination with small values of β_j' so that ξ_m can be bounded by $\kappa \|A\|$. See Scott (1978) for the details. Thus the product of the two terms is bounded by $\kappa^2 \|A\|$ and this leads to the choice of $\kappa = \sqrt{\epsilon}$.

Although $\kappa = \sqrt{\epsilon}$ is much smaller than $\kappa = 0.01$ a threshold vector will satisfy

$$\beta_{ji} \leq \sqrt{\epsilon} \|A\|$$

which means that in most cases the Ritz vector will have converged to half the working accuracy while the Ritz value will have converged to working accuracy. Hence very few Ritz vectors will ever cross the threshold which explains the computational efficiency of the scheme.

There is another reason for choosing $\kappa = \sqrt{\epsilon}$. The Ritz vectors computed by the algorithm will not be orthornormal to working accuracy. Instead they will satisfy

$$|\underline{y}_m^* \underline{y}_k| \le \kappa$$

Since the true eigenvectors are orthogonal, it is necessary to orthogonalize the Ritz vectors (in order of decreasing accuracy). The Rayleigh quotients of the new vectors will be perturbed by $O(\kappa^2 \|A\|)$ and choosing $\kappa = \sqrt{\epsilon}$ insures that this perturbation can be ignored.

Further insight into the behaviour of selective orthogonalization can be obtained by taking the terms representing the orthogonalizations and considering them as perturbations to the matrix T_j. The new matrix T_j' will be tridiagonal except for the addition of non-zero spikes in certain columns which are multiples of the eigenvectors of T which generated the Ritz vector used in the orthogonalization. For example if $\underline{y}_m^{(j)} = Q_j \underline{s}_m^{(j)}$ is the first threshold vector then

$$AQ_j - Q_j T_j = (\underline{r}_j + \underline{y}_m^{(j)} \xi_m^{(j)}) \underline{e}_j^*$$

can be rewritten as

$$\underline{AQ}_j - Q_j[T_j + \underline{s}_m^{(j)} \underline{e}_j^* \xi_m^{(j)}] = \underline{r}_j \underline{e}_j^*.$$

The choice of $\kappa = \sqrt{\epsilon}$ insures that the perturbation $\underline{s}_m^{(j)} \beta_m^{(j)}$ can be ignored in computing the eigensystem of later T's except for the required orthonormalization of the computed Ritz vectors. See Parlett (1980b) for more details.

The current implementation (see Scott (1979b)) employs selective orthogonalization only for those eigenpairs desired by the user. Orthogonality is allowed to dissipate in the direction of undesired eigenvectors. In theory this invalidates the a priori error estimates but the desired Ritz vectors still converge and their residual norms provide a posteriori error estimates. For problems where eigenpairs are desired from only one end of the spectrum this halves the number of β_{ji} which must be computed at each step. Only in very extreme examples would the convergence of the Ritz pairs be seriously degraded by the failure to purge the undesired Ritz vectors. This also removes the uncertainty in the amount of storage needed by the algorithm.

The current implementation employs block Lanczos with all its advantages discussed in section 2. In fact the ability to detect multiple eigenvalues simultaneously is so important that a blocksize of one is recommended only when only one eigenvalue is desired. The quantities computed by block Lanczos satisfy

$$AQ_j - Q_jT_j = R_jE_j^*$$

where $Q_j = (P_1, P_2, P_3, \ldots, P_j)$, each P_i is an n×p orthonormal matrix, T_j is a jp×jp block tridiagonal (band) matrix

$$T_j = \begin{bmatrix} A_1 & B_1^* & & & \\ B_1 & A_2 & B_2^* & & \\ & B_2 & \ddots & \ddots & \\ & & \ddots & \ddots & B_{j-1}^* \\ & & & B_{j-1} & A_j \end{bmatrix},$$

$R_j = P_{j+1}B_j$, and $E_j^* = [O\ I]$ is a p×jp matrix. The analogue of equation (3.1) for block Lanczos is

$$\|A\underline{y}_i - \underline{y}_i\theta_i\| = \|B_j\underline{s}_{ji}\| \equiv \beta_{ji},$$

where $\underline{s}_{ji}$ is a p-vector of the last p components of the vector $\underline{g}_i$. The analogue of equation (3.2) is

$$\underline{y}_i^* R_j \underline{s}_{ji} = \gamma_{ji} \tag{4.3}$$

where $|\gamma_{ji}| \simeq \epsilon\,\|A\|$. Since $R_j = P_{j+1}B_j$ equation (4.3) can be rewritten as

$$[\underline{y}_i^* P_{j+1}][B_j\underline{s}_{ji}] = \gamma_{ji}. \tag{4.4}$$

The vector $B_j\underline{s}_{ji}$ still contains all the needed information. It is $\|B_j\underline{s}_{ji}\|$ which measures convergence but it is a small element of $B_j\underline{s}_{ji}$ which signals loss of orthogonality. Thus

convergence ⇒ loss of orthogonality

but the arrow no longer goes both ways as the following example shows.

Let $R_0 = (\underline{q}, A\underline{q})$ for an arbitrary vector $\underline{q}$. The columns of P_1 form an orthonormal basis of span (R_0) so that span (AP_1) = span $(A\underline{q}, A^2\underline{q})$. Thus R_1 will have rank 1 and one column of P_2 would be zero. Numerically the second column of P_2 would be normalized noise and serious loss of orthogonality would occur even though no convergence had occurred.

The Lanczos algorithm with selective orthogonalization provides a black box method for computing a few eigenvalues and eigenvectors of a symmetric operator which is ideally suited for large sparse problems. The details of the current implementation are in Scott (1979b).

5. CURRENT RESEARCH

Current research in the Lanczos algorithm has been focused on using it to solve other related problems. This section will survey current research areas.

A problem which arises in some applications is to compute <u>all</u> the distinct eigenvalues of a large sparse symmetric matrix. The Lanczos algorithm (with no reorthogonalization) was used by Cullum and Willoughby with a clever method of identifying and discarding repeated copies (Cullum and Willoughby (1978, 1980)). The main difficulty is to determine when to stop the process. $j \geq n$ is clearly needed but $j \geq 10n$ is sometimes required to resolve clusters of eigenvalues and the cost of computing eigenvalues of such large tridiagonal matrices is non-trivial. More recently Parlett and Reid have employed an efficient method of monitoring the convergence of the algorithm which allows termination to occur as soon as possible (Parlett and Reid (1980)).

In some applications (especially generalized eigenvalue problems) it is possible to factor the matrix A (or the shifted matrix $A - \sigma I$) and run the Lanczos algorithm on A^{-1} (or $(A-\sigma I)^{-1}$). The inversion of the operator causes a transformation of the spectrum. The eigenvalues far from the shift σ are compressed near zero so that the eigenvalues near σ become the extreme ones with large relative separations. Thus by the Kaniel-Paige-Saad theory discussed in section 2, convergence of the algorithm to the eigenvalues near σ is quite rapid. If only one factorization is possible then the shift σ should be chosen near the centre of the desired eigenvalues and Lanczos with selective orthogonalization should be used. (One of the codes described in Scott (1979b) is designed for this context). Some problems are so difficult that they can be solved only with an inversion and in any case the convergence rate of the eigenvalue nearest σ is <u>always</u> improved by such an inversion.

If several factorizations can be computed then the questions are how long should the Lanczos algorithm be run with one factorization before computing a new factorization, what should the new shift parameter be, and how should the new starting vector (or vectors) be chosen? One convenient result is that the factorization of $A - \sigma I$ gives a count of the number of eigenvalues of A less than σ (equal to the number of negative pivots encountered). This means that eigenvalues cannot be missed by the algorithm without a warning message.

A preliminary code for this problem is described by Ericsson and Ruhe (1979). The code uses simple Lanczos with full reorthogonalization. The length of a Lanczos run is determined by storage and cost considerations. The new shift is chosen so that the last converged eigenvalue lies halfway between the new and old shift. The starting vectors are chosen randomly and are then orthogonalized against the converged vectors. The code employs a clever modification which improves the accuracy of the eigenvectors.

The author has a preliminary code which uses block Lanczos with no reorthogonalization. The end of a Lanczos run is signalled by storage considerations or significant loss of orthogonality ($\beta_{ji} < 10n\varepsilon\|A\|$ say). The new shift bisects the p and (p + 1)st unconverged eigenvalues (p is the blocksize) and choses the p starting vectors as linear combinations of the unconverged eigenvectors. It is recommended that the blocksize be chosen as $p = \sqrt{\text{mband}}$ where mband is the bandwidth of the matrix A. This choice is made to try to equalize to some extent the cost of a Lanczos step and the cost of a factorization.

Both codes are capable of solving generalized problems $A-\lambda M$ including the important special case when M is diagonal and only positive semi-definite.

No direct comparisons of the codes have been made but I would expect that as the bandwidth of A increased (and with it the cost of a factorization) the better and better the block version would perform. Ericsson and Ruhe make no use of the unconverged Ritz vectors due to difficulties with multiple eigenvalues. Since block Lanczos essentially solves the multiple eigenvalue difficulty my code is able to retain more information at each iteration. However a multiplicity larger than the blocksize can cause my code to return erroneous results so more work is needed.

There is a close relationship between the conjugate gradients algorithm for solving $A\underline{x} = \underline{b}$ (with initial guess $\underline{x}_0 = \underline{0}$) and the Lanczos algorithm with $\underline{r}_0 = \underline{b}$. In theory the conjugate gradient solution at step j is

$$\underline{x}_j = A_j^+ \underline{b}$$

where A_j is the projection of A onto the subspace $K_j(\underline{r}_0) = \text{span}(\underline{b}, A\underline{b}, A^2\underline{b}, \ldots, A^{j-1}\underline{b})$. Using quantities computed by the Lanczos algorithm,

$$\underline{x}_j = Q_j T_j^{-1} Q_j^* \underline{b},$$

where $Q_j^*\underline{b} = \underline{e}_1 \beta_0$ by the special choice of $\underline{r}_0$. Due to rounding errors both conjugate gradients and the Lanczos algorithm diverge completely from their theoretical paths but the relation

$$\underline{x}_j = Q_j T_j^{-1} \underline{e}_1 \beta_0 \tag{5.1}$$

continues to hold.

Equation (5.1) gives an easy explanation of the potential instability of conjugate gradients for indefinite matrices. If A is indefinite T_j can be singular and the formula breaks down. Numerically this corresponds to the solution $\underline{x}_j$ growing very large at one step and then shrinking at the next causing serious loss of accuracy due to cancellation. (T_j and T_{j+1} cannot both be ill conditioned unless A itself is).

This relationship between conjugate gradients and Lanczos was exploited in Paige and Saunders (1975) to produce a stable algorithm, SYMMLQ, for solving symmetric indefinite systems of equations. Since SYMMLQ is less efficient than simple conjugate gradients it was not recommended for positive definite A.

At the cost of storing the Lanczos vectors ($\underline{q}$'s) in secondary store, the Lanczos formulation offers several important advantages.

1. Indefinite A causes no appreciable extra expense. It is only necessary to test the condition of the small tridiagonal matrix T_j and take one more Lanczos step if T_j is ill conditioned.
2. Selective orthogonalization can be used to maintain robust linear independence among the Lanczos vectors. This will decrease the number of steps needed by the algorithm.
3. If the solution of more than one right hand side is needed

then the subspace generated for one right hand side can be used to find a good starting guess for $\underline{b}_2$ as

$$\underline{x}_0 = Q_j T_j^{-1} Q_j^* \underline{b}_2 .$$

It is possible (even likely) that this $\underline{x}_0$ will already be sufficiently accurate and no further modification will be needed. (If A has multiple or near multiple eigenvalues it will be necessary to use block Lanczos in working on $\underline{b}_1$ to retain the high likelihood that the initial guess $\underline{x}_0$ for $\underline{b}_2$ is sufficiently accurate).

It is important to point out that preconditioning schemes can be incorporated into the Lanczos formulation without difficulty. See Parlett (1980a) for a more detailed discussion of using (simple) Lanczos for solving linear equations.

Another area of interest for the Lanczos algorithm is in solving λ-matrix problems of the form

$$(\lambda^k A_k + \lambda^{k-1} A_{k-1} + \dots \lambda_0)\underline{x} = \underline{0}. \tag{5.2}$$

Such problems can always be rewritten as linear problems $A - \lambda M$ where A and M are $kn \times kn$ matrices but it would be nice to have some method of finding eigenvalues of (5.2) working with only $n \times n$ matrices.

A method given in Scott (1979c) uses the Lanczos algorithm to solve linear problems $A - \lambda M$ for one eigenvalue at a time without factorizing any matrix. Extensions of this idea are being investigated for λ-matrices of degree bigger than 1. In particular, a method of solving quadratic λ-matrix problems using the Lanczos algorithm and factorizing only one $n \times n$ matrix is given in Scott (1980).

6. CONCLUSIONS

The Lanczos algorithm is a powerful technique for computing eigenvalues of sparse symmetric matrices. Its behaviour in finite precision arithmetic is well understood and effective means now exist for correcting its known deficiencies. If the formation of matrix-vector products is the only available information then the Lanczos algorithm is the best method for computing eigenvalues.

Inversion of the spectrum via a sparse factorization greatly improves the condition of the desired eigenvalues and Lanczos applied to the inverted problem is again the best solution method. Only when a complete factorization is impossible will other methods, based on some approximate inversion, prove superior to the Lanczos algorithm.

REFERENCES

Cullum, J. and Donath, W.E. (1974) A block generalization of the the symmetric S-step Lanczos algorithm, Report # RC 4845 (# 21570), IBM Thomas J. Watson Research Center, Yorktown Heights, New York.

Cullum, J. and Willoughby, R.A. (1979) Lanczos and the computation in specified intervals of the spectrum of large, sparse, real symmetric matrices, in Sparse Matrix Proceedings 1978 (I.S. Duff and G.W. Stewart eds.) SIAM Press, pp. 220-255.

Cullum, J. (1980) Computing eigenvectors (and eigenvalues) of large, symmetric matrices using Lanczos tridiagonalization, in Numerical Analysis, (G.A. Watson ed) Springer-Verlag, pp. 46-63.

Ericsson, T. and Ruhe, A. (1980) The spectral transformation Lanczos method in the numerical solution of large, sparse, generalized, symmetric eigenvalue problems, Math. Comp. 34, to appear.

Golub, G.H. (1973) Some uses of the Lanczos algorithm in numerical linear algebra, in Topics in Numerical Analysis (J.J.H. Miller ed), Academic Press, pp. 173-184.

Golub, G.H., Underwood, R. and Wilkinson, J.H. (1972) The Lanczos algorithm for the symmetric $Ax = \lambda Bx$ problem, Technical Report STAN-CS-72-270, Computer Science Dept., Stanford University.

Jensen, P.S. (1972) The solution of large symmetric eigenproblem by sectioning, SIAM J. Numer. Anal. 9, pp. 534-545.

Kahan, W. and Parlett, B.N. (1976) How far should you go with the Lanczos algorithm?, in Sparse Matrix Computations, (J.R. Bunch and D.J. Rose eds.) Academic Press, pp. 131-144.

Kaniel, S. (1966) Estimates for some computational techniques in linear algebra, Math. Comp. 20, pp. 369-378.

Lanczos, C. (1950) An iteration method for the solution of the eigenvalue problem of linear differential and integral operators, J. Res. Nat. Bur. Standards, Sect. B. 45, pp. 225-280.

Paige, C.C. (1971) The computation of eigenvalues and eigenvectors of very large sparse matrices, PhD Thesis, University of London.

Paige, C.C. (1972) Computational variants of the Lanczos method for the eigenproblem, J. Inst. Maths. Applics. 10, pp. 373-381.

Paige, C.C. (1976) Error analysis of the Lanczos algorithm for tridiagonalizing a symmetric matrix, J. Inst. Maths. Applics. 18, pp. 341-349.

Paige, C.C. and Saunders, M.A. (1975) Solution of sparse indefinite systems of linear equations, SIAM J. Numer. Anal. 12, pp. 617-629.

Parlett, B.N. (1980a) A new look at the Lanczos algorithm for solving symmetric systems of linear equations, Lin. Alg. and its Applics. 29, pp. 323-346.

Parlett, B.N. (1980b) The symmetric eigenvalue problem, Prentice-Hall.

Parlett, B.N. and Reid, J.K. (1980) Tracking the progress of the Lanczos algorithm for large symmetric eigenvalue problems, Report CSS # 83, Computer Sciences and Systems Div., AERE, Harwell, England.

Parlett, B.N. and Scott, D.S. (1979) The Lanczos algorithm with selective orthogonalization, Math. Comp. 33, pp. 217-238.

Ruhe, A. (1979) Implementation aspects of band Lanczos algorithm for computation of eigenvalues of large sparse matrices, Math. Comp. 33, pp. 680-687.

Saad, Y. (1980) Error bounds on the interior Rayleigh-Ritz approximations from Krylov subspaces, SIAM J. Numer. Anal. 17, to appear.

Scott, D. (1978) Analysis of the symmetric Lanczos algorithm, PhD dissertation, Department of Mathematics, University of California, Berkeley.

Scott, D. (1979a) How to make the Lanczos algorithm converge slowly, Math. Comp. 33, pp. 239-247.

Scott, D. (1979b) Block Lanczos software for symmetric eigenvalue problems, Technical Report ORNL/CSD-48, Union Carbide Corporation, Nuclear Division, Oak Ridge, TN.

Scott, D. (1979c) Solving sparse symmetric generalized eigenvalue problems without factorization, Technical Report

ORNL/CSD-49, Union Carbide Corporation, Nuclear Division, Oak Ridge, TN. SIAM J. Numer. Anal. to appear.

Scott, D. (1980) Solving sparse quadratic λ-matrix problems, Technical Report ORNL/CSD-69, Union Carbide Corporation, Nuclear Division, Oak Ridge, TN.

Underwood, R. (1975) An iterative block Lanczos method for the solution of large sparse symmetric eigenproblems, PhD dissertation, STAN-CS-75-496, Stanford University.

DISCUSSION

PROFESSOR A. JENNINGS (Belfast). In my experience simultaneous iteration will work as well as the Lanczos algorithm if the eigenvalues are well separated.

SCOTT. I agree although in many practical situations we do not have good separation of eigenvalues.

DR. A.M. ERISMAN (Boeing). Could you give me some idea of the sizes of problems you have run your Lanczos codes on and the relative speed of your codes?

SCOTT. We are currently solving problems of order greater than 2000 and have found that when we replaced the simultaneous iteration routines in the SAP package (Structural Analysis Program) with our Lanczos routines we can get an improvement of a factor of 10 in computing speeds. I don't have actual times with me but they will, of course, be very dependent on the machine being used. Certainly, the Lanczos algorithm is much less sensitive to poor starting vectors.

DR. J.K. REID (Harwell). We have found that a single inverse iteration procedure is very successful in obtaining the smallest eigenvalue of a plasma physics problem where that eigenvalue is simple but close to a large set of very close eigenvalues.

SCOTT. Certainly for these problems arising from the study of the stability in plasmas in Tokamaks by a finite element discretization one must factor the matrix. Typically one isolated eigenvalue is near zero (either positive or negative) and the rest form an essentially continuous spectrum of positive eigenvalues whose upper limit is around 10^{15}. Thus one needs to factor $(K-\sigma M)$ to increase the separation and although you are only seeking one eigenvector you must use an accurate estimation of σ. Currently, the plasma codes (for example the ERATO code from Lausanne)use coarse grids to get such an estimate.

UNKNOWN. What happens if you can't afford a matrix factorization?

SCOTT. It depends on the spectrum. For example if you wish to find the eigenpair corresponding to eigenvalue 1 for the distribution, 1, 2, 3 ... 1000 then about 200 iterations would be required whereas if the distribution is of the form 1 1/2 1/3 ... 1/10000 only 6 iterations would be needed. Typical structures problems have a distribution of the form 2^k where the small eigenvalues are required. Often in excess of 30,000 iterations will be required if factorization is not performed.

LARGE MATRIX TECHNIQUES IN QUANTUM CHEMISTRY AND ATOMIC PHYSICS*

R.K. Nesbet†

(Fachbereich Physik, Universität Kaiserslautern, 6750 Kaiserslautern, W. Germany)

ABSTRACT

The matrix eigenvalue problem

$$(H - \varepsilon I)\ \underline{c} = \underline{0}$$

arises in variational approximations to the Schrödinger equation. The N×N real symmetric matrix H can be very large ($N = 10^3 - 10^5$) with about 0.1 N^2 non-zero elements. In general, the spectrum of H is bounded below, only a few (less than 10) lowest eigenvalues are required, and H is diagonally dominant. Accurate eigenvectors are required. Iterative methods are favoured, because viable methods must have computation times that vary as N^2 rather than N^3. A new iterative method, based on a partitioned modification of the Jacobi algorithm, is proposed for the large matrix eigenvalue problem. In other applications, the matrix expression

$$A^T (H - \varepsilon I)^{-1} B$$

must be evaluated. Here A and B are rectangular matrices, and H has the properties mentioned above. Methods currently in use for these problems will be discussed.

1. INTRODUCTION

Bound states in quantum mechanics are characterized as eigenstates of a linear Hamiltonian operator, whose energy eigenvalue

---

*Work supported in part by the US Office of Naval Research and by the Deutsche Forschungsgemeinschaft (SFB 91)

†Permanent address: IBM Research Laboratory, San Jose, California 95193, USA

spectrum is bounded below but not above. Variational calculations require the evaluation of the lower eigenvalues of a Hermitian matrix H. The dimension can be very large ($N=10^3-10^5$), with perhaps $N^2/10$ non-zero elements. The eigenvalue spectrum is bounded below, and states of physical interest correspond to the lowest eigenvalues. The matrix H, while not sparse in any strict sense, is diagonally dominant. Accurate eigenvectors are required.

These properties favour iterative methods. Viable methods must have computation times that vary as N^2 rather than N^3. Particular methods that have been found in practice to be successful are described here. Methods for the evaluation of expressions involving the inverse $(H-\varepsilon I)^{-1}$ are also discussed. Extensive bibliographies and reviews of these subjects are to be found in a recent report (NRCC (1978)).

Relaxation methods appropriate for the lowest eigenvalue are discussed in section 2, and methods for higher eigenvalues in section 3. A new algorithm for the latter problem, presented in section 4, uses the classical Jacobi method to block-diagonalize a partitioned matrix. Applications involving matrix inversion are discussed in section 5.

2. RELAXATION METHODS - LOWEST EIGENVALUE

The linear eigenvalue problem

$$H\underline{c}_k = \underline{c}_k \lambda_k \quad k=1,\ldots N \tag{1}$$

or the generalized eigenvalue problem

$$H\underline{c}_k = S\underline{c}_k \lambda_k \quad k=1,\ldots N \tag{2}$$

is to be solved for a few of the lowest eigenvalues and for the corresponding eigenvectors. The matrices H and S are real and symmetric, S is positive definite, but the dimension N can be very large.

Boys (1950) used a simple relaxation method based on the method of Seidel (1874). For a single eigenvector, (2) can be written as

$$[H-(\lambda+\Delta\lambda)S](\underline{c}+\underline{\Delta c}) = \underline{0} \tag{3}$$

where $\underline{c}$ is an approximate eigenvector and λ is an approximate eigenvalue. The residual vector is

$$(H-\lambda S)\underline{c} = \underline{q} \tag{4}$$

Given a value of λ, the ith equation of the linear system is satisfied exactly if $\underline{\Delta c}$ has

$$\Delta c_i = -(H_{ii}-\lambda S_{ii})^{-1} q_i \tag{5}$$

as its only non-zero coefficient. If one coefficient c_1 dominates the eigenvector, it can be fixed at unity. The updated eigenvalue is determined from the equation of index 1,

$$\Delta\lambda = \{(H-\lambda S)(\underline{c}+\underline{\Delta c})\}_1/\{S(\underline{c}+\underline{\Delta c})\}_1 \tag{6}$$

Boys (1950) pointed out that a variationally stable estimate of $\Delta\lambda$ should be obtained from the Rayleigh quotient,

$$\lambda = N/D, \tag{7}$$

where, for current values of the vector $\underline{c}$,

$$N = \underline{c}^T H\underline{c} \tag{8}$$

$$D = \underline{c}^T S\underline{c}, \tag{9}$$

in the generalized eigenvalue problem. Iterative schemes based on use of the Rayleigh quotient were considered by Löwdin (1962).

These two approaches were combined by Nesbet (1965) in a variant of an algorithm published earlier by Cooper (1948). If a single eigenvector component is updated by (5), then

$$\Delta N = (2\underline{c}^T H + \underline{\Delta c}^T H)\underline{\Delta c}\ , \tag{10}$$

$$\Delta D = (2\underline{c}^T S + \underline{\Delta c}^T S)\underline{\Delta c}\ . \tag{11}$$

Since

$$\lambda+\Delta\lambda = (N+\Delta N)/(D+\Delta D) \tag{12}$$

then

$$\Delta\lambda = (\Delta N-\lambda\Delta D)/(D+\Delta D), \tag{13}$$

which reduces to

$$\Delta\lambda = q_i\Delta c_i/(D+\Delta D). \tag{14}$$

This simple formula is used to update the Rayleigh quotient as each successive Δc_i is computed. (5) and (14) imply monotonic convergence to the lowest (and highest) eigenvalue of H if the current λ is such that all denominators in (5) have the same sign.

This algorithm has been widely used in quantum chemical calculations. It is rapidly convergent in most applications, easily programmed, and efficient in its data-handling aspects. The dominant computational step is the evaluation of the residual vector $\underline{q}$. As given by (4), each component of $\underline{q}$ is a scalar product, requiring only a single row of H.

In electronic wave function calculations, the matrix elements H_{ij} can be represented as linear combinations of a much smaller number of basic integrals h_A, with coefficients $X_{A,ij}$ that can be computed easily as required. When $N^2 \gg 10^8$, the direct use of H_{ij} becomes impractical. In the direct CI method (Roos(1972)), the formula

$$(H\underline{c})_i = \Sigma_A h_A \Sigma_j X_{A,ij} c_j \tag{15}$$

is used to evaluate the residual q_i required in (5) and (14). This requires storage and retrieval only of the elements h_A.

Shavitt (1970) modified the relaxation algorithm so that symmetric matrices H and S, stored as lower triangular arrays of independent elements, could be accessed sequentially. This is done (for H) by accumulating the partial sums

$$t_i = \sum_{j>i} H_{ij} c_j = \sum_{j>i} H_{ji} c_j \tag{16}$$

for each i, as the rows j of H with j>i are read.

(5) is a linear approximation to the solution of a 2 × 2 secular equation in a basis consisting of $\underline{c}$ and the unit vector $\underline{e}_i$. More generally, an arbitrary perturbing vector $\underline{d}$ can be used, with coefficient α determined from the secular equation, equivalent to making the Rayleigh quotient stationary. Such optimal relaxation methods have been considered by Fadeev and Fadeeva (1963) and by Shavitt et al (1973). In an obvious notation

$$(H_{dc} \equiv \underline{d}^T H \underline{c}, \text{ for example}),$$

equation (12) gives

$$\left.\begin{aligned} H_{cc}-\lambda S_{cc} &= 0 \\ \text{and we define } q_d \text{ by} \\ H_{dc}-\lambda S_{dc} &= q_d . \end{aligned}\right\} \quad (17)$$

The secular equation for $\Delta\lambda$ gives

$$-\Delta\lambda S_{cc}(H_{dd}-\lambda S_{dd}-\Delta\lambda S_{dd})=(q_d-\Delta\lambda S_{dc})^2 \quad (18)$$

The coefficient of vector $\underline{d}$ is given by

$$\alpha=-(H_{dd}-\lambda S_{dd}-\Delta\lambda S_{dd})^{-1}(q_d-\Delta\lambda S_{dc}) \quad (19)$$

When terms of $O(q_d^2)$ are neglected, if d is the vector $\underline{e}_i$, (19) reduces to (5). It should be noted that (13) is valid without approximation.

For large matrices, the extra work involved in solving (18) is negligible compared with the computation of q_d. Hence, in principle, the optimal relaxation algorithm should be used in preference to (5) and (14). In practice, because $\Delta\lambda$ vanishes quadratically, similar convergence rates are achieved unless the starting vector has been badly chosen.

Schwarz (1974) and Ruhe (1974) have considered over-relaxation as a refinement of these methods.

In the optimal relaxation algorithm, the vector $\underline{d}$ can be chosen arbitrarily. Davidson (1975) has devised an algorithm based on use of the vector $\underline{\Delta c}$, whose components are given by Δc_i of (5). Since this method is used primarily for non-extreme eigenvalues, it will be described in the following section. Test calculations (Davidson (1975)) indicate that for the lowest eigenvalue alone, the optimal relaxation algorithm, using co-ordinate vectors, converges in somewhat fewer iterations than does Davidson's method. For matrices of the kind considered, the use of the Krylov sequence of successive vectors $\underline{c}$, $H\underline{c}$, $H^2\underline{c}$, ... leads to slow convergence for the lowest eigenvalue. Davidson (1975) points out that the gradient method (Hestenes and Karush (1951a, 1951b)) and the method of Lanczos (1950) involve essentially the same sequence of vectors. Test calculations showed very poor convergence for the Lanczos method (Davidson (1975)) and for the gradient method (Shavitt et al (1973)).

3. METHODS FOR HIGHER EIGENVALUES

The relaxation method has been proven to be very effective for obtaining the lowest eigenvalue of a large matrix. Convergence to higher eigenvalues cannot be guaranteed. Shavitt et al (1973) have applied two procedures to obtain higher eigenvalues, in the context of the optimized relaxation method, equations (18) and (19). In the first method (Hotelling (1933)), each known eigenvalue is shifted upwards by adding the appropriate dyadic to H_{ij}. For the generalized eigenvalue problem, this is

$$\Delta H_{ij} = (S\underline{c}_k)_i \, \mu_k (\underline{c}_k^T S)_j . \tag{20}$$

This root-shifting method was used earlier by Cooper (1948). The parameter μ_k must be large enough so that all values of $\lambda_k + \mu_k$ are greater than the next eigenvalue to be determined, which then becomes the lowest of the modified eigenvalue problem.

The scalar products $\underline{c}_k^T S \underline{c}_k$ are updated each time the new approximate eigenvector is changed. If the shift parameter μ_k is too large, diagonal dominance can be lost, and the method is slowly convergent.

The second method considered by Shavitt et al (1973) is specific orthogonalization of the coordinate vectors $\underline{e}_i$ to the known eigenvectors, at each updating step. Both methods were found to require comparable computation times in trial calculations (Shavitt et al(1973)).

These methods require successive evaluation of all eigenvalues below the one desired, and have convergence difficulties for nearly degenerate eigenvalues. In order to achieve more stable convergence in difficult cases, Davidson (1975) developed a method in which several vectors are considered together, so that a particular eigenvector can be selected after diagonalizing a small matrix. When combined with a root-homing procedure, based on maintaining maximum overlap at each iteration with an approximate eigenvector obtained in a preliminary calculation (Butscher and Kammer (1976)), Davidson's method can be used to select a particular upper eigensolution out of a set with closely-spaced eigenvalues.

The version of this method to be described here is due to Liu (1978), who showed that computations could be organized so that several eigenvectors can be obtained simultaneously. This is generally more efficient for closely-spaced eigenvalues than the original procedure, which obtained selected eigenvectors in succession. To simplify the notation, only the standard eigenvalue problem (1) will be considered here.

In Davidson's method, the original basis $\{\underline{e}_i\}_N$ of H_{ij} is condensed iteratively into a much smaller working basis of orthonormal vectors $\{\underline{b}_a\}_M$, with $M \ll N$. Eigenvectors $\{\underline{c}_k\}_L$ with $L \leq M$, are constructed by diagonalizing H within the basis $\{\underline{b}_a\}$.

The vector $H\underline{b}$ must be computed for each working vector $\underline{b}$. For large matrices this is the dominant computational step. The matrix of scalar products

$$G_{ab} = \underline{b}_a^T H \underline{b}_b \tag{21}$$

is computed and diagonalized by a standard small-matrix method to give approximate eigenvectors of H in the form

$$\underline{c}_k = \Sigma_a \underline{b}_a \alpha_{ak} \tag{22}$$

The residual vectors are

$$\underline{q}_k = (H-\lambda_k I)\underline{c}_k \tag{23}$$

$$= \Sigma_a (H\underline{b}_a - \lambda_k \underline{b}_a)\alpha_{ak}. \tag{24}$$

Because of their structure, the residual vectors do not have to be stored separately.

Instead of updating the vectors $\underline{c}_k$ element by element, a simultaneous update vector $\Delta\underline{c}_k$ is defined, with elements given by (5). Because this procedure neglects the interaction between different components of the eigenvector, $\underline{c} + \Delta\underline{c}$ is not the exact solution of the component equations. Instead, the vectors $\Delta\underline{c}$ are used at each iterative step as new basis vectors, augmenting the set $\{\underline{b}\}$. In practice (Liu (1978)), the vectors $\Delta\underline{c}$ can be orthogonalized to the set $\{\underline{b}\}$ and discarded if the resulting norm is less than a certain criterion. The remaining vectors are orthonormalized and included in an extended basis $\{\underline{b}\}$. If this set grows to inconvenient size, it can be contracted to the current set of vectors $\{\underline{c}\}$ selected by an eigenvalue or overlap criterion before subsequent iterations. Trial calculations (Davidson (1975), Liu (1978)) show convergence in a number of full iterations only moderately larger than that required by the successive relaxation method for a single eigenvector.

4. A PARTITIONED JACOBI METHOD FOR LARGE MATRICES

The classical Jacobi method is a direct transformation of the matrix H to diagonal form by successive 2×2 orthogonal transformations. The method is convergent, because it reduces the total norm of off-diagonal matrix elements at each elementary step. When used for complete diagonalization, the Jacobi method is slower than methods based on reduction to tridiagonal form. However, because it deals with individual elements of the transformed matrix, it can be used to reduce H to block-diagonal form, with computational effort proportional to the number of off-diagonal elements to be eliminated.

Because it requires transformation of the entire matrix, the standard Jacobi method is not suitable for large matrices. However, as will be shown here, a modified algorithm can be devised that avoids transformation of the larger sector of a partitioned matrix, while diagonalizing the smaller sector and reducing the rectangular array of coupling matrix elements to zero.

Preliminary test calculations with this method indicate that for the lowest eigenvalue of a large matrix it converges in times comparable to the successive relaxation method, and that it converges equally efficiently in applications requiring any specified number of eigenvalues, the computation time being proportional to this number. Thus this method appears to be competitive with those described above.

In the partitioned Jacobi method, the approximate eigenvectors $\{\underline{c}_k\}_L$ are used directly as basis vectors. As in Davidson's method, the vectors $\{H\underline{c}_k\}_L$ must also be computed, or updated concurrently with the vectors c. For each coordinate vector $\underline{e}_i$, updating is done by solving the 2×2 eigenvalue problem in the basis $\underline{c}_k$, $\underline{e}_i$ for each $\underline{c}_k$ independently of the others. This step is equivalent, for a single eigenvector, to use of the optimal relaxation algorithm. Orthonormality is retained in the Jacobi algorithm, if the vector $\underline{e}_i$ is orthogonalized to the current set of vectors $\underline{c}$. The resulting normalized vector is

$$\underline{n}_i = \frac{1}{d_i}[\underline{e}_i - \Sigma_k (\underline{e}_i, \underline{c}_k)\underline{c}_k], \ i=L+1,\ldots N \qquad (25)$$

Given a specific index i, the row of matrix elements

$$(\underline{n}_i, H\underline{c}_k) = \frac{1}{d_i}[H_{ik} - \Sigma_{k'} (\underline{c}_{k'})_i H_{k'k}], \ k=1,\ldots L \qquad (26)$$

where

$$H_{ik} = (\underline{e}_i, H\underline{c}_k) \tag{27}$$

and

$$H_{k'k} = (\underline{c}_{k'}, H\underline{c}_k), \tag{28}$$

are scanned successively, as in the standard Jacobi algorithm, and are transformed to zero if larger than a current criterion. The criterion is reduced after each complete iteration through the range of index i.

The original matrix elements

$$H_{ij} = (\underline{e}_i, H\underline{e}_j), \quad i,j = L+1, \ldots N \tag{29}$$

are not transformed, but the vector $H\underline{c}_k$, whose elements are H_{ik} of (27), must be transformed as $\underline{c}_k$ is updated. Because only one row of H_{ij} is required for operations involving $\underline{e}_i$, the matrix can be scanned by rows. In addition to (26), a specific formula required for row i of H_{ij} is

$$(\underline{n}_i, H\underline{n}_i) = \frac{1}{d_i^2}[H_{ii} - 2\Sigma_{k'} (\underline{c}_{k'})_i H_{ik'} + \Sigma_{k'} \Sigma_{k''} (\underline{c}_{k'})_i H_{k'k''} (\underline{c}_{k''})_i]. \tag{30}$$

The matrix element $(\underline{n}_i, H\underline{c}_k)$ is eliminated by a 2 × 2 orthogonal transformation. The eigenvector corresponding to the lower eigenvalue defines the updated vector

$$\underline{c}_k + \underline{\Delta c}_k = \underline{c}_k \alpha + \underline{n}_i \beta \tag{31}$$

$$= \Sigma_{k'} \underline{c}_{k'} [\alpha \delta_{k'k} - \frac{\beta}{d_i} (\underline{c}_{k'})_i] + \frac{\beta}{d_i} \underline{e}_i \tag{32}$$

where $\delta_{k'k}$ is the Kronecker delta and

$$\alpha^2 + \beta^2 = 1. \tag{33}$$

It can easily be verified that this transformation retains the orthonormality of the vectors $\underline{c}$. The updated vector $H\underline{c}_k$ is

$$H(\underline{c}_k+\Delta\underline{c}_k) = \Sigma_{k'} H\underline{c}_{k'} [\alpha\delta_{k'k} - \frac{\beta}{d_i} (\underline{c}_{k'})_i] + \frac{\beta}{d_i} H\underline{e}_i \quad (34)$$

The elements $H_{ij} = H_{ji}$ of $H\underline{e}_i$ are accessible as the current row of H.

This algorithm has been programmed, following exactly the logic of a standard Jacobi subroutine. Another version has been designed for use with H_{ij} stored as a lower triangular array of independent elements. A vector of partial sums, defined as in (16), must be saved for each vector $H\underline{c}_k$. Each vector component $(H\underline{c}_k)_i$ is updated only when $H\underline{e}_i$ is accessed, so that the full residual partial sum is available. Because H_{ik} cannot be kept current, it must be recomputed from the vectors $H\underline{c}_k$ at the beginning of each full iteration. This second version has been found to converge satisfactorily in test calculations.

5. THE MATRIX INVERSION PROBLEM

The matrix expression

$$M^T (H-\varepsilon I)^{-1} M, \quad (35)$$

where H is a large Hermitian matrix, occurs in linear expansion methods for bound state or scattering solutions of the Schrödinger equation. Applications include partitioning methods for the eigenvalue problem, variational methods in electron scattering theory, and the Stieltjes imaging theory of oscillator strength distributions and photo-ionization. Examples of such applications and relevant computational methods have recently been reviewed (Nesbet (1978)).

The optimal method for (35) would be an N^2 iterative procedure, valid for $H-\varepsilon I$ not positive definite, that would converge in much less than N steps.

Diagonalization of H is the most straightforward method, but it is inefficient or impossible for large matrices, because all N eigenvalues are required. The matrix H is real and symmetric but not in general sparse. If results are needed for many values of ε, diagonalization is advantageous because it has only to be carried out once.

A modified Cholesky method (Nesbet (1971)) has been used for (35). Although $H-\varepsilon I$ is not positive definite, it can be represented in the form

$$H - \varepsilon I = T\Sigma T^T, \tag{36}$$

where T is a lower triangular matrix and Σ is a diagonal matrix whose elements are either +1 or -1. If M is a rectangular matrix and

$$B = T^{-1}M \tag{37}$$

then

$$M^T(H-\varepsilon I)^{-1}M = B^T\Sigma B. \tag{38}$$

Use of (38) eliminates the back-substitution step that would be required if $(H-\varepsilon I)^{-1}M$ were evaluated directly. Algorithms including the out-of-core data-handling necessary for large matrices have been described. In practice the method is numerically stable.

The modified Cholesky method can also be used to compute inverse moment expressions such as

$$\mu_k = C\underline{d}^T(H-\varepsilon I)^{-k}\underline{d}, \tag{39}$$

but back-substitution cannot be avoided.

Making use of (36), the steps are

$$\begin{aligned} \underline{b}_0 &= \underline{d} \\ \underline{b}_1 &= T^{-1}\underline{b}_0 \\ \underline{b}_2 &= (T^T)^{-1}\Sigma \underline{b}_1 \\ &\cdots \end{aligned} \tag{40}$$

Then

$$\left.\begin{aligned} \underline{d}^T(H-\varepsilon I)^{-k}\underline{d} &= \underline{b}_k^T\Sigma\underline{b}_k \quad k \text{ odd} \\ &= \underline{b}_k^T\,\underline{b}_k \quad k \text{ even.} \end{aligned}\right\} \tag{41}$$

REFERENCES

Boys, S.F. (1950) Electronic wave functions, II, A calculation for the ground state of the beryllium atom, Proc. Roy. Soc. (London), A201, pp. 125-137.

Butscher, W. and Kammer, W.E. (1976) Modification of Davidson's method for the calculation of eigenvalues and eigenvectors of large real symmetric matrices, Root homing procedure, J. Comput. Phys. 20, pp. 313-325.

Cooper, J.L.B. (1948) The solution of natural frequency equations by relaxation methods, Quart. Appl. Math. 6, pp. 179-183.

Davidson, E.R. (1975) The iterative calculation of a few of the lowest eigenvalues and corresponding eigenvectors of large real-symmetric matrices, J. Comput. Phys. 17, pp.87-94.

Fadeev, D.K. and Fadeeva, V.N. (1963) Computational methods of linear algebra, section 61, Freeman, San Francisco.

Hestenes, M.R. and Karush, W. (1951a) A method of gradients for the calculation of the characteristic roots and vectors of a real symmetric matrix, J. Res. Nat. Bur. Standards 47, pp. 45-61.

Hestenes, M.R. and Karush, W. (1951b) Solutions of $A\underline{x}=\lambda B\underline{x}$, J. Res. Nat. Bur. Standards 47, pp. 471-478.

Hotelling, H. (1933) Analysis of a complex of statistical variables into principal components, J. Educ. Psychology 24, pp. 417-441 and pp. 498-520.

Lanczos, C. (1950) An iteration method for the solution of the eigenvalue problem of linear differential and integral operators, J. Res. Nat. Bur. Standards 45, pp. 255-281.

Liu, B. (1978) The simultaneous expansion method for the iterative solution of several of the lowest eigenvalues and corresponding eigenvectors of large real-symmetric matrices, In NRCC Workshop (1978), pp. 49-53.

Löwdin, P.-O. (1962) Studies in perturbation theory, IV, Solution of eigenvalue problem by projection operator formalism, J. Math. Phys. 3, pp. 969-982.

Nesbet, R.K. (1965) Algorithm for diagonalization of large matrices, J. Chem. Phys. 43, pp. 311-312.

Nesbet, R.K. (1971) Evaluation of $m^{\dagger}(h-\varepsilon)^{-1}m$ for a large Hermitian matrix h, J. Comput. Phys. 8, pp. 483-488.

Nesbet, R.K. (1978) Linear equation systems in bound state and scattering problems, In NRCC Workshop (1978), pp. 79-88.

NRCC Workshop (1978) Numerical algorithms in chemistry: Algebraic methods, LBL-8158, Lawrence Berkeley Laboratories, California.

Roos, B. (1972) A new method for large-scale CI calculations, Chem. Phys. Lett. 15, pp. 153-159.

Ruhe, A. (1974) SOR-methods for the eigenvalue problem, Math. Comp. 28, pp. 695-710.

Schwarz, H.R. (1974) The eigenvalue problem $(A-\lambda B)x=0$ for symmetric matrices of high order, Comput. Meth. Appl. Mech. Eng. 3, pp. 11-28.

Seidel, L. (1874) Über ein Verfahren, die Gleichungen, auf welche die Methode der kleinsten Quadrate führt, sowie lineäre Gleichungen überhaupt, durch successive Annäherung aufzulösen, Abhandl. Akad. Wiss. München. Math. Phys. Kl 11, pp. 81-108.

Shavitt, I. (1970) Modification of Nesbet's algorithm for the iterative evaluation of eigenvalues and eigenvectors of large matrices, J. Comput. Phys. 6, pp. 124-130.

Shavitt, I., Bender, C.F., Pipano, A. and Hosteny, R.P. (1973) The iterative calculation of several of the lowest or highest eigenvalues and corresponding eigenvectors of very large symmetric matrices, J. Comput. Phys. 11, pp. 90-108.

DISCUSSION

J.K. REID (Harwell). Am I right in thinking that your method for obtaining non-extreme eigenvalues approximately solves a system of linear equations at each step?

NESBET. The eigenvectors are updated by the approximate solution of inhomogeneous linear equations in the coordinate relaxation method. Davidson's simultaneous update is a little more complicated. Assuming that the matrix is diagonally dominant, Davidson's method approximates the Krylov sequence of the inverse matrix.

D.S. SCOTT (Oak Ridge). So you can think of it as preconditioning with A^{-1} and hence the amount of diagonal dominance should have an impact on the convergence.

NESBET. Certainly. Diagonal dominance is crucial for the convergence of these iterative methods.

DR. A.M. ERISMAN (Boeing). Is your matrix in the inversion section (section 5) the same as that in the earlier part of the paper. If so, how are you then able to compute the factors? Also, since the matrix is not positive definite is it not possible that Cholesky's method could break down?

NESBET. It is the same but usually smaller. We use out-of-core decomposition techniques for the inversion problem. These would be prohibitive if used in typical applications of the iterative eigensolution procedures. I agree that Cholesky's method is not guaranteed to be stable on our matrices since they are not positive definite. However, we monitor the decomposition and have never had any failures. An exact zero eigenvalue of a leading submatrix is highly improbable in 64 bit arithmetic.

PROFESSOR A. JENNINGS (Belfast). From my memory of the fine detail in Davidson's method, I believe failure can occur because a divisor can be zero. Does this cause you any problems? Does your method avoid this?

NESBET. In practice, this does not happen and indeed it is always possible to avoid a zero divisor. No such problem occurs with the partitioned Jacobi method or with the optimized relaxation method.

SPARSE MATRIX TECHNIQUES IN ENGINEERING PRACTICE

P.J. Beresford

(Atkins Research and Development,
Epsom, Surrey)

ABSTRACT

This paper describes an assortment of problems in engineering mechanics and some techniques used for their solution. Particular attention is paid to methods of implementing the linear algebra involved in the calculations, especially with respect to the sparsity introduced by the finite element discretisation of the problems. The selection of equation solver or eigen-solution technique is seen to be dependent on the particular engineering problem being tackled. Economy of solution is the goal - how to achieve it is the game.

1. INTRODUCTION

Many problems in engineering mechanics are amenable to some discretisation process, such as the finite element method, and can furthermore be presented as a set of linear equations. This is true even for many non-linear problems, since they can often be split into sequences of linear steps. The provision of suitable software tools to solve such problems is a very costly business. Several program packages have been produced which enable the engineering analyst to describe his problem and to obtain its solution in a suitable form. (See, for example, Henrywood and Sheffield (1976)). The program user generally sees the input data and the ensuing results, but the processing between these two positions is usually hidden. If the intermediate process is the solution of a set of linear equations, which solution procedure should be used? If it is an eigensolution, which eigenvalue extraction scheme? These are questions which the software provider must ask. It is laudable and important that the program be easy to use from the analyst's point of view, that the data input and presentation of results be appropriate to his needs; but if the results are not reliable, or the cost is exorbitant, and this because of an inappropriate solution scheme, then the program as a whole will not be used.

2. LINEAR EQUATION SOLVING

It is immediately apparent that many engineering problems involve the solution of a set of linear equations at some stage in the analysis. Static analyses with prescribed forces or displacements or with temperature changes which produce small deformations (hence the linear assumption) are all of this type. Less obvious are analyses which involve large deformations or nonlinear behaviour, or timestepping schemes, but often these can be represented as linear steps, and the same solution procedures can be used.

The general problem to be solved is of the form:

$$A\ X = B$$

where the solution is sought for X given A and B.

A is square, $n \times n$

X and B are rectangular, $n \times m$, where m

is often unity, in which case X and B are vectors of length n. However, the case of $m > 1$ is common and this fact alone has a great influence on the choice of solution procedure.

There are two categories of equation solver - the direct methods and the iterative methods. The choice of category can depend upon many factors, but broadly, if several solution vectors are required (the case $m > 1$) direct methods can use most of the results of the effort already expended in calculating the first vector in order to evaluate the remaining vectors. Iterative schemes gain no such advantage. On the other hand, if only one solution vector is required, and this is only one of a sequence of vectors in a non-linear problem, then the solution vector at the end of one linearised step can be a good starting vector for the next step, and an iterative scheme can be attractive.

The coefficient matrix A constructed via a finite element discretisation is generally sparse, since each row of the matrix corresponds to a variable at a node, and each node is only connected to its neighbours. The effects of this sparsity and how it can be exploited to provide efficient solutions, will be examined in terms of a direct method - the Cholesky decomposition. This paper will also only be concerned with symmetric positive definite matrices, since the vast majority of engineering analyses are of this type.

3. MATRIX STORAGE SCHEMES

It can be shown that direct methods for the solution of linear equations are equivalent to the factorisation of the coefficient matrix. The Cholesky factorisation (or decomposition as it is usually called) produces a lower triangular matrix L such that $LL^T = A$. Once this triangular matrix has been found, the solution for X can be performed in two steps

(i) solve $L Y = B$ for Y

(ii) solve $L^T X = Y$ for X

both of these steps being straightforward since only one unknown at a time is met in the first equation by performing forward substitution, and only one at a time in the second by backward substitution. The initial evaluation of L is also a simple matter (see, for example, Jennings (1977)).

If only a small number of equations is involved, then L can be stored as a full triangular matrix and the coding of the factorisation and substitution algorithms is straightforward. However, when larger problems are encountered which do not fit into the machine storage or which involve redundant operations with a significant number of zero values, then other storage schemes become advantageous.

The first departure from full triangular storage arises from the observation that many matrices have a banded structure, in that for every non-zero element a_{ij} of a matrix A we can calculate the difference $|i-j|$, and we call the largest of these the half-bandwidth. This can be much smaller than the order of the matrix. It is only necessary to store the elements of the matrix within the band.

If the pattern of non-zero matrix elements is observed further, it is seen that the bandwidth of each row of the matrix is not affected by the Cholesky factorisation process, although many elements within the band which are zero in A become non-zero in L. This feature is exploited in the variable bandwidth storage scheme. Since for symmetric matrices only half of the matrix need be stored, there is a choice of either storing the lower triangle or its transpose. If the upper triangle is stored, variable bandwidth schemes are often termed "skyline" schemes, since the columns of the matrix retain their lengths.

The storage saving achieved by adopting such schemes may still not be sufficient to store the larger matrices in the memory of the machine being used. It becomes necessary to split the matrices into several blocks, which can be brought into memory as required from a suitable large storage device. Nowadays, it is usual to use some form of direct access to such devices. This is termed blocked variable bandwidth storage.

The final development is to cater for the problem which has difficulty storing even one block in memory - simply block the matrix the other way as well, and we have submatrix storage. The submatrix size can be chosen to suit the machine, and if possible, the problem as well. Null submatrices need not be stored unless they become non-null during the factorisation process.

These storage schemes are depicted in Figure 1.

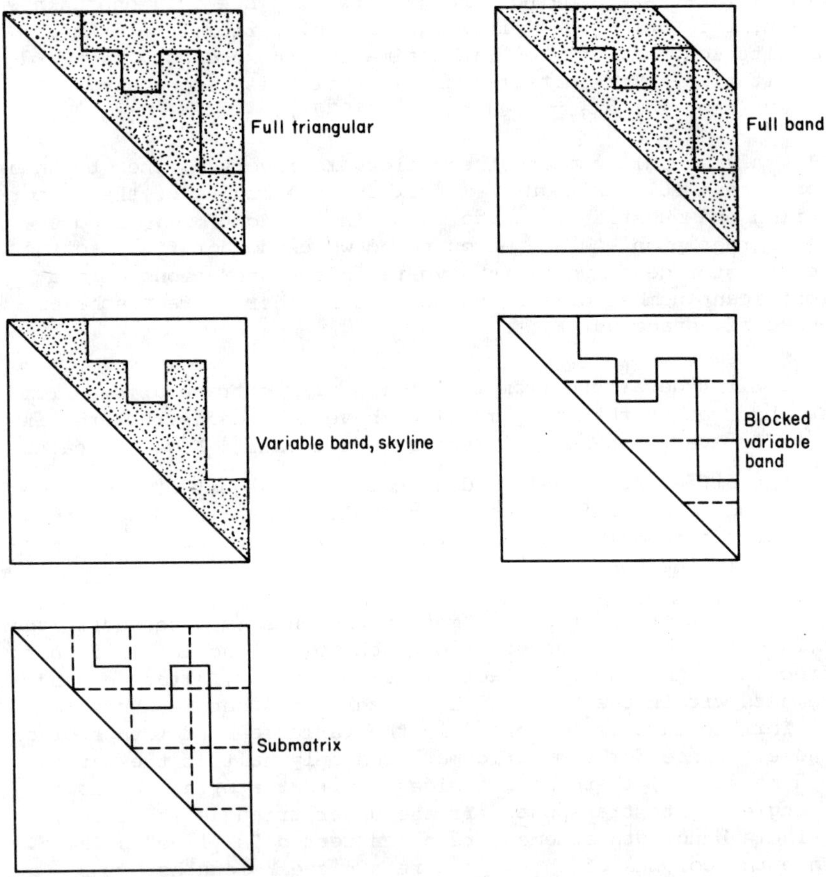

Figure 1 Storage schemes for symmetric matrices

4. BANDWIDTH REDUCTION

The cost of solving a set of linear equations depends on both the number of arithmetic operations being performed and on the amount of data transfer between main memory and any storage devices being used. Cost means different things to different people. For some, with their own dedicated machine, it might mean elapsed time; for others, it might mean degradation of response for other users; for the commercial bureau user, it means money, calculated by some complicated formula, which will undoubtedly be different on another bureau. When an equation solution procedure is being tuned to perform well on a particular machine, many factors must be taken into account. It is not always obvious how much memory should be used, or how much data should be transferred at a time.

One feature can almost invariably reduce costs, however, and that is to reduce the size of the problem. Taking account of any symmetry will help here, for example by only representing half of the structure. Whether this is the case or not, the order of the equations is not usually important, so it is often possible to reorder them so that the bandwidth of the matrix is significantly reduced. Fewer arithmetic operations will be required, less storage will be used, and less data transfer will take place. Several bandwidth reduction algorithms have been devised, perhaps the most commonly used being the reverse Cuthill-McKee (see Cuthill and McKee (1969), and George (1971)). Similar results can be obtained for less computational effort, using the scheme reported by Gibbs, Poole and Stockmeyer (1976).

5. SUBSTRUCTURES

It is no coincidence that coefficient matrices arising from the discretisation of large structural engineering problems are sparse, or indeed that a submatrix storage scheme will reflect the connectivity of the actual structure. A natural extension of the submatrix scheme is to use a substructure technique. An example will illustrate the principle. Figure 2 depicts an aircraft (the Dash 7 built by De Havilland of Canada), which was built in several variations. An analysis of the fuselage was required for airworthiness certification. Each section of the fuselage was modelled as a separate substructure, and several variants of the design were analysed for little more than the cost of a single analysis.

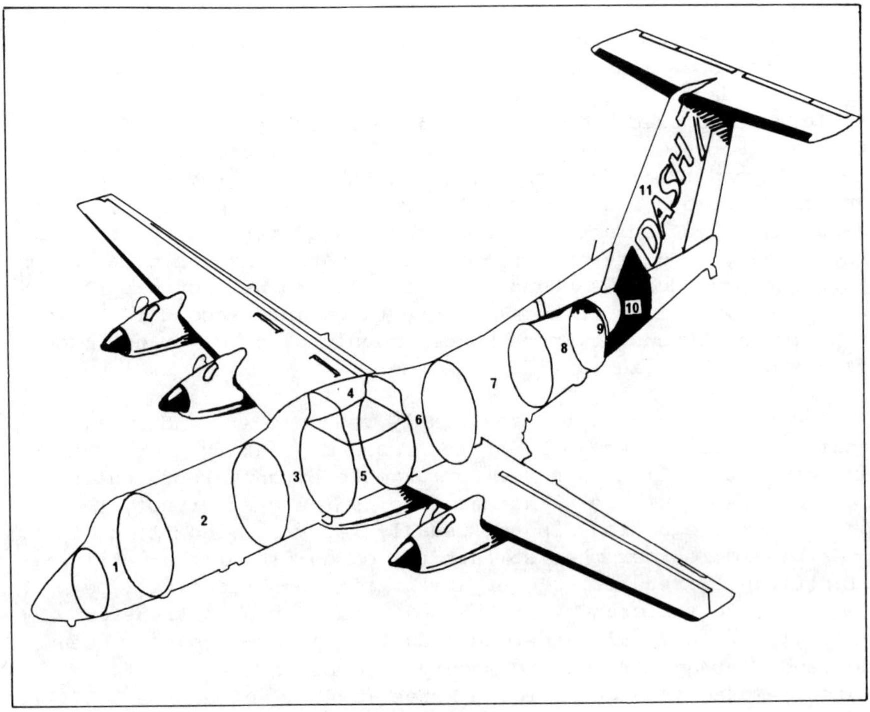

Figure 2 Analysis of aircraft fuselage using substructures

If the equations to be solved for one substructure are partitioned between those variables internal to the substructure (subscript 1), and those linked to other substructures (subscript 2), we can write

$$\begin{bmatrix} A_{11} & A_{12} \\ A_{21} & A_{22} \end{bmatrix} \begin{bmatrix} X_1 \\ X_2 \end{bmatrix} = \begin{bmatrix} B_1 \\ B_2 \end{bmatrix}$$

where $A_{12} = A_{21}^T$

and the second equation can be expanded to be

$$\left[A_{22} - A_{12}^T A_{11}^{-1} A_{12}\right] X_2 = B_2 - A_{12}^T A_{11}^{-1} B_1$$

or

$$C X_2 = D$$

The coefficient matrix C and its corresponding right-hand side matrix D can be calculated for each substructure in turn, and assembled in the usual manner to produce a "structure" matrix of link variables. This set of equations can be solved and the internal variables of each substructure can be evaluated by substitution in the usual way.

The calculation of the substructure matrices C and D does not of course involve any matrix inversion. If A_{11} is assumed to be non-singular, then it can be factored as $L_{11} L_{11}^T$, and the second equation becomes:

$$\left[A_{22} - A_{12}^T L_{11}^{-T} L_{11}^{-1} A_{12}\right] X_2 = B_2 - A_{12}^T L_{11}^{-T} L_{11}^{-1} B_1$$

If $Y = L_{11}^{-1} A_{12}$ then C becomes:

$$A_{22} - Y^T Y$$

and computations become a sequence of forward and backward substitutions and matrix multiplications. Explicit matrix transposition is not required if routines are written specially for the necessary operations.

6. CONSTRAINT EQUATIONS

Sometimes the discretisation of a problem is not simply a matter of one equation for each variable at a node. There are occasions when several variables are constrained to act together in some manner. For example, (Figure 3), an offshore-oil platform might have several piles driven down its legs into the sea bed. These piles will not be rigidly connected to the legs, but pass through several guides. The transverse motion of the legs and the piles at each of these guides will be identical, yet the piles will be free to move independently of the legs in the longitudinal directions. The transverse variables must be constrained to be equal. Another example is the soil-structure interaction of a power station on a rigid basemat, (Figure 4). The difference in stiffness between the idealised basemat and the neighbouring soil is so large that the overall coefficient matrix becomes ill-conditioned. However, if the rigidity is modelled by constraining the motions of the basement nodes, so that it remains flat, then the problem can be solved satisfactorily.

The straightforward method for introducing constraints into the solution procedure is by using linear transformations. If we require the solution of $\min_{\underline{x}} (\frac{1}{2} \underline{x}^T A\underline{x} - \underline{b}^T\underline{x})$ subject to $\underline{x} = T\underline{y}$

where A is n×n, $\underline{x}$ and $\underline{b}$ are vectors of length n, $\underline{y}$ is a vector of length m, and T is n×m (m < n) then, after substitution, the

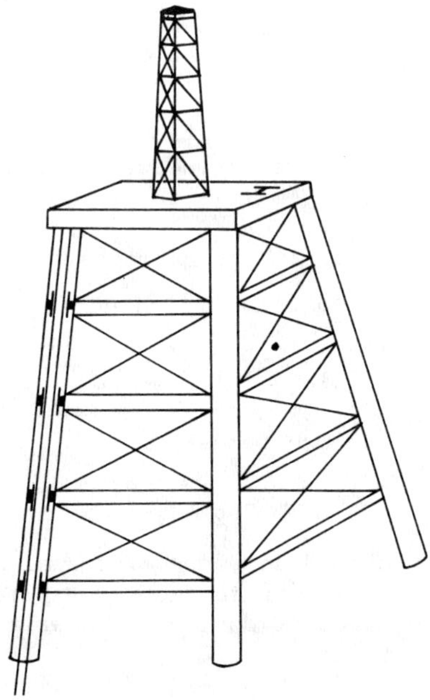

Figure 3 Pile guides in offshore platforms

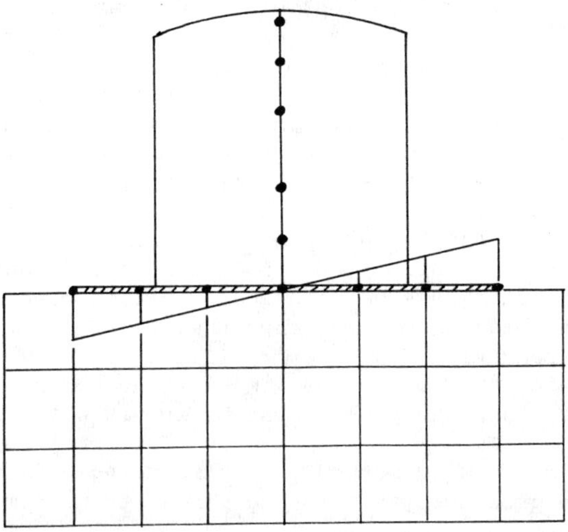

Figure 4 Soil-structure interaction with rigid basemat

solution to our minimization problem yields the equations

$$T^T \, A \, T \, \underline{y} = T^T \, \underline{b} \quad .$$

If the matrix A is partitioned with the dependent equations in the first rows, and if T can be partitioned $\begin{bmatrix} t \\ I \end{bmatrix}$, t being (n-m) × m and I being m×m, then the equations to be solved can be written

$$\begin{bmatrix} t^T I \end{bmatrix} \begin{bmatrix} A_{11} & A_{12} \\ A_{21} & A_{22} \end{bmatrix} \begin{bmatrix} t \\ I \end{bmatrix} \underline{y} = \begin{bmatrix} t^T I \end{bmatrix} \begin{pmatrix} \underline{b}_1 \\ \underline{b}_2 \end{pmatrix}$$

which expands to:

$$\left[t^T \, A_{11} \, t + t^T \, A_{12} + A_{21} \, t + A_{22}\right] \underline{y} = t^T \, \underline{b}_1 + \underline{b}_2$$

If the number of dependent equations (n-m) is small, then the amount of work involved in these transformations is not large.

Another technique which can be used if the original coefficient matrix is not singular, is the Lagrange multiplier method. Here the coefficient matrix is expanded to include the con straints as extra equations.

If the solution is sought to A $\underline{x}$ = $\underline{b}$ subject to C$\underline{x}$ - $\underline{d}$ then we can write:

$$\begin{bmatrix} A & C^T \\ C & O \end{bmatrix} \begin{pmatrix} \underline{x} \\ \underline{\lambda} \end{pmatrix} = \begin{pmatrix} \underline{b} \\ \underline{d} \end{pmatrix}$$

The first is written to retain symmetry in the coefficient matrix, and introduces a further set of unknowns, $\underline{\lambda}$.

This equation gives:

$$C^T \, \underline{\lambda} = \underline{b} - A \, \underline{x}$$

and $\underline{x}$ can be eliminated by premultiplying firstly by A^{-1} and then by C, yielding:

$$C \, A^{-1} \, C^T \, \underline{\lambda} = C \, A^{-1} \, \underline{b} - \underline{d}$$

If a Cholesky decomposition is performed on A yielding Cholesky factors $L^T L$, a solution can be found for $\underline{\lambda}$ using:

$$\left[C\ L^{-T}\ L^{-1}\ C^T\right]\underline{\lambda}, = C\ L^{-T}\ L^{-1}\ \underline{b} - \underline{d}$$

This small set of equations is solved for $\underline{\lambda}$, then $\underline{x}$ is found from:

$$A\ \underline{x} = \underline{b} - C^T\ \underline{\lambda}$$

This is the standard problem with a modified right-hand side.

7. EIGENVALUE PROBLEMS

Many engineering problems are dynamic in nature, such as wave loading on sea structures, earthquake excitation of buildings, vibration in vehicles. Often the physical processes involved are dependent on their history, and linear solution techniques are not applicable. However, many practical problems involve small deformations which allow a reasonable linearisation, and in such cases it can be advantageous to transform the equations to some normal coordinates and perform the bulk of the computation in the new system. Not only can lengthy transient calculations be performed on a number of independent single equations, but the number of equations can often be reduced. For structural vibration problems it is usual for only the lower frequency modes to be significant in calculating the response; typically fewer than fifty modes for a structure with several thousand variables (see paper by Jennings (1980) at this conference).

As ever, the choice of eigensolution technique to perform the transformation to normal coordinates depends on what the problem requires. A different technique will be used to find all the eigenvalues of a small matrix than to find only the first few eigenvalues and vectors of a large matrix.

An example of the former type of problem is the dynamic analysis of piping systems in power stations. Typical problems involve fewer than 200 dynamic equations, so it is cheap enough to use an eigensolution technique such as the Householder-QR method with inverse iteration for the vectors.

The solution is sought to the undamped problem:

$$K\ X = M\ X\ \Lambda \tag{1}$$

where K is the stiffness matrix

M the mass matrix

X the eigenvectors

and Λ the diagonal matrix of eigenvalues

For a problem of this size, it is convenient to find all the eigenvalues, and then to calculate only the eigenvectors felt to be significant in contributing to the response. Often the order of the eigenvalue problem can be much smaller than the corresponding structural matrices, since reasonable assumptions can be made that many of the equations will have little effect on the eigenvalues or vectors. For example, in these same piping systems, the rotational inertia terms are usually neglected in the eigensolution, the assumption being made that the overall structural response is dominated by the translational terms.

Thus, if we consider a subset of m eigenpairs and partition the equation obtained from (1) as

$$\begin{bmatrix} K_{11} & K_{12} \\ K_{21} & K_{22} \end{bmatrix} \begin{bmatrix} X_1 \\ X_2 \end{bmatrix} = \begin{bmatrix} 0 & 0 \\ 0 & M_{22} \end{bmatrix} \begin{bmatrix} X_1 \\ X_2 \end{bmatrix} \Lambda' \tag{2}$$

(where subscript 1 is associated with zero mass, X_2 is square m×m and Λ' is also m×m), then elimination of X_1 yields

$$\left[K_{22} - K_{21} K_{11}^{-1} K_{12}\right] X_2 = M_{22} X_2 \Lambda'$$

We can find the eigenvalues Λ' and eigenvectors X_2 of this reduced system. Since the right hand side of (2) can now be evaluated without knowing X_1 we can find the components in X_1 by solving

$$\begin{bmatrix} K_{11} & K_{12} \\ K_{21} & K_{22} \end{bmatrix} \begin{bmatrix} Y_1 \\ Y_2 \end{bmatrix} = \begin{bmatrix} 0 \\ M_{22} X_2 \Lambda' \end{bmatrix}$$

for Y_1 and Y_2. Then $X_1 = Y_1$ (the desired terms) and $X_2 = Y_2$ (unchanged).

8. SIMULTANEOUS ITERATION METHODS FOR EIGENVALUE PROBLEMS

For large problems which require the calculation of only a few eigenvalues and vectors, simultaneous iteration methods have proved popular. These methods make use of the invariance of eigenvectors or of sets of eigenvectors under multiplication by the matrix A and their implementation involves A only in such multiplications. Thus they are ideal for large sparse problems. We can write our eigenproblem as determining a vector $\underline{x}$ and a scalar λ such that

$$A \underline{x} = \lambda \underline{x}$$

One method for finding such a vector/scalar pair is the <u>power method</u>. This simply takes an arbitrary vector $\underline{x}_0$, and multiplies it by the matrix A to obtain a new vector $\underline{x}_1$, and after repeated multiplication by A,

$$\underline{x}_{i+1} = A \underline{x}_i$$

the vector $\underline{x}_{i+1}$ converges with a scalar multiplier λ over the vector $\underline{x}_i$.

It is usual to normalise the vector after each multiplication to prevent overflow.

For a symmetric matrix of order n, there will be n eigenvalues (not necessarily all distinct) and n corresponding eigenvectors (that is, n eigenpairs). The <u>simultaneous iteration method</u> is an extension of the power method, finding a number of eigenpairs simultaneously. It is sometimes called the <u>subspace iteration method</u> (conveniently also with the initials SI) because we are working with invariant subspaces, most of the calculation being performed with a reduced number of equations, just enough for the number of eigenpairs required.

For undamped structural analysis, the eigenproblem is as before:

Find the eigenvalues Λ and the eigenvectors X such that $KX = MX\Lambda$ where K is the stiffness matrix and M is the mass matrix.

Note that K and M are of size $n \times n$
X is $n \times m$

and Λ is $m \times m$ and diagonal,

where n is the number of equations in the system

and m is the number of frequencies required.

We can transform this to the standard eigenvalue problem

$$Y = D\ Y\ \Lambda$$

where $Y = L^T X$, $LL^T = K$, $D = L^{-1}\ M\ L^{-T}$;

Y is $n \times m$, D is $n \times n$.

The S.I. scheme chooses a set of starting vectors Y_O, forms $A = Y_O^T\ D\ Y_O$ (the projection of D onto an $m \times m$ subspace), performs the spectral decomposition of A into its eigenvalues Λ_O and eigenvectors V_O, and then produces the improved vectors $Y_1 = D\ Y_O\ V_O$. Convergence is achieved by repeated application of this Rayleigh-Ritz procedure.

In order to find the first m eigenpairs of our $n \times n$ system we only need to solve the eigenproblem for the very much smaller $m \times m$ system represented by the matrix A. This procedure is similar to the power method but converges to several vectors simultaneously.

8.1 Efficient computation scheme

The major steps of the simultaneous iteration scheme outlined above are:

(i) form initial vectors Y_O

(ii) form matrix $A = Y_O^T\ D\ Y_O = Y_O^T\ L^{-1}\ M\ L^{-T}\ Y_O$

(iii) solve for eigenvalues Λ_O and eigenvectors V_O of A

(iv) form $Y_1 = D\ Y_O\ V_O$

(v) normalise Y_1 so that $Y_1^T\ Y_1 = I$

(vi) test for convergence of eigenvalues Λ_O, set $Y_O = Y_1$, and if necessary go back to (ii)

The lengthy matrix manipulations in steps (ii) and (iv) need not be performed in a brute-force fashion but can be implemented by using substitution techniques as we now illustrate.

For the calculation of the matrix A in step (ii), we can take the Cholesky decomposition of the mass matrix, $M = L_m\ L_m^T$

Then $$A = Y_o^T L^{-1} L_m L_m^T L^{-T} Y_o = Z^T Z$$

$$\text{where } Z = L_m^T L^{-T} Y_o$$

Z can be calculated in two stages:

$$X = L^{-T} Y_o \quad \text{by back substitution}$$

$$Z = L_m^T X \quad \text{by multiplication}$$

Then form $A = Z^T Z$ and find the full eigensolution using a JACOBI routine.

As the estimates Y_o become better, the matrix A becomes more diagonally dominant, and the JACOBI scheme becomes more efficient.

Again, stage (iv) can be split into several sections.

$$\begin{aligned} \text{We are trying to form } Y_1 &= D\, Y_o\, V_o \\ &= L^{-1} L_m L_m^T L^{-T} Y_o V_o \\ &= L^{-1} L_m Z V_o \end{aligned}$$

So, multiply V_o by Z, then premultiply by L_m, and finally perform a forward substitution to get the final result.

Once convergence has been achieved in step (vi), the eigenvectors X in the original coordinate system are found by performing the first calculation of step (ii) as outlined above.

9. CONCLUSIONS

It has been seen that, in taking account of the structure of the matrices involved in engineering analyses, substantial economies in both storage requirements and computational effort can be achieved. The basic matrix manipulation tools of factorisation, substitution and multiplication arise time and time again, and the solutions to particular classes of problem can usually be obtained efficiently by suitable restatement and reorganisation.

ACKNOWLEDGEMENTS

The ideas presented in this paper have, of course, been gleaned from many sources; thanks are particularly due to

Mr. D. Hitchings of the Department of Aeronautics, Imperial College, London, for many fruitful discussions, and to Dr. I.S. Duff of AERE, Harwell for clarifying some points of detail.

REFERENCES

Cuthill, E. and McKee, J. (1969) Reducing the bandwidth of sparse symmetric matrices, Proc. 24th ACM National Conference, New York, Brandon Systems Press, pp. 157-172.

George, A. (1971) Computer implementation of the finite element method, Report STAN-CS-71-208, Stanford University, California.

Gibbs, N.E., Poole, W.G. Jr., and Stockmeyer, P.K. (1976) An algorithm for reducing the bandwidth and profile of a sparse matrix, SIAM J. Numer. Anal, 13, pp. 236-250.

Henrywood, R.K. and Sheffield, J.D. (1976) ASAS - A general-purpose finite element analysis system, Comput. Aided Des. 8, pp. 2-8.

Jennings, A. (1977) Matrix Computation for Engineers and Scientists, Wiley.

Jennings, A. (1980) Eigenvalue methods and the analysis of structural vibration, these proceedings.

THE EXPLOITATION OF SPARSITY BY ITERATIVE METHODS

D.A.H. Jacobs

(Central Electricity Research Laboratories, Leatherhead)

ABSTRACT

The majority of iterative methods for solving large systems of algebraic equations exploit the sparsity of those equations in order to be efficient.

This paper commences with a historical review of some 'classical' iterative methods from Point Jacobi to the Strongly Implicit Procedure. The way in which each of these methods exploits the sparsity is highlighted. In general the greater the assumptions about the sparsity made by these iterative methods, the faster is the rate of convergence.

This trend to greater restrictions on the sparsity pattern is not present in the recently developed preconditioned methods linked to a solution procedure which is often the conjugate gradient method. Ways in which the sparsity is exploited in some of the algorithms developed to date are described.

It is shown that a general basis of many iterative procedures is that of iterated defect correction. This form is then developed to extend the application of iterative and direct methods to a much wider set of sparsity patterns which need not fit the requirements of a particular solution method. In this way, for example, an iterative solution scheme for a five diagonal matrix can be used to solve systems other than those which are five diagonal. A number of examples illustrating the range and power of this method are described.

1. INTRODUCTION

The requirement to solve systems of algebraic equations

$$A\underline{x} = \underline{b} \tag{1}$$

where A is a square matrix of order N, $\underline{b}$ is a given vector and the value of the vector $\underline{x}$ is sought, arises as part of the mathematical modelling of many physical systems. In general, non-linear systems are solved by linearizing or by Newton iteration, in both cases a sequence of linear systems must be solved. Consideration will therefore be given to solving linear systems.

There are two general classes of methods for solving linear systems of algebraic equations, direct and iterative. There have been extensive developments in both classes in recent years which have produced significant improvements in computational efficiency. With few exceptions the methods all exploit the sparsity of the algebraic system. In general the advances in both direct and iterative methods have made an increasing use of special features of the problem being solved - sometimes manifested solely in the form of the sparsity, and sometimes demanding such requirements as the equality of many coefficients.

A number of the more recent developments make explicit use of the fact that the algebraic system is derived from a partial differential equation, and they are frequently not directly applicable to more general systems. In the class of direct methods, the use of fast Fourier transform techniques and the marching methods have provided major improvements in solution efficiency. The method of nested dissection is also worthy of note since for certain classes of problems it can be proved to be close to the optimal implementation employing a Gaussian elimination type of method. Major advances have not been confined to direct methods. The Strongly Implicit Procedure and more recently the use of preconditioning together with conjugate gradients have both provided significant increases in the power of iterative methods.

Iterative methods have historically exploited the sparsity of the algebraic system with two major objectives. The first is to reduce the amount of storage required during the solution of the system of equations, and the second is to reduce the amount of computation needed to obtain a solution of adequate accuracy. If the matrix A was full, a direct method of solution based on Gaussian elimination would need $O(N^3)$ computational operations, and storage of $O(N^2)$. In contrast an iterative method based on calculating the residual of the kth approximation to the solution

$$\underline{r}_k = \underline{b} - A\underline{x}_k$$

and relating it to the change $\underline{\delta x}_k = \underline{x}_{k+1} - \underline{x}_k$ in a simple manner involves $O(N^2)$ operations per iteration. Hence if rapid convergence can be achieved the total amount of computation may be much less than for a direct method.

We are primarily interested in systems with a sparse matrix A. For example, systems derived from partial differential equations usually have a very small number of non-zeros in each row of the matrix. In addition the matrix may have a very 'nice' structure (for example five diagonal) for an appropriate ordering of the unknowns and many of the algebraic equations may be of an identical form. In this case the residual vector at each iteration can be calculated in $O(\tau)$ computations where τ is the number of non-zeros in the matrix A.

In this paper a number of the available iterative methods are considered and the way in which they exploit special features of the problem and in particular the sparsity structure of the system of equations is highlighted. The paper is divided into three parts.

First, in Section 2, some 'classical' iterative methods from the Point-Jacobi iterative scheme to the Strongly Implicit Procedure are briefly considered. The methods use an increasing degree of 'implicitness' to provide increased rates of convergence and they do this by 'extracting' increasing numbers of the non-zeros of the matrix A to form the basis of the iteration update formula.

The more recently developed family of preconditioned conjugate gradient methods is considered in Section 3. In 1977 Meijerink and van der Vorst used preconditioning to transform the algebraic systems derived from a discretization model to a form which would be better suited to solution by the 'theoretically' direct conjugate gradient method and would exploit its rapid residual-reducing features. The method proves to be very powerful and less restrictive than some of the other iterative methods. The preconditionings generally exploit the sparsity pattern of the coefficient matrix, and the ease of calculating the residual from a sparse system is also exploited in the implementations. A number of alternative preconditionings are mentioned, as are extensions of the basic method to a much wider class of problems.

In Section 4 the method of iterated defect correction is described together with several of its applications. Iterated defect correction is a general form, of which many of the classical iterative methods are examples. In addition a number of other iterative methods can be derived from this class, each of which in general exploits the sparsity of the coefficient matrix. Iterated defect correction provides a general method by which the advanced iterative and direct methods which are generally restricted in their application to a small range of problems can be used in an iterative mode to solve a much wider class of problems.

2. CLASSICAL ITERATIVE METHODS

2.1 Sparsity and Iterative Methods

Iterative methods for solving systems of algebraic equations have been comprehensively reviewed by Fox (1977). Since that time the notable advance has been in the use of preconditioned conjugate gradient methods - details of which are given in Section 3. In this section we analyse some of the 'classical' schemes and place particular emphasis on how the methods exploit the sparsity of the algebraic system.

The basic form of most of the 'classical' iterative methods is a splitting of the coefficient matrix A into the sum of two matrices B and C so that

$$A = B + C. \tag{2}$$

The splitting is chosen so that the system

$$B\underline{y} = \underline{c} \tag{3}$$

is 'easily' solved. The iteration procedure is then based on the repeated use of the update equation

$$B\underline{x}_{k+1} = \underline{b} - C\underline{x}_k \tag{4}$$

where k is the iteration index. Because of the choice of B, this system can be solved very economically for each new approximation to the solution. Some of the methods do indeed work in terms of the approximate solutions $\underline{x}_k$ and the matrices B and C. An alternative, but equivalent, formulation is used by other methods: this is obtained by subtracting $B\underline{x}_k$ from both sides of the iterative update equation (4) to obtain

$$B\delta\underline{x}_k = \underline{r}_k \tag{5}$$

where

$$\delta\underline{x}_k = \underline{x}_{k+1} - \underline{x}_k$$

is the change vector and

$$\underline{r}_k = \underline{b} - A\underline{x}_k$$

is the residual vector of the kth approximation to the solution. It is clear from the form of the iteration that if it converges, which implies that $\delta\underline{x}_k \to 0$, then $\underline{r}_k \to 0$ and hence $\underline{x}_k$ converges to the solution of the original system. The requirement that $\delta \underline{x}_k \to 0$ as $k \to \infty$ demands that the matrix $B^{-1}C$ has eigenvalues whose absolute magnitudes are all less than unity.

All the 'classical' methods considered in this section exploit the sparsity of the coefficient matrix A. They do this in two ways. First the methods are generally so straightforward to code that the user at least provides the code which calculates the terms of the vector $\underline{b} - C\underline{x}_k$ or $\underline{b} - A\underline{x}_k$ dependent on the form of the iterative procedure used. Thus the sparsity pattern of the matrix can be fully exploited.

The second way in which the sparsity is exploited is in the selection of the matrix B which has to be such that the system $B\underline{y} = \underline{c}$ can be solved very efficiently. The various iterative procedures all exploit the sparsity present in the matrix A by selecting the matrix B to have much of, or even more, sparsity than the matrix A.

A few of the well-known splittings are shown in Figure 1 for a five diagonal system of equations obtained, for example, from a finite difference approximation to Poisson's equation in a rectangle.

2.2 The Point-Jacobi Scheme

For the Point-Jacobi scheme the matrix B consists of the main diagonal terms of the matrix A. The operation count per iteration is $O(\tau)$.

2.3 The Gauss-Seidel Scheme

The Gauss-Seidel scheme uses the latest values of the components of $\underline{x}$ as soon as they have been determined. In this case the matrix B consists of all the non-zero elements of the matrix A which lie on or below the main diagonal, the remaining elements being zero. Hence the sparsity present in the lower triangular part of A is maintained in the matrix B. The convergence of the Gauss-Seidel method is generally faster than that of the Point Jacobi method, for many problems it is twice as fast. The amount of computation per iteration is again $O(\tau)$.

2.4 The Successive Over-Relaxation Scheme (SOR)

SOR combines the Gauss-Seidel scheme with an acceleration procedure to enhance the rate of convergence. The exploitation of sparsity is therefore identical to that in Gauss-Seidel. However, as is well known, the rate of convergence may be significantly better but is dependent on the choice of the relaxation parameter. The optimization of the parameter may be both diffi-

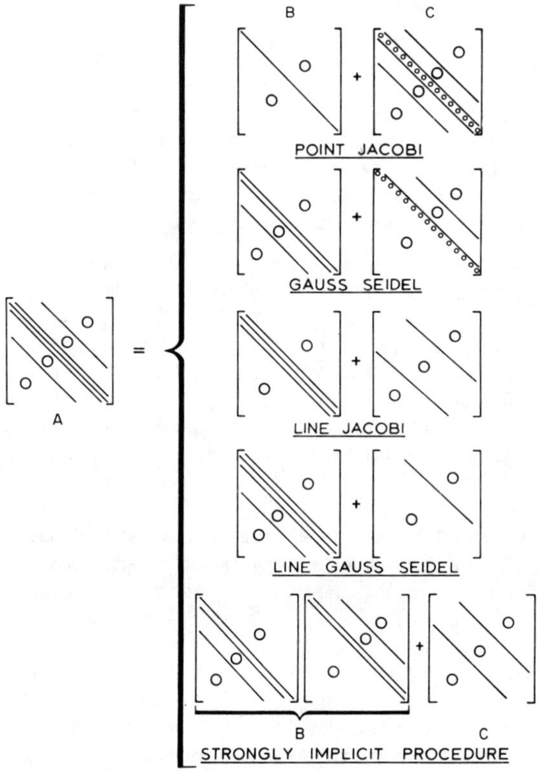

Figure 1 Schematic factorizations for five iterative schemes illustrated for an equation of Poisson type

cult and critical.

2.5 Line Iterative Schemes

'Single line' iterative methods use the relative ease with which a tridiagonal matrix (with non-zero elements on only the main three diagonals) can be inverted. In the Line Jacobi scheme the matrix B comprises the non-zero elements of A which lie on or immediately adjacent to the main diagonal; all the remaining elements are zero. For the line Gauss-Seidel scheme, B comprises all the non-zero elements of A which lie on or below the main diagonal together with those immediately above the main diagonal. The successive line over-relaxation (SLOR) schemes and alternating direction implicit (ADI) methods are also based on the ease with which a tridiagonal system can be solved exactly.

Multi-line iterative methods (for example, Parter (1959)) put even 'more' of the non-zeros of the matrix A into the factor matrix B, but the solution of the system $B\underline{y} = \underline{c}$ then becomes more demanding both in terms of the amount of computation and of the sparsity structure of the matrix A.

The form of all these matrices B means that the system $B\underline{y} = \underline{c}$ can be solved in a straightforward manner. In each case much of the sparsity structure in the matrix A is extracted and put in the matrix B to provide an efficient solution procedure.

In general most, if not all, of the methods considered so far are sufficiently easy to comprehend and implement that users generally integrate the iterative procedure with the problem. In this way the sparsity structure and even the repeated nature of many equations can be exploited to the extent demanded by the application.

2.6 The Strongly Implicit Procedure (SIP)

SIP (Stone (1968)) is an implicit method which resembles a direct method because a new estimate for the components of the solution vector are simultaneously derived from all the equations of the system, that is each new value for a component depends on the new values of all the other components.

Referring to Figure 1, SIP places in B all the non-zero elements of the matrix A together with some further non-zero elements which are added in order that the resulting matrix is easily factorized into the product of two special matrices L (lower triangular) and U (upper triangular). Each has non-zero elements only on those diagonals in the same positions as the non-zero diagonals of the corresponding triangular parts of the matrix A. Since L and U are triangular, the system

$$LU\underline{y} = \underline{c}$$

is easily solved.

For the case of Laplace's equation being solved by finite differences when a five diagonal system of equations is obtained, the matrices L and U will each have only three non-zero diagonals as shown in Figure 1. The product of two such matrices has seven non-zero diagonals, five in positions coincident with those of A and two additional ones adjacent to the outermost diagonals of A. In determining the coefficients of the matrices L and U there are five values to be obtained on each row, one value is redundant and an efficient exploitation of this sets the main diagonal elements of the upper triangular matrix U to unity. With the five values on each row of the matrix A given,

L an U are uniquely determined if the matrix B is made identical to the matrix A on the non-zero diagonals of the latter. There is scope for reducing the size of the terms of the two extra diagonals of B not present in A and this is done by relating the seven diagonal form to a difference molecule and representing fractions of the extra terms using a Taylor series as linear combinations of the five primary points. The fraction which is cancelled is varied on successive pairs of iterations to provide rapid convergence, the pair consisting of a sweep of the nodes in one direction followed by a sweep with one index swept in the reverse order.

The amount of computation to solve the system of equations (3) when B is of the form described above, namely the product of two matrices with non-zeros confined to a few diagonals, is $O(\tau)$, although the multiplying factor is greater than that for the other iterative methods described above. However the rate of convergence obtained has generally been found to be significantly better. There are also a number of other advantages. SOR and some of the other methods require properties such as diagonal dominance or positive definiteness of the systems of equations, and sometimes symmetry, to work, or at least for the theoretical convergence properties of the methods to be valid. Such restrictions have not been found to be necessary for SIP. An acceleration parameter is used in SIP but it has been found that its value is far less critical than, say, the relaxation parameter for SOR, indeed a standard parameter proves adequate, if not ideal, for a wide range of problems. Overall the method has proved to be very robust and efficient.

It does have some drawbacks, however. Firstly since SIP makes considerable use of the structure of the sparsity of the coefficient matrix A, different sparsity structures demand new algorithms, although of course, it is this exploitation of structure which contributes to the efficiency of the method. Extensions can be made to the original method as described by Stone (1968) which applied to five diagonal systems. Weinstein, Stone and Kwan (1969) extended the method to seven diagonal systems and Jacobs (1974) extended it to thirteen diagonal systems. The method of iterated defect correction described in Section 4 also enables the use of SIP to be considerably extended.

A second drawback of the method is that SIP is inherently more complex than the other iterative methods considered so far making it less easy for a user to implement. For this reason we developed eight years ago a number of subroutines based on SIP for solving the systems of equations which frequently arise in finite difference modelling. These have been used on a diverse range of problems, including a number involving the solution of non-linear equations. The significant power in terms of the efficiency, and the ease of use assisted by the alleviation of the need to optimize acceleration parameters for the vast majority of the problems have been the principal attributes. However for certain large problems, for example a Poisson's equation on a 70 by 70 mesh, SIP has produced very slow convergence.

However, in general, it is less easy to exploit the sparsity present in a system if library routines are employed. In the light of the experience of users of the original programs and with a view to easing and widening their application, a revised suite of programs (Jacobs (1980a)) has recently been produced. There are two iterative routines for each of the four basic molecule types considered - five point (for example, the Laplacian in two dimensions using three point replacements of second derivatives), seven point (for example as for the five point but in three dimensions), nine point (for example for a high accuracy replacement of the Laplacian in two dimensions) and thirteen point (for example fourth order operators such as the biharmonic in two dimensions). The basic routine performs one SIP iteration to determine the update vector $\underline{\delta x}$ of equation (5) given the current residual and the coefficients of the matrix on which the iteration matrix B is to be based. The user must provide the driving routine to calculate the residual, to update the approximate solution and check for convergence. In this he can exploit the sparsity to the full. The coefficients of the matrix supplied with which to calculate the LU factors which form the factor matrix B need not be identical to those of the matrix A and so defect correction can be used as described in Section 4. These single iteration routines are also designed for use with non-linear problems. Single, double precision and complex versions are provided.

The single-iteration routines described above form the kernel of four standard 'driver' routines which are designed to control the iterative procedure for all 'normal' applications. The input includes the coefficients of the original system of equations(1) which in this case must conform to the multidiagonal structure specified, the source terms, an initial approximation to the solution and convergence criteria. The routines organize the iteration: calculate the current residual, call the basic routine, update the approximation; and repeat this cycle until convergence is achieved.

Six specific problem solving programs have also been provided which make use of the above iterative modules as appropriate. There are five programs for solving Poisson's equation in two and three dimensional rectangular regions, two dimensional circles and annuli and three dimensional hollow and solid cylindrical regions, and one for solving the biharmonic equation in a two dimensional rectangle. For a number of these programs additional facilities are provided to enable the user to define internal boundaries to the overall region on which, for example, the potential can be specified, or within which the conductivity (permeability, diffusivity, etc) takes a different constant value to that in the remainder of the region. Thus these programs are suitable for solving Poisson's equation for a wide range of geometries in two and three dimensions using finite differences.

This suite of programs provides users with a range of iterative solution packages and problem solving programs which either provide the necessary problem solving tools or at least form the basis for the development of special purpose applications packages. Indeed the problem solving packages have been written in a form which makes their adaptation for solving other problems as easy as possible. Thus a solution package for the Helmholtz equation can be easily developed for one of the regions for which a Poisson solver is available.

2.7 Summary

To conclude this section it is worth summarizing the common underlying features of the methods considered so far.

The iterative procedures have all been based on the use of the iterative update formula $B\underline{\delta x}_k = \underline{r}_k$ where $\underline{r}_k$ is the residual of the kth approximation to the solution vector and $\underline{\delta x}_k$ is the change vector. The methods exploit the sparsity in two ways - they first provide an efficient means of calculating $\underline{r}_k$ which will generally need $O(\tau)$ computations and secondly they generate a matrix B which

(i) provides an easy means of calculating $\underline{\delta x}_k$ because of its form.

(ii) is close to the matrix A, or more specifically is such that $B^{-1}C$ has eigenvalues no greater, and preferably less, than one in modulus,

(iii) is sparse so that the system of equations (5) can be solved very economically.

Of course in general the methods considered all use a matrix B which is similar to the matrix A, the matrix being obtained by extracting just some of the matrix A, or by adding, as in the case of SIP, some extra coefficients.

3. PRECONDITIONED CONJUGATE GRADIENT METHODS

3.1 Introduction

The class of iterative methods described in this section are of an inherently different form to those described in Section 2. One of the ways in which they exploit sparsity is also rather different.

The section is divided into two parts. Part A contains a description of the conjugate gradient method and the basis of preconditioning. The original method of Meijerink and van der Vorst (1977) for preconditioning symmetric M-matrices is then described. The way in which sparsity is employed in this factorization is explained. Then, in Part B, a number of recent extensions to the method are described for a wide range of different types of systems - for example non-symmetric and complex. The sparsity exploitation employed in these is generally analogous to that used in the original method.

PART A: CONJUGATE GRADIENTS AND PRECONDITIONING

3.2 The Conjugate Gradient Method

The conjugate gradient method (Hestenes and Stiefel (1952)) is a gradient method belonging to the same family as the method of steepest descent.

Given a system of equations $A\underline{x} = \underline{b}$ for which the solution $\underline{x}$ is required, and given an approximation $\underline{x}_0$ to the solution, a gradient method seeks the minimum of a suitable measure of the residual $\underline{r} = \underline{b} - A\underline{x}$ by deriving a sequence of estimates $\underline{x}_k$ with residual $\underline{r}_k$. The measure is chosen so that the minimum occurs when $\underline{r}_k = \underline{0}$, that is at the exact solution. An obvious measure to use is the L_2 norm, namely $h^2 = \underline{r}_k^T\underline{r}_k$. However if the matrix A is positive definite and symmetric, then so is the matrix A^{-1}, and a 'better' measure to take for many applications is $h^2 = \underline{r}_k^T A^{-1} \underline{r}_k$. It is a 'better' measure when gradient methods are used for the minimization since it results in less computation and also better properties of the algorithm (see also Section 3.6). Since A^{-1} is a positive definite symmetric matrix, the minimum of h^2 occurs when $\underline{r}_k = \underline{0}$. For each estimate $\underline{x}_k$ of the solution a direction $\underline{p}_k$ is selected and the new estimate $\underline{x}_{k+1}$ is the point on the line $\underline{x}_k + \alpha\underline{p}_k$ which minimizes h^2.

The various gradient methods are distinguished by their choice of $\underline{p}_k$. The method of steepest descent selects $\underline{p}_k$ as the direction along which h^2 changes most rapidly, that is $\underline{p}_k = \underline{r}_k$. With this choice, an infinite number of steps are generally required to obtain the exact solution.

The method of conjugate gradients selects $\underline{p}_k$ to be as close as possible to the direction of steepest descent with the proviso that $\underline{p}_k$ is conjugate to all the previous directions of search. Since the vectors $\underline{p}_k$ form a set of conjugate vectors, a set of N of them will form a basis of the N dimensional space. Since h^2 has been separately minimized along each of these conjugate directions, the conjugate gradient algorithm will terminate, given exact arithmetic, after N steps. The algorithm for real symmetric positive definite systems is a special case of the general algorithm given in sub-section 3.8.

Conjugate gradients would be a direct method if exact arithmetic were used. However, in practice, when finite precision storage and arithmetic operations are used, it behaves like an iterative method. It is for this reason, and because the amount of computation required to perform N steps would in general be greater than that required by other direct methods, such as Gaussian elimination, that the conjugate gradient method has not been greatly used as a direct method. However Reid (1971) and others proposed and demonstrated its use as an iterative method, and that particular paper can be seen as a foundation stage in the developments that have since taken place. The iterative nature employed is not that more than N steps are used, but that, if only limited accuracy is required, this may be obtained in many fewer than N steps.

There were two further developments in the evolution of preconditioned conjugate gradient methods which are worthy of note at this stage. The first was the increased understanding of the properties and behaviour of the conjugate gradient method. This in part arose from the analogy of the method with the Lanczos (1950) algorithm for finding eigenvalues of a matrix from which many theoretical results carry over to conjugate gradients. Secondly, papers such as those of Reid (1971) and Jennings (1977) showed that if the eigenvalues are clustered in a small number of groups, then the conjugate gradient method will significantly reduce the residual in a number of steps equal to the number of groups, which may be very much smaller than the order of the system. The theoretical result of which these are manifestations, is that conjugate gradients requires, using exact arithmetic, one step for each distinct eigenvalue. However, for the systems of algebraic equations derived from finite difference or finite element models of partial differential equations, the eigenvalues of the coefficient matrix are generally evenly distributed. Thus conjugate gradients is not ideally suited for such systems, at least in their 'raw' state.

3.3 Preconditioning for SOR

The other development of note was the use of preconditioning methods applied to systems of equations prior to using iterative methods. Evans (1967) used a preconditioning method together with SOR to generate a system for which the choice of the acceleration parameter is not as critical as it is with SOR. Thereby better rates of convergence can be obtained given the same effort in parameter optimization.

3.4 Preconditioning for conjugate gradients

Meijerink and van der Vorst (1977) were the first to consider using preconditioning followed by conjugate gradients. They used preconditioning to generate a transformed system of equations which has a much better eigenvalue spectrum from the point of view of subsequent solution by conjugate gradients, in particular the eigenvalues were more closely clustered together. They derived concrete existence and convergence results in the case of M-matrices and described an algorithm for symmetric M-matrices. (An M-matrix is a non-singular matrix with all its off-main diagonal terms non-positive and with $A^{-1} \geq 0$. Finite difference and finite element models frequently, but by no means always, give rise to M-matrices.) Other researchers (for example Kershaw (1978)) have since exploited the power of the resulting method by applying it in a modified form 'pragmatically' to a wider class of systems with, in general, great success.

The preconditionings, of which many possible forms have been tried, all exploit the sparsity of the original system of equations in one form or another. For example, Meijerink and van der Vorst (1977) considered a five diagonal system of equations with a symmetric M-matrix. They sought an approximate Cholesky factorization of the coefficient matrix

$$A \doteq LL^T \tag{6}$$

where the matrix LL^T is called the preconditioning matrix.

The sparsity pattern of the factor was selected to be either the same as that of the original matrix A in the lower triangle, or the same with the addition of three extra diagonals, resulting in the so called ICCG(0) and ICCG(3) methods respectively

(Incomplete Cholesky factorization with Conjugate Gradients). The product LL^T was made to have the same element values on the non-zero diagonals corresponding to those of the matrix A as the corresponding elements of A. The analogous equalities, but increased in number, are used with the ICCG(3) method for which an additional three non-zero diagonals are included in the factor matrix L, two adjacent to the outermost diagonal of the band, and one adjacent to the inner diagonals. These particular diagonals are added since the coefficients of the factors generally decrease away from the diagonals in the same position as those present in the matrix A. Their inclusion provides a better approximation to the exact Cholesky factors. More recently Meijerink and van der Vorst (1978 and 1979) have explored factors with yet more additional diagonals. The improvement in approximating the factors, which is problem dependent, must be balanced by the extra work involved in obtaining them, and the possible additional storage requirements.

The approximate factorization obtained in the manner described above is used to precondition the original system of equations. The system (1) can be rewritten, provided L of equation (6) is non-singular, in the equivalent form

$$L^{-1} A L^{-T} L^T \underline{x} = L^{-1} \underline{b} \tag{7}$$

that is as

$$D\underline{z} = \underline{d} \tag{8a}$$

where

$$D = L^{-1} A L^{-T} , \tag{8b}$$

$$\underline{z} = L^T \underline{x} \tag{8c}$$

and

$$\underline{d} = L^{-1} \underline{b}. \tag{8d}$$

By the very choice of the factor L, we know that the matrix D is very close to the identity and hence will have a great many eigenvalues much more closely clustered than those of the original matrix A. Conjugate gradients applied to the preconditioned systems, for a great number of problems reduces the residual by a sufficient amount for most purposes in the first few steps, and is generally more economic than most other iterative methods.

The sparsity of the coefficient matrix, A, of the original system is exploited in two ways in the algorithm described. During the conjugate gradient algorithm the calculation of the product of the matrix A with a given vector can make full use of the sparsity. Secondly, the approximate factors used to precondition the system of equations are selected to have much of the sparsity of the original system. In calculating the product of the factor matrix L with a vector, as is necessary during the algorithm, full advantage can be taken of the sparsity pattern.

An additional attraction of the preconditioned conjugate gradient method is the ease with which most of the algorithm can be vectorized, thus making it ideal for the increasing number of specialized 'parallel' computers which are being used to solve a great many of the largest models of differential equations.

The methods described by Meijerink and van der Vorst (1977) were restricted to systems characterized by M-matrices. However Kershaw (1978) successfully applied the same basic idea to a wider class of problems characterized by general symmetric positive definite matrices. With an M-matrix Meijerink and van der Vorst (1977) had proved that the main diagonal terms of the approximate factor are always positive for the inexact factorization they used (as they are for the complete Cholesky factorization). Similar results are not available for more general matrices. Since it is necessary that the matrix factor is non-singular in order that the preconditioning (equation (7)) is invertible, the main diagonal elements must all be non-zero and for the minimization property, it needs to be positive definite. Kershaw (1978) recommended forcing all values of the main diagonal coefficients to be positive, setting any elements which were calculated to be non-positive during the incomplete factorization to a 'suitable' positive value. Such modifications, it is argued, will only have a very limited effect on the convergence rate of conjugate gradients provided they are few in number - and any deleterious effect they do have will be removed by the first few steps of the conjugate gradient algorithm. Kershaw describes very favourable experience with this approach and this has been echoed by other users, including ourselves.

PART B: RECENT DEVELOPMENTS IN PRECONDITIONED CONJUGATE GRADIENT METHODS

3.5 Types of preconditioning

Munksgaard (1980) developed an analogous algorithm which made no assumptions about the sparsity structure of the matrix A, and incorporates a very flexible approach to fill-ins, thereby allowing the user to control the level of sparsity in the approximate factor. In his method, a tolerance level is set so that during the calculation of the approximate Cholesky factor, fill-ins are ignored only if they are smaller in magnitude than the tolerance factor times the main diagonal element in their row and column. Like Kershaw (1978), Munksgaard modified the diagonal coefficients of the approximate factor to ensure positive definiteness of the preconditioning matrix. The calculations of the approximate preconditioning factor is based on the work of Gustafsson (1977) who ensured the equality of the row sums of the preconditioning matrix and of the original matrix, with a possible main diagonal enhancement. Munksgaard reports results on a wide range of problems and compares the computation times with both conjugate gradients and the Yale sparse matrix package. It is an interesting flexible approach which must have considerable potential for more general sparse systems than those derived from regular finite differences and finite elements.

Since the publication of the original preconditioned conjugate gradient method, a number of other preconditionings have been tried. The preconditionings of Meijerink and van der Vorst (1977) did not ensure that the row-sum of the preconditioning matrix, $L^T L$, was equal to that of the original matrix. The equality of row sums can be shown (for example Gustafsson (1977) or Wong (1978)) to be a very desirable feature in ensuring that the eigenvalues of the two matrices are very close, and hence that the preconditioned matrix (matrix D in equation (8a)) is close to the identity.

Wong (1978) considered three different types of preconditioning for symmetric positive definite systems of a kind similar to ICCG(0). Each ensures the equality of row-sums and a number of other combinations of requirements were enforced, for example equality of main diagonal terms, etc. He concluded from the results of a number of experiments that the 'best' choice in terms of convergence of the methods he tried was to force the equality of row sums and the equality of non-zero off-diagonal coefficients. It is interesting that this scheme also requires only the main diagonal element of the factor matrix L to be stored, the other elements are very simply related to the equivalent elements of the matrix A. Hence the sparsity is more 'fully' exploited for this preconditioning than for the others he considered.

It is this author's opinion that a general purpose code for diagonally structured matrices should be based on an algorithm employing the non-zero diagonal structure of the original coefficient matrix A for the approximate factorization. We also believe the same is true for more generally structured sparse matrices and are developing packages (Markham (1981)) using packed storage of the elements implementing the preconditioning described above.

We have also considered the use of a preconditioning based on the SIP factorization with correction for systems of equations resulting from finite difference models which may be non-symmetric and non-positive definite.

For this preconditioning, the row sums are preserved but all the diagonal elements of L and U need to be stored as they are different from those of A. A large number of tests (Jacobs (1980b)) has shown the method to be never better than that based on equality of row-sums and off-diagonal elements, and in some cases, it is considerably less good. For systems which are originally symmetric, the preconditioned system may become unsymmetric, and the methods described in the next subsection may be required. Because of the convergence results and the additional storage required,we are using a preconditioning based on the equality of row sums and off-diagonal elements. This preconditioning is used for both symmetric and non-symmetric systems. For the latter systems, as described in the next subsection, the bi-conjugate gradient method is used to solve the preconditioned system.

3.6 Extensions to Non-Symmetric Systems

The basic conjugate gradient method is applicable only to symmetric positive definite systems. The minimization property is not valid for indefinite systems. Kershaw (1978) recommended the use of conjugate gradients to solve the normal equations formed after preconditioning the original non-symmetric system. However the eigenvalues which are clustered by the incomplete LU decomposition preconditioning are then spread out by forming the normal equations, indeed the condition number of the latter system is the square of that of the preconditioned system.

An alternative approach described by Wong (1978) is to use the bi-conjugate gradient method developed by Fletcher (1976). This is an extension of the method proposed by Lanczos (1950) for determining the eigenvalues of an unsymmetric matrix. The amount of computation the algorithm involves is certainly greater than that used by conjugate gradients, indeed it is about the same as that required for solving the normal equations. On symmetric systems it is identical to conjugate gradients, although using more computation because some calculations are unnecessarily repeated. The experience of both Wong and ourselves with this method has been very favourable, producing rates of convergence only slightly slower than conjugate gradients applied to a similarly sized and conditioned system.

The actual algorithm will not be detailed here since it is a special case of the complex bi-conjugate gradient algorithm described in subsection 3.8.

A paper just published (Axelsson (1980)), which has been noted at the conference, describes a conjugate gradient method applicable to non-symmetric systems. It is not known whether the strictly monotonic decay of the error has been proved for this method.

3.7 Extensions to singular systems

In endeavouring to produce a method applicable to a wide class of applications, the problem of solving singular systems of algebraic equations arises. Such systems can be obtained, for example, when seeking a solution of Poisson's equation in a region with boundary conditions only on the derivative (see Jacobs (1980c)).

In the case of preconditioned conjugate gradient methods we have tried several different means of tackling such problems. The first was analogous to that used in the SIP problem solving packages described earlier in which a rescaling of the solution is used when solving singular problems which only have a unique solution modulo an additive constant. Such a rescaling, bearing in mind that, for the transformed variable $\underline{z}$ of equation (8a) the non-uniqueness is in the form of a multiple of the additive vector $L^t\underline{1}$ where $\underline{1}$ is the vector with each component equal to one, failed to provide a convergent algorithm.

The second method we considered was to seek out 'a redundant equation' and to eliminate it from the system thereby reducing the order by one. (Note that, since the problem being solved may not be within a rectangular region if Cartesian coordinates are being used, there may be a number of identity equations uncoupled to the rest of the system. The last equation of the set need not therefore be necessarily a redundant one.) For a general solution scheme such a method is not, we believe, viable.

We finally settled on simply using an extension of the pragmatic approach of Kershaw (1978) which he recommended for symmetric positive definite systems which are not M-matrices. Namely, during the determination of the incomplete factors, any main diagonal terms which are found to be less than a tolerance factor times the corresponding term of the matrix A, taking due account of signs, are reset to the value of the corresponding term of the matrix A. If the main diagonal of A is zero, then a typical value of the other elements in the row is used in its place. This ensures that the preconditioning matrix LU is non-singular. However, the matrix D of equation (8a) which is to be solved by the conjugate gradient method is still singular and may also be non-symmetric. The basic method of conjugate gradients has proved to be unsuitable. The bi-conjugate gradient algorithm (see section 3.6), which was developed for both non-symmetric and indefinite systems, has worked very well in these cases (Jacobs (1980b)). The rate of convergence for such problems is only slightly less fast than for problems with the same number of algebraic equations but which are non-singular. The possibility arises of a direction of search being parallel to the eigenvector corresponding to the zero eigenvalue. This is most unlikely in practice, and it has never occurred in our studies.

3.8 Extensions to Complex Systems

In a number of application areas systems of algebraic equations with complex coefficients and vectors have to be solved. It is interesting to note that the SIP method works well for such problems.

When applying preconditioned conjugate gradient methods to such problems there are several approaches that could be used. The variables could be split into their real and imaginary components to generate a real system of twice the size, that is of order 2N. However for many of the problems of this type, a very strong coupling between the real and imaginary components of each variable would exist and this generally makes iterative solutions much less efficient. It is preferable to work with the complex quantities.

The preconditioning using an analogous incomplete Cholesky or LU factorization as appropriate can be performed making the necessary checks to ensure that the factor matrices are not singular.

The preconditioned system obtained will consist of complex equations and the conjugate gradient method cannot be applied since it is only suitable for real, symmetric, positive definite systems. In section 3.6, where non-symmetric systems were considered, it was shown that such systems could be solved by forming the normal equations although the use of bi-conjugate gradients resulted in a better method. For a complex system one could form the equivalent of the normal equations by multiplying the system by the conjugate transpose of the preconditioned coefficient matrix; however as described in section 3.6 this increases the ill-conditioning of the system and may negate much of the effect of preconditioning.

The 'best' approach is to develop a version of conjugate gradients applicable to complex systems of equations. The method of bi-conjugate gradients mentioned in section 3.6 was developed by Fletcher (1976) for non-symmetric, indefinite systems. The author (Jacobs (1980d)), has developed a complex-bi-conjugate gradient algorithm which extends the application of conjugate gradients to complex, indefinite systems. The algorithm is as follows:

The Complex Bi-Conjugate Gradient Algorithm

To solve: $A\underline{x} = \underline{b}$

Given A - a non-singular matrix with complex elements of order n

$\underline{b}$ - the vector of right-hand sides of order n

$\underline{x}_o$ - an initial estimate of the solution

Solution Procedure:

Calculate the initial residual $\underline{r}_o = \underline{b} - A\underline{x}_o$

and set the initial bi-residual $\bar{\underline{r}}_o = \underline{r}_o^*$

Set the first direction vector $\underline{p}_o = \underline{r}_o$

and the first bi-direction vector $\bar{\underline{p}}_o = \bar{\underline{r}}_o$

Then for each i = 0,1,2,...

$$\text{Calculate the coefficient } \alpha_i = \begin{cases} \text{either} & \dfrac{(\bar{\underline{r}}_i, \underline{r}_i)}{(\bar{\underline{p}}_i, A\underline{p}_i)} \\[2ex] \text{or} & \dfrac{(\bar{\underline{r}}_i, \underline{p}_i)}{(\bar{\underline{p}}_i, A\underline{p}_i)} \\[2ex] \text{or} & \dfrac{(\bar{\underline{p}}_i, \underline{r}_i)}{(\bar{\underline{p}}_i, A\underline{p}_i)} \end{cases}$$

Calculate the new estimate $\underline{x}_{i+1} = \underline{x}_i + \alpha_i \underline{p}_i$

$$\text{Determine the new residual } \underline{r}_{i+1} = \begin{cases} \text{either} & \underline{b} - A\underline{x}_{i+1} \\ \text{or} & \underline{r}_i - \alpha_i A\underline{p}_i \end{cases}$$

Determine the new bi-residual $\bar{\underline{r}}_{i+1} = \bar{r}_i - \alpha_i^* A^{H} \bar{\underline{p}}_i$

$$\text{Determine the coefficient } \beta_i = \begin{cases} \text{either} & \dfrac{(\bar{\underline{r}}_{i+1}, \underline{r}_{i+1})}{(\bar{\underline{r}}_i, \underline{r}_i)} \\ \text{or} & -\dfrac{(\bar{\underline{r}}_{i+1}, A\underline{p}_i)}{(\bar{\underline{p}}_i, A\underline{p}_i)} \\ \text{or} & -\dfrac{(A^{H}\bar{\underline{p}}_i, \underline{r}_{i+1})}{(\bar{\underline{p}}_i, A\underline{p}_i)} \end{cases}$$

Set the new direction vector $\underline{p}_{i+1} = \underline{r}_{i+1} + \beta_i \underline{p}_i$

and the new bi-direction vector $\bar{\underline{p}}_{i+1} = \bar{\underline{r}}_{i+1} + \beta_i^* \bar{\underline{p}}_i$

and continue until $\underline{r}_{i+1} = \underline{0}$.

Note that ^{H} means the transpose of the vector or matrix of complex conjugate elements and that the inner product $(\underline{x},\underline{y}) = \underline{x}^{H}\underline{y}$, and * denotes the complex conjugate.

It is of interest to note that this algorithm incorporates the bi-conjugate gradient algorithm as a special case when the matrix and vectors are real, and the conjugate gradient method when the matrix is further restricted to being symmetric and positive definite. Thus the complex-bi-conjugate gradient algorithm is a generalized conjugate gradient method. The paper (Jacobs (1980d)) contains a full proof of the algorithm.

It has been used successfully together with a preconditioning by Biddlecombe (1980) on complex systems of equations arising from electromagnetic applications.

3.9 Convergence rates

Our own work, as described in the foregoing subsections, has been directed at developing preconditioned conjugate gradient methods for use as general purpose solution procedures for systems of equations derived from a wide spectrum of models, for example finite difference and finite elements. Methods have been developed for non-symmetric, indefinite and complex systems. They have been tested on a range of problems to identify the best general purpose version. Our test problems, all taken from partial differential equations, have included systems ranging from 100 equations to over 40,000 equations and a set of problems constructed to produce a system of equations the condition number of which could be varied by changing just one parameter in the specification of the physical problem. A number of the problems have also been solved using the SIP programs for comparison. The results (see Jacobs (1980c)), in summary, show that for small 'well behaved' problems, that is, up to about 2500 equations and reasonably well conditioned, SIP and preconditioned conjugate methods both provide very good rates of convergence. However, for larger problems, and for problems which result in systems of ill-conditioned equations, the preconditioned conjugate gradient method continues to provide similar rates of convergence whereas SIP is far less good and fails to converge for some very large problems.

4. ITERATED DEFECT CORRECTION

4.1 Introduction

The iterative methods considered in the previous two sections provide means of solving systems of algebraic equations, both groups exploiting the sparsity of the coefficient matrix, although doing so in different ways.

Iterated defect correction (also known as iterated deferred correction or difference correction (Fox (1977), Stetter (1978) and as the iterative replacement scheme (Jacobs (1975))), is a generic name given to a family of iterative methods to which those of Section 2 belong as special cases. The basis of the method is as follows: given the requirement to solve the algebraic system $A\underline{x} = \underline{b}$ the method of 'iterated defect correction' employs a matrix B which is in some sense 'comparable' to that of A but has the additional property that the system $B\underline{y} = \underline{c}$ is very easily and/or economically solved. The original system is then cast into the form $B\underline{x} = \underline{b} - (A - B)\underline{x}$ where (A-B) is the 'defect matrix', being the difference between the actual matrix of the system and that used for the purposes of generating an iterative scheme. As with the 'classical' methods of section 2 the iterative update formula can then be written as

$$B\,\delta\underline{x}_k = \underline{b} - A\underline{x}_k = \underline{r}_k,$$

namely equation (5).

The choice of the matrix B will depend on the type of solution procedure to be used for solving the system (5). However an overriding requirement, if defect correction is to converge, is that the matrix $B^{-1}(A-B)$ has no eigenvalues greater than one in modulus, a condition also required of the iterative methods described in section 2. In the following subsections a number of special cases of this general iterative technique are detailed. The general objective of all such methods is to obtain a solution to the system of equations (1) by using an iterative procedure of the form (5), the solution, albeit perhaps approximate, of this latter system must therefore be obtained with much less effort than that required to solve the original system, since it will have to be solved perhaps many times to obtain convergence. This is generally achieved by fully exploiting the sparsity of the matrix A in the choice of non-zero elements of the matrix B, and often omitting large groups of elements to ensure that the system (5) is easily solved.

4.2. 'Classical' iterative methods

A number of the 'classical' iterative methods have already been surveyed in section 2 where the particular form chosen for the matrix B by a number of different methods was described.

4.3 Exploitation of fast direct methods

Two notable advances of the past few years in the area of solution methods for large algebraic systems are the development of so called 'fast direct methods', namely those based on Fourier analysis or cyclic reduction, and the novel methods which have been termed 'marching'. These are reviewed in the paper by Detyna (1980) in these proceedings.

The point of major interest in this section is that the methods provide means of obtaining the solution to the system of N equations derived from a two dimensional problem in only $O(N \log_2 \log_2 N)$ computations, that is at least comparable with an iterative method. For each there are generally restrictions on their use. However their power makes it desirable that they be used on as many problems as possible. Iterated defect correction is one means of broadening their range of application.

For example if Poisson's equation is to be solved in a rectangular region within which a number of subregions are defined with different conductivities, then the fast direct methods are generally not applicable. However, if the matrix B is generated to be that representing the system of equations for a uniform region, then each iteration of the system (5) can be solved using fast direct methods instead of using an iterative procedure throughout. Concus and Golub (1973) describe one such application. Many others are possible, using either Fourier methods or marching methods as appropriate. It is an area of application which has not been fully explored.

4.4 Higher order approximations

Another use is that for which the method of iterated deferred correction was first employed and on which Pereyra (1966, 1970) has made considerable advances for both one and two dimensional problems. Suppose a particular differential equation is to be solved, and both low and high accuracy discrete formulations can be obtained using the same mesh nodes. Take for example the case of Poisson's equation on a uniform square finite difference mesh. Then a five point difference replacement of the Laplacian gives $O(h^2)$ accuracy, whereas the nine point replacement gives $O(h^6)$ accuracy. The amount of extra work involved in solving the more accurate system will depend on the solution method being employed - in some cases a new version of the solution procedure may have to be derived (for example with the Strongly Implicit Procedure or preconditioned conjugate gradients), for other methods, for example the fast direct methods, it may not be possible to solve directly the more complex system. Iterated defect correction provides a means of applying the best available techniques.

In the example mentioned above, two systems of algebraic equations are obtained $A_5\underline{x} = \underline{b}_5$ and $A_9\underline{x} = \underline{b}_9$. Both matrices are of the same order, but A_9 is less sparse than A_5. The author (Jacobs (1975)) used iterated defect correction to solve this system using the iterative formula $A_5'\delta\underline{x}_k = \underline{r}_k = \underline{b}_9 - A_9\underline{x}_k$ where the matrix A_5' has been scaled to make the main diagonal coefficients the same as the corresponding ones in the matrix A_9. Then by using a five diagonal equation solver (in this case based on SIP), a solution satisfying the more accurate nine-point molecule (that is accurate to $O(h^6)$) was obtained using 43 iterations as compared to 34 had the residual been calculated from the five diagonal system which would have produced a solution with a truncation error of $O(h^2)$.

Another application of this type has been to the solution of Poisson's equation on a non-uniform finite difference mesh. The commonly used five point difference molecule has a truncation error of $O(h)$ on a non-uniform grid, and at least one further point in each coordinate direction is required in the difference molecule to obtain second order accuracy. (An alternative means of obtaining the increased accuracy would be to reduce the mesh size and/or use a uniform mesh, both of which are likely to result in an increased number of nodes and therefore greater expense in obtaining the solution.) The author (Jacobs (1975)) showed that there was no systematic way of selecting the extra nodes so that the difference equations are diagonally dominant. Experiments showed that the 'best' method was that with the extra nodes selected so that the coefficients of the four nodes adjacent to the central one in the difference molecule had the same sign. Physically this corresponds to choosing the extra nodes on the side with the larger mesh increment adjacent to the central node. After suitably scaling the coefficients, an algorithm was obtained which converged for moderate mesh non-uniformities in fewer iterations than the same routine solving the less accurate five diagonal equations.

4.5 Complex molecules and boundary conditions

In much the same way as described in the previous subsections, iterated defect correction can be employed to obtain solutions to a wide class of problems which result in systems of algebraic equations which do not lend themselves to solution by a pre-existing method or algorithm. In each case if a 'suitable' matrix 'comparable' to the given system can be 'extracted' from the actual coefficient matrix, then iterated defect correction can be employed. Possible examples, over and above those already considered, include the solution of higher order differential equations (for example, the biharmonic equation) which yield large complex difference molecules and for which iterated defect correction based on a simpler subset of the coefficients, for example a five diagonal set, could be used (and indeed has been (Jacobs (1975))).

Boundary conditions of the same accuracy as the other finite difference equations cannot always be expressed in the same form as those for internal nodes. This is especially common when a boundary is being considered which is not defined entirely by mesh lines. In this case the iterative scheme is derived using finite difference replacements for the boundary conditions of the same general form as that for internal nodal points, although there will be a reduction in accuracy. But the residual is calculated using a more accurate difference replacement of the boundary conditions.

4.6 Multi-grid mesh rebalancing

The applications discussed in the previous subsections have all employed iterated defect correction with an iteration matrix of the same order as the original matrix. Multiple-grid methods, (see Brandt (1977a,b), Hackbush (1979) for original papers and Reid (1979) and Nakamura (1977) for brief surveys) also known as coarse mesh rebalancing, make use of a sequence of different approximations to a differential equation, where each approximation is derived on a separate mesh, but the meshes overlap. This can be viewed as a different form of sequenced iterated defect correction. The system of equations for which the solution is sought is $A_1\underline{x} = \underline{q}_1$. A family of other systems can be generated with a decreasing number of unknowns - that is on coarser meshes.

Given the current residual of the approximate solution, $\underline{x}_1^c$ namely

$$\underline{r}_1 = \underline{q}_1 - A_1\underline{x}_1^c$$

multiple grid methods interpolate $\underline{r}_1$ onto the second grid to form $\underline{r}_2$, solve

$$A_2\underline{\delta}_2 = \underline{r}_2$$

and then interpolate $\underline{\delta}_2$ back onto the original grid to form a correction $\underline{\delta}_1$ to $\underline{x}_1^c$. The equation $A_2\underline{\delta}_2 = \underline{r}_2$ is solved either directly or iteratively with possible use of a third grid for a similar stage of defect correction. The idea can be nested to any depth and may employ a direct solution on the coarsest grid.

The raison d'être for the method is that the convergence of most iterative methods is limited by the speed at which they can decay the components of the residual which have long wavelengths compared to the grid spacing. By determining the iterative update $\underline{\delta}_1$ through a sequence of problems, each component of the residual is 'solved' on a mesh comparable in size to its wavelength.

The power of the method has been demonstrated on a number of problems derived from both finite differences and finite elements (see Nakamura (1977) for references). The interpolation used for the residuals and the corrections between the different meshes is of major importance to the overall efficiency of the scheme and their selection is still very much an 'art'.

These are interesting methods which exploit both the sparsity of the original system and its relationship with subsystems of smaller dimension, and the convergence properties of iterative methods. Their actual success in practice is very dependent on tuning for each application and they have therefore been most used on very large and complicated problems.

4.7 Generating sparsity from full systems

Another form of iterated defect correction can be applied to the solution of some classes of systems of equations with a virtually full coefficient matrix, for example systems which arise from many of the integral methods and for which little of the foregoing sections is directly applicable. However, for these matrices, the dominant, or at least relatively important, features of the problem may be characterized by just a few of the coefficients of the matrix, perhaps the diagonal with a few additional entries that is a much more sparse coefficient matrix. If this reduced matrix system is amenable to solution by one of the efficient techniques, for example fast direct or iterative methods, then the vast computational saving they offer over a direct method applied to the full matrix can be exploited by using iterated defect correction. Of course the standard iterative methods are already of this form and it is being advocated here that the use of the newer iterative methods and even some of the fast direct methods should be explored - the potential saving could be very large.

5. CONCLUSIONS

Iterative methods have been developed historically to provide means of solving large problems. The meaning of 'large' increases with time as the tools available become more powerful. In general a major contribution to the efficiency of iterative methods is obtained from their exploitation of sparsity.

Significant advances in both iterative and direct methods for solving large systems of equations are being made. Frequently these make use of sparsity, and many also exploit other additional special features of classes of problems.

Iterative and direct methods are also becoming much more 'alike'. Consider, for example, the preconditioned conjugate gradient method which is 'almost' a direct method - indeed it would be if infinite precision computation were available and if a sufficient number of steps were used. In addition the greatly increased efficiencies being achieved by some of the direct methods and the improved rates of convergence of many of the new iterative methods means that the choice of method is now more problem dependent than ever before. Not only is the size of problem critical when selecting a solution method, but the form of the problem - for example special features such as separability of the differential equation - are also highly relevant. In this paper not only has the direct application of specialized methods been considered, but also the case has been made for exploiting these methods to the full on wider classes of problems by employing iterated defect correction. A number of these extensions are based on sparsity exploitation.

The preconditioned conjugate gradient methods are undoubtedly the biggest advance in iterative methods that has been made in the last decade or so. With the extensions which have been developed, for example those to singular and to complex systems surveyed in this paper, they provide the best general purpose iterative method available for very large and for very complicated problems.

For solving very large systems of algebraic equations derived from three dimensional problems, iterative methods remain the only viable solution technique. The greatly improved convergence rates achieved by preconditioned conjugate gradient methods means that the solution of such problems is more attainable now than ever before.

At a conference such as this, it is appropriate to speculate on likely future developments. Solution methods can evolve in an orderly structured fashion for many years, for example Point Jacobi through to SIP, providing a steady improvement in efficiency - each exploiting the sparsity or special features of the problem to a greater degree to achieve better convergence. However, there can also be very sudden, dramatic developments, not along the current mainstream. The preconditioned conjugate gradient method is one example and the marching method is another. What many of these developments have in common is the exploitation of some feature of the problem or of an algorithm so as to enable the 'best' combination of particular methods to be achieved. Thus, because the solution of large algebraic systems is occupying more people and more computers than ever before, and because new computer architectures are becoming available which will strongly influence the performance of solution methods, there are some exciting times ahead!

ACKNOWLEDGEMENTS

This paper has been written at the Central Electricity Research Laboratories. I am grateful to Geoff Markham for his comments on an early draft. The paper is published by permission of the Central Electricity Generating Board.

REFERENCES

Axelsson, O. (1980) Conjugate gradient type methods for unsymmetric and inconsistent systems of linear equations, Lin. Alg. and its Applics. 29, pp. 1-16.

Biddlecombe, C. (1980) ICCBCG - A symmetric complex sparse matrix solver, Note CAG/80-14, SRC Rutherford Laboratories

Brandt, A. (1977a) Multi-level adaptive techniques (MLAT) for partial differential equations: ideas and software, in Mathematical Software III, (J.R. Rice, ed.), Academic Press, pp. 277-318.

Brandt, A. (1977b) Multi-level adaptive solutions to boundary-value problems, Math. Comp. 31, pp. 333-390.

Concus, P. and Golub, G.H. (1973) Use of fast direct methods for the efficient numerical solution of nonseparable elliptic equations, SIAM J. Numer. Anal. 10, pp. 1103-1120.

Detyna, E. (1980) Rapid Elliptic Solvers, these proceedings.

Evans, D.J. (1967) The use of preconditioning in iterative methods for solving linear equations with symmetric positive definite matrices, J. Inst. Maths. Applics. 4, pp. 295-314.

Fletcher, R. (1976) Conjugate gradient methods for indefinite systems, in Numerical Analysis, Dundee (1975), Lecture Note 506, (G.A. Watson ed.), Springer Verlag, pp. 73-89.

Fox, L. (1977) Finite-difference methods for elliptic boundary value problems, in 'The State of the Art in Numerical Analysis' (D.A.H. Jacobs ed.), Academic Press, London pp. 799-881.

Gustafsson, I. (1977) A class of first order factorization methods, Research Report 77.04R, Chalmers University of Technology, Göteborg, Sweden.

Hackbush, W. (1979) On the computation of approximate eigenvalues and eigenfunctions of elliptic operators by means of a multi-grid method, SIAM J. Numer. Anal. 16, pp. 201-215.

Hestenes, M.R. and Stiefel, E. (1952) Methods of conjugate gradients for solving linear systems, J. Res. Nat. Bur. Standards, 49, pp. 409-436.

Jacobs, D.A.H. (1974) The strongly implicit procedure for biharmonic problems, J. Comput. Phys. 13, pp. 303 315.

Jacobs, D.A.H. (1975) Iterative replacement schemes for complicated banded difference equations, Note No. RD/L/N 169/75, CERL, Leatherhead.

Jacobs, D.A.H. (1980a) A summary of subroutines and packages employing the strongly implicit procedure for solving elliptic and parabolic partial differential equations, Note No. RD/L/N 55/80, CERL, Leatherhead.

Jacobs, D.A.H. (1980b) Preconditioned conjugate gradient methods for solving systems of algebraic equations, Note No. RD/L/N193/80, CERL, Leatherhead.

Jacobs, D.A.H. (1980c) Solving Poisson's equation subject to Neumann boundary conditions, Note No. RD/L/N 116/80, CERL, Leatherhead.

Jacobs, D.A.H. (1980d) Generalizations of the conjugate gradient method for solving non-symmetric and complex systems of algebraic equations, Note No. RD/L/N 70/80, CERL, Leatherhead.

Jennings, A. (1977) Influence of the eigenvalue spectrum on the convergence rate of the conjugate gradient method, J. Inst. Maths. Applics. 20, pp. 61-72.

Kershaw, D.S. (1978) The incomplete Cholesky-conjugate gradient method for the iterative solution of systems of linear equations, J. Comput. Phys. 26, pp. 43-65.

Lanczos, C. (1950) An iteration method for the solution of the eigenvalue problem of linear differential and integral operators, J. Res. Nat. Bur. Standards 45, pp. 225-282.

Lorenz, E.N. (1976) A rapid procedure for inverting del-square with certain computers, Monthly Weather Rev. 104, pp. 961-966.

Markham, G. (1981) A user's guide to CGS, BCGS and CCGS: Sparse code packages using preconditioned conjugate gradient methods for solving systems of algebraic equations, CERL Laboratory Program, to be published.

Meijerink, J.A. and van der Vorst, H.A. (1977) An iterative solution method for systems of which the coefficient matrix is a symmetric M-matrix, Math. Comp. 31, pp. 148-162.

Meijerink, J.A. and van der Vorst, H.A. (1978) Guidelines for the usage of incomplete decompositions in solving sets of linear equations as occur in practical problems, Private communication.

Meijerink, J.A. and van der Vorst, H.A. (1979) Incomplete decompositions as preconditioning for the conjugate gradient algorithm in Conjugate Gradient Methods and Similar Techniques (I.S. Duff ed.) AERE Report R9636, UKAEA Harwell.

Munksgaard, N. (1980) Solving sparse symmetric sets of linear equations by preconditioned conjugate gradients, TOMS 6, pp. 206-219.

Nakamura, S. (1977) Computational Methods in Engineering and Science, Wiley.

Parter, S.V. (1959) On two line iterative methods for the Laplace and biharmonic difference equations, Numer. Math. 1, pp. 240-252.

Pereyra, V. (1966) On improving an approximate solution of a functional equation by deferred corrections, Numer. Math. 8, pp. 376-391.

Pereyra, V. (1970) Highly accurate numerical solution of quasi-linear elliptic boundary-value problems in n-dimensions, Math. Comp. 24, pp. 771-783.

Reid, J.K. (1971) On the method of conjugate gradients for the solution of large sparse systems of equations, in Large Sparse Sets of Linear Equations (J.K. Reid ed.) Academic Press, London pp. 231-254.

Reid, J.K. (1979) Finite-difference equations and their iterative solution, in A Survey of Numerical Methods for Partial Differential Equations (I. Gladwell and R. Wait eds) Oxford University Press.

Stetter, H.J. (1978) The defect correction principle and discretization methods, Numer. Math. 29, pp. 425-443.

Stone, H.L. (1968) Iterative solution of implicit approximations of multi-dimensional partial differential equations, SIAM J. Numer. Anal. 5, pp. 530-558.

Weinstein, H.G. Stone, H.L. and Kwan, T.V. (1969) Iterative procedure for solution of systems of parabolic and elliptic equations in three dimensions, Ind. and Eng. Chem. Fund. 8, pp. 281-287.

Wong, Y.S. (1978) Iterative methods for problems in numerical analysis, D. Phil Thesis, Oxford.

DISCUSSION

PROFESSOR J.M. McNAMEE (York, Ontario). You made some comments on the relative merits of iterative and direct solvers. Could you summarize these please?

JACOBS. Certainly. The preferred method is both problem and environment dependent. For 2 dimensional PDE's we would normally use a direct method if storage allows. For most 3 dimensional problems I feel that only iterative methods are viable. For non-linear problems I believe that direct methods are not ideally suited for use at each step of the non-linear iteration except perhaps for the first step or two, unless they are very efficient, for example the fast direct methods.

PROFESSOR A. GEORGE (Waterloo). When you use the term "fast" what do you mean by it?

JACOBS. It depends on the problem and problem size. A "fast" direct method for many two dimensional problems may be one that is more efficient than a banded Cholesky solver, thus it may include nested dissection and would certainly include the methods discussed by Detyna at this conference.

PROFESSOR A.JENNINGS (Belfast). Have you much experience of the iterative techniques you discussed being applied outside the area of partial differential equations?

JACOBS. No, all my experience and that of users of our codes is on problems in partial differential equations.

PROFESSOR M.J.D. POWELL (Cambridge). When solving non-linear problems, there is always the possibility of using conjugate gradients directly. Could you comment on this?

JACOBS. I have not tried that idea at all. It is the preconditioning which has made conjugate gradients a viable method for our problems.

VARIATIONAL METHODS IN OIL RESERVOIR MODELLING

J. Rae

(Theoretical Physics Division, AERE, Harwell)

ABSTRACT

The lecture gives an introduction to the physical processes which form the basis for modelling the behaviour of oil fields. The governing mathematical equations are set up and some of their properties discussed in two cases of importance, the "parabolic" case when capillary pressure dispersion effects are large and the more interesting "hyperbolic" case when they are negligible. Some indications are given as to the numerical methods which are employed, especially to the recent rise of interest in variational methods.

1. INTRODUCTION

My intention in preparing this lecture is to interest mathematicians in a set of equations which, at least in this country, have far too long been neglected. The equations are those which model the behaviour of fluids, water oils and gases, as they move under external influences in the porous rock of underground reservoirs. Not only can these models be of real economic importance but they lead to a rich variety of mathematical and numerical problems worthy of wider attention. In the United States of America, which of course dominates most technical aspects of the petroleum industry, these problems have led to many of the advances in the last twenty-five years in numerical methods for solving partial differential equations. Even so, many interesting problems remain or have arisen recently.

Any attempt to model a real oil reservoir leads to some form of numerical calculation and, as the methods used are often sparse matrix techniques, it is an appropriate topic to discuss at this Conference. I will not, however, be saying much about the techniques themselves - they are well covered in other lectures - but rather trying to introduce a field in which they can usefully be applied. I should also emphasize that only the simplest cases are mentioned here. The subject has a history of some decades and has thrown up many and varied problems impossible to cover in one short lecture.

The scheme of this lecture is quite straightforward. There are three main sections which, with no attempt at a comprehensive review, treat in turn the basic equations which form the mathematical model of oil and water in two dimensional horizontal reservoirs, some mathematical properties of these equations, and a brief outline of some of the methods used and problems encountered when one attempts to solve the equations by numerical methods. Variational methods are given some prominence to reflect the attention recently given to them and the bias of the author rather than to imply that they are the most widely used.

2. BASIC EQUATIONS FOR OIL AND WATER IN A POROUS MEDIUM

2.1 Introduction

On a fundamental level the mathematical description of flow in a porous medium encounters very difficult statistical problems. Although much has been done to clarify these difficulties (Beran (1968), Scheidegger (1960), and Bear (1972)) most flow modelling has been based on a simpler, more empirical, approach which uses experimentally determined relations among a few fundamental variables. Such an approach is quite appropriate, and perhaps even necessary, in modelling oil and water in underground reservoirs where the detailed nature of the porous rock and of the fluids is difficult and sometimes impossible to obtain. By using empirical relations a mathematical model of flow in a porous medium can be built from three simple tenets of basic physics, namely equilibrium thermodynamics, conservation of matter and Newton's laws of motion.

2.2. Thermodynamics

From the equilibrium thermodynamics of simple systems, for example two locally homogeneous and isotropic fluids which do not interact chemically, it is known that pressure, density and temperature are not independent quantities but are related by an equation of state (Zemansky (1957)). For most purposes reservoirs can be assumed to be isothermal so that the equation of state reduces to a relation between density ρ and pressure p. There are several cases of interest. In the simplest, the fluids are taken as incompressible, that is ρ is a constant independent of p. We adopt this case in the rest of this paper. For compressible fluids a physicist would use a compressibility $\frac{1}{\rho}\frac{\partial \rho}{\partial p}$ (often essentially constant) and a reservoir engineer a formation volume factor, the ratio of a volume of liquid in reservoir conditions to its volume at reference surface conditions. If gas is present in the reservoir the thermodynamical picture becomes much more complicated.

2.3 Mass conservation

The simplest description of a porous medium is in terms of its porosity ϕ, the fraction of its bulk volume not occupied by solid material. In describing flow through such a medium we will always be concerned with the case where there are many small capillaries or linked pores throughout the medium. In particular, when we refer to a flow velocity $\underline{u}$ at a "point" x, y, z we mean the average velocity in a cell centred at x, y, z which is very small compared to the scale of variation of the medium's properties but nonetheless contains many pores.

In view of these remarks the continuity equations which express the conservation of mass of oil or water can be written in terms of water and oil saturations S_W and S_O, the fraction of fluid which is each phase, as

$$\phi \rho_O \frac{\partial S_O}{\partial t} = Q_O - \rho_O \underline{\nabla} . \underline{u}_O \tag{1}$$

$$\phi \rho_W \frac{\partial S_W}{\partial t} = Q_W - \rho_W \underline{\nabla} . \underline{u}_W \tag{2}$$

where t is time

ϕ is rock porosity

ρ_O is the (constant) density of oil

Q_O is the source (+ve) or sink (-ve) mass rate density for oil

$\underline{u}_O$ is the oil flow velocity.

The corresponding quantities for water have the subscript W. The left-hand side of these equations represents the rate of change of mass of fluid in a unit volume of porous medium, Q accounts for sources or sinks in that volume and the divergence terms measure the rate of flow out of the volume.

2.4 Darcy's law

Flow in a porous medium under any particular pressure distribution depends not only on porosity but on the number and manner of linkings of pores. An empirical law describing this was first established by Darcy (1856) and for a single-phase one-dimensional flow it says that the flow velocity is in proportion directly to the local pressure gradient and inversely to the fluid viscosity. The constant of proportionality, which has the dimensions of area, is known as the rock permeability. In general its value will vary when measured in different directions in a porous bed so permeability is a second-order tensor. However, in practice for sedimentary rock beds this tensor is nearly diagonal for some natural coordinate system which follows the bedding planes. As we will consider here only areal reservoirs and neglect the vertical dimension we further assume the permeability is isotropic in the x y plane.

When several phases are present together in a porous medium it is found experimentally that the flow velocity of any phase i is related to the corresponding pressure gradient by a law of the Darcy form

$$\underline{u}_i = - \frac{K_i}{\mu_i} \nabla p_i \tag{3}$$

with μ_i the fluid viscosity. Because the fluids interfere with each other K_i is not equal to the single phase permeability K but is empirically related to it by an equation

$$K_i = K\, k_{ri} \tag{4}$$

where k_{ri} is the relative permeability of fluid i, a function of the saturations of the various phases present (Collins (1961), Dake (1978)). When just oil and water are present the relative permeabilities are usually expressed as a function of water saturation, typically of the form shown in Figure 1.

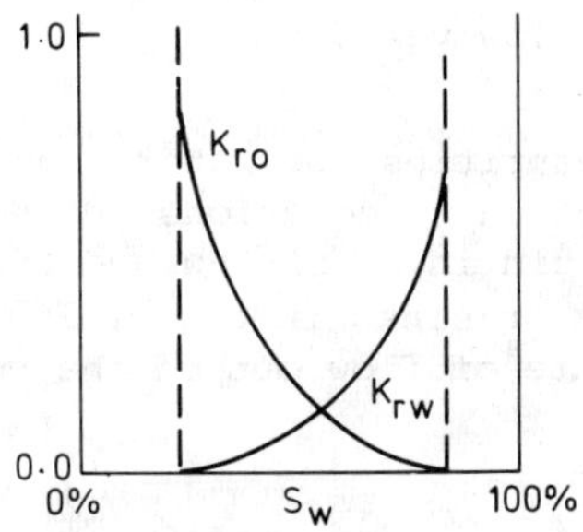

Figure 1 Relative permeabilities as a function of water saturation

The cut-off at about 20% represents a typical immobile connate water saturation and the upper cut-off residual oil.

The modified Darcy's law for the two-phase system may now be written

$$\underset{\sim}{u}_O = - \frac{Kk_{rO}}{\mu_O} \nabla P_O \tag{5}$$

$$\underset{\sim}{u}_W = - \frac{K\ k_{rW}}{\mu_W} \nabla P_W \tag{6}$$

Since we are considering only thin horizontal reservoirs we are neglecting all effects of gravity. If this has to be included the pressures appearing above have to be replaced by flow potentials, the pressures minus their hydrostatic parts.

2.5 Further relations

The equations written above may be put together in the form

$$\phi \frac{\partial S_O}{\partial t} = \frac{Q_O}{\rho_O} + \nabla .\ T_O \nabla P_O \tag{7}$$

$$\phi \frac{\partial S_W}{\partial t} = \frac{Q_W}{\rho_W} + \nabla .\ T_W \nabla P_W \tag{8}$$

where $$T_O = \frac{K\ k_{rO}}{\mu_O} \tag{9}$$

is the oil mobility, and similarly for T_W. The two equations (7) (8) relate the four field variables S_O, S_W, P_O, P_W and have to be supplemented to complete the sytem.

Two immiscible fluid phases, such as oil and water, present together in a system of pores, display physical effects not found in a single phase. One fluid, generally but not always the water, wets the rock medium more than the other and is referred to as the wetting phase. In a single capillary pore, the meniscus interface between the phases is convex as looked at from the wetting side and there is a positive jump in pressure P_C, the capillary pressure, in moving from the wetting to non-wetting side. It is found experimentally that in a given porous medium the capillary pressure is a function of the local saturation (Scheidegger (1960), Collins (1961), Dake (1978)). With just oil and water present the capillary pressure is usually taken as a measured function of the water saturation

$$P_O - P_W = P_C (S_W) \tag{10}$$

and has a form typified in Figure 2.

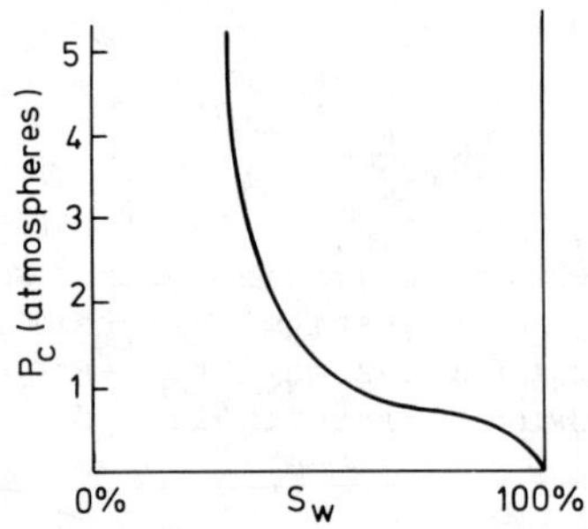

Figure 2 Capillary pressure as a function of water saturation

For some oil and rock combinations, however, P_C may be almost constant and this has important mathematical consequences as will be seen later.

Equation (10) is a relation between P_O and P_W and the set of equations is completed by the obvious relation, following from the definition in 2.3,

$$S_O + S_W = 1 \tag{11}$$

Thus (7) (8) (10) (11) are the equations to be solved.

Further useful forms of the equations can be set up (Peaceman (1977)) by adding (1) and (2) and introducing the total velocity $\underline{u} = \underline{u}_O + \underline{u}_W$

$$\nabla . \underline{u} = \frac{Q_O}{\rho_O} + \frac{Q_W}{\rho_W} \tag{12}$$

We will further assume that, as happens normally, sources and sinks inject and extract fluids in proportion to their local mobilities so

$$\frac{Q_W}{\rho_W} = \frac{T_W}{T_O + T_W}\left(\frac{Q_W}{\rho_W} + \frac{Q_O}{\rho_O}\right) \tag{13}$$

As we also have

$$\underline{u} = - (T_O + T_W) \nabla P_W - T_O \nabla P_c \tag{14}$$

equation (8) can now be rewritten in the useful form

$$\phi \frac{\partial S_W}{\partial t} = - \underline{u}. \nabla \left(\frac{T_W}{T_W + T_O} \right) - \nabla. \left(\frac{T_O T_W}{T_O + T_W} \frac{\partial P_c}{\partial S_W} \nabla S_W \right) \tag{15}$$

The equation (15) indicates some of the difficulties inherent in this system of equations. Even if the total velocity $\underline{u}$ is considered as a known function of x, y, t, and of course it usually has to be determined self consistently from equations involving S_W, the equation (15) displays two distinct characters. If P_c is a constant the dispersion terms drop out and the equation becomes a nonlinear hyperbolic equation for S_W. The fractional flow $T_W/(T_W + T_O)$ is however not usually a convex or concave function of S_W, which puts the equation out of the class of well studied hyperbolic equations. We refer to this as the hyperbolic case and return to it later. If P_c is not constant it is a decreasing function of S_W, so the equation has the form of a nonlinear parabolic equation. We will call this the parabolic case. Even here though, there are possible complications as the dispersion coefficient can vanish at certain values of S_W, for example by T_W vanishing at connate water saturation as in Figure 1. This can lead to dispersive solutions which have a continuous saturation S_W with infinite spatial gradients at certain points in space (Robinson et al(1980)) as indicated in Figure 3 which shows a water region spreading radially by dispersion but maintaining a sharp corner on the profile

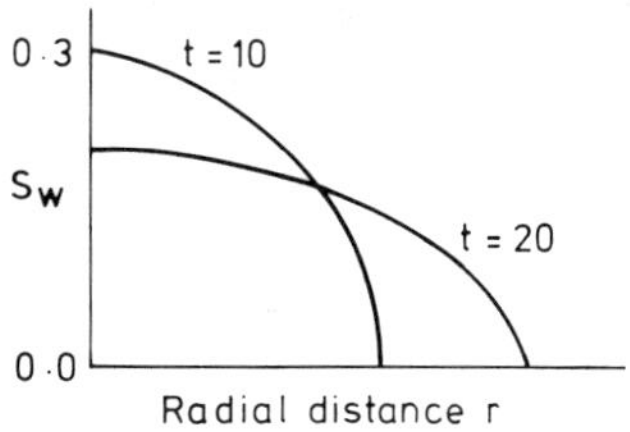

Figure 3 Water saturation profile spreading by nonlinear dispersion

3. MATHEMATICAL ASPECTS

3.1 *Weak formulation*

As shown by the example in section 2.5 even in the parabolic case it is appropriate to look for solutions in a weak sense, rather than for classical solutions. For general information on weak solutions see, for example, Courant and Hilbert (1962), Morrey (1966).

When equations (12), (14) and (15) are integrated with suitable test functions $\Phi(x)$, $\Psi(x)$ an appropriate weak formulation can be written in terms of oil pressure P and water saturation S

$$\int_{\Omega} d\tau \; \phi(x) \; \{\nabla.(T_W + T_O)\nabla P - \nabla.T_W \nabla P_c + \frac{Q_O}{\rho_O} + \frac{Q_W}{\rho_W}\} = 0 \qquad (16)$$

$$\phi \frac{\partial}{\partial t} \int_{\Omega} d\tau \; S(t)\Psi(x) = - \int_{\Omega} d\tau \; \underline{u}\cdot\nabla \left(\frac{T_W}{T_O+T_W}\right) \Psi(x)$$

$$+ \int_{\Omega} d\tau \frac{T_O T_W}{T_O+T_W} P_c' \nabla S . \nabla\Psi$$

$$- \int_{\partial\Omega} d\ell \frac{T_O T_W}{T_O+T_W} P_c' \frac{\partial S}{\partial n} \Psi \qquad (17)$$

Here P_c' is dP_c/dS, $\partial S/\partial n$ is the normal derivative on the boundary of the region Ω, and the last term in (17) is integrated round that boundary.

The equations (16) and (17) are to hold for all choices of tests functions Φ, Ψ from suitable spaces. For solutions P, S which are smooth enough, these last equations are equivalent to the "classical" differential equations (12), (15) but in general they admit a wider class of solutions. As the equations hold as the test functions are varied in their space the formulation above leads to the "variational" methods now widely used for numerical solution of the problem. It is important to realise that we are not necessarily assuming here that the governing equations (12), (15) can be derived from a variational or minimum principle in the traditional sense (Courant and Hilbert (1962)) although this possiblility is included. Another weak form commonly used is to allow the test functions a time dependence and integrate over an arbitrary time interval in (16), (17) as well as over the space region Ω.

3.2 *Solutions in the parabolic case*

Even for the parabolic case where $P_c' \not\equiv 0$ and known solutions are well behaved there seems to be no published existence theorem for solutions to equations (16), (17)†. They slip through the net of known results, for example Friedman (1964) or Ladyzhenskaya (1968) and other references therein, partly by their nonlinearity, partly by their awkward mixed elliptic/parabolic form and partly because P_c' is a function of S which can vanish for certain values. Even the simpler case of equation (17) alone with $\underline{u}$ taken as a known solenoidal field has not been covered. It seems to the author that methods recently used (Temam (1979), Girault and Raviart (1979)) to show existence of solutions to the Navier-Stokes equations can be adapted to the present case, but a detailed watertight proof has not yet been constructed. Once existence is assumed or known there is little problem, for the parabolic case, in extending existing contraction arguments (Oleinik (1964), Quinn (1971)) to show uniqueness.

Because of the nonlinearity it is difficult to find solvable examples of these problems even when simple velocity fields $\underline{u}$ are imposed. Some solutions to nonlinear diffusion equations are known. Ames (1965) and Robinson et al (1980) have applied one to the present context.

3.3 *Solutions in the hyperbolic case*

The equations for the hyperbolic case are just (16) (17) again with the terms with P_c' removed. Orthodox arguments for the existence of solutions would generally establish them as limits of solutions to (16), (17) as P_c' tends to zero in a suitable way. Thus we lack an existence theorem in this case also. From what is known about hyperbolic systems in general (see for example Courant and Hilbert (1962), Lax (1971), and Dafermos (1974)) we must expect there to be many discontinuous solutions which consist of piecewise smooth regions separated by jumps or shocks. Certain conditions have to be satisfied at the shocks and enable us, hopefully, to pick out the unique allowable solution which is correct physically. This correct solution will, of course, be that reached from the parabolic equations as the physical dispersion term P_c' is taken to zero in a suitable way. In the oil-water context of this paper, the

†See the note added at the end of this paper.

shocks or jumps are in the water saturation S. Thus as a waterflood is pushed through an oil rich reservoir one typically has little water ahead of the flood, a sudden jump in water saturation at a well-defined interface which moves, and a region of flooded reservoir behind.

The first of the jump conditions to hold at a shock, the Rankine-Hugoniot conditions, can be obtained in a completely standard way (Lax (1971), Dafermos (1974)) by integrating equations (16), (17) over a small time interval containing the time in question and over a region $\bar{\Omega}$ containing the hypothesized shock line. In this way equation (16) gives the condition

$$\left[(T_O + T_W) \nabla P\right]^+ \cdot \underline{n} = \left[(T_O + T_W) \nabla P\right]^- \cdot \underline{n} \qquad (18)$$

where $\underline{n}$ is the unit normal to the shock line (or surface in three-dimensional case) and + and - indicate limit values obtained on approaching the shock from opposite sides. Condition (18) expresses nothing but the continuity of the component of total velocity $\underline{u}$ normal to the shock. In a region with no saturation jump the $T_O + T_W$ factors are equal and the pressure gradient (in any direction) is continuous; discontinuities of pressure gradient lie on the shocks, where S is discontinuous. The condition which arises from equation (17) is

$$C\ (S^+ - S^-) = \left[(T_W \nabla P)^+ - (T_W \nabla P)^-\right].\underline{n} \qquad (19)$$

where C is the speed the shock moves normal to itself. The right hand side of this condition is the difference in water fluxes on the two sides of the shock; the condition says that the shock moves at such a speed as to conserve the amount of water.

Unfortunately it is known, from much simpler hyperbolic equations, that satisfying the equation in the smooth regions and the Rankine-Hugoniot jump conditions at the shocks does not give a unique solution. Perhaps the simplest example of this is the Hopf equation

$$\frac{\partial S}{\partial t} + \frac{S\,\partial S}{\partial x} = 0 \qquad (20)$$

which, for the initial conditions $S_O = 0$ if $x < 0$, $S_O = 1$ otherwise, has the two weak solutions

$$\text{A)} \quad S(x,t) = \begin{cases} 0 & \text{if } x < 0 \\ \frac{x}{t} & \text{if } 0 \leq x \leq t \\ 1 & \text{if } x > t \end{cases} \tag{21}$$

$$\text{B)} \quad S(x,t) = \begin{cases} 0 & \text{if } x \leq \frac{t}{2} \\ 1 & x > \frac{t}{2} \end{cases} \tag{22}$$

which satisfy both the equation and the jump condition corresponding to (19)

$$\tfrac{1}{2}(S^+ - S^-) = \tfrac{1}{2}(S^2)^+ - \tfrac{1}{2}(S^2)^- \tag{23}$$

For uniqueness a further shock condition has to be imposed, a condition which for historical reasons has become known as an entropy condition. For hyperbolic equations of the form

$$\frac{\partial S}{\partial t} + \frac{\partial}{\partial x} g(S) = 0 \tag{24}$$

where the function g is convex (or concave) the requirements are well understood. Various proposed entropy conditions have been shown to be equivalent (Lax (1957)); for convex positive g the shock moves to the right and the simplest form of the condition says that the speed to the left of any point must not be less than the speed to the right. In general terms it is a stability condition; a shock can only maintain itself if the material fluxes to it are appropriate and the shock will reform after perturbation. For the Hopf equation above it shows that the solution A is stable in this sense while B is unstable and unphysical. For the convex case the theory has been extended to more complicated sets of equations than (24).

As remarked earlier the fractional flow $T_W/(T_O + T_W)$ appearing in equation (15) is not in general convex or concave but has both characters as indicated in Figure 4.

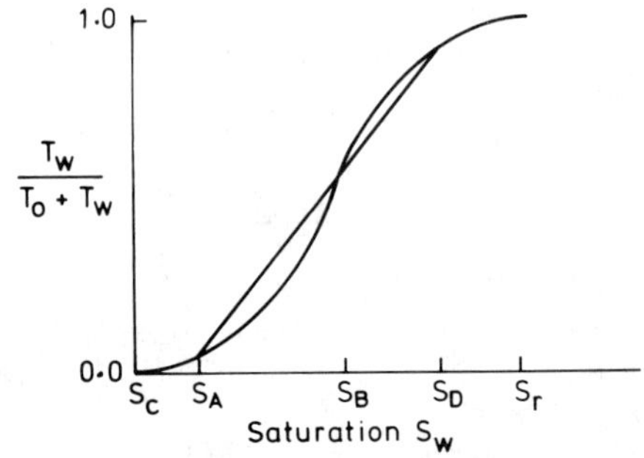

Figure 4 A typical fractional flow curve

Very little theory exists for this more general case and the literature is rather confused. It turns out however that there is enough to cover the two-phase case of water flooding which is considered here. The proper entropy condition seems to have been given first by Oleinik (1963) but is probably better understood from Hopf (1969) or, especially, Quinn (1971). For equation (24) with a function g as in Figure 4 and S^-, S^+ indicating the saturation limits from the left and right in x the condition can be written

$$\frac{g^- - g}{S^- - S} \geq \frac{g^- - g^+}{S^- - S^+} \geq \frac{g - g^+}{S - S^+} \tag{25}$$

for any S between S^+ and S^-. Geometrically, this says that the chord slope on Figure 4 between shock limits S^+, S^- must be greater than the chord slope from S^+ to any intermediate S and less than that from S^- to any intermediate S. In fact either inequality in (25) implies the other. Thus the condition says that a shock from S_A to S_B in Figure 4 is allowable but that S_A to S_D would not be; it would develop into a smoothly falling saturation followed by a smaller jump. In physical terms the condition is again a kind of stability criterion. From the Rankine-Hugoniot relation applied to equation (24) we recognise the middle term of (25) as the shock speed. So the Oleinik condition says that the speed of the physical shock lies between the speeds of any two partial shocks got by stepping down the saturation jump in two stages. The upper partial would catch up and the lower be caught up, so the physical shock is in this sense stable. When the g curve is convex the new entropy condition reduces to the old, in this case $S^- > S^+$. It applies more generally than to equation (24), even by the method of Quinn (1971) to the system (12), (15), and is the uniqueness condition we need.

Solvable cases of the hyperbolic system (12), (15) are quite difficult to find and simplifications are needed. A useful set of tractable examples arises when the two fluids in the reservoir have equal mobilities and each completely fills part of the reservoir (so relative permeability is always unity). In these cases the pressure field satisfies the Laplace equation across the reservoir and can be solved by any of the standard methods. The shock interfaces then move with the common fluid velocity $\underline{u}$ along trajectories governed by the now known pressure gradient. Many interesting examples can be constructed in this way (Bear (1972), Collins (1961), Duxbury and Rae (1979)). Finally, mention must be made of another "solvable" case, the Buckley-Leverett (1942) solution of equation (24) for the case of an oil filled reservoir subject to a water drive from one side. This has become almost a classic test problem for numerical methods in reservoir studies. The solution of (24) and its boundary conditions is obtained by changing variables from S and t and finding x(S, t). When this is inverted it leads to a double valued solution of the form shown in Figure 5. Buckley and Leverett (1942) removed the double valuedness by postulating a discontinuity at $x_F(t)$ chosen so that the two shaded areas in Figure 5 are equal. This leads to the solution shown. The shaded areas have to be equal to satisfy mass conservation and this imposes the requirement

$$\int_0^{S_F} \frac{dg}{ds}\, ds = C\, S_F$$

where S_F is the saturation just behind the front and C is again the shock speed. It is, of course, also given by $C = \left.\frac{dg}{ds}\right|_{S_F}$ and it is easy to see that these two relations lead to a speed which is the slope of the line from the origin tangent to the g curve. This method of finding C is known as the Welge construction (Welge (1952)) and can now be seen as justified by being a special case of the Oleinik condition.

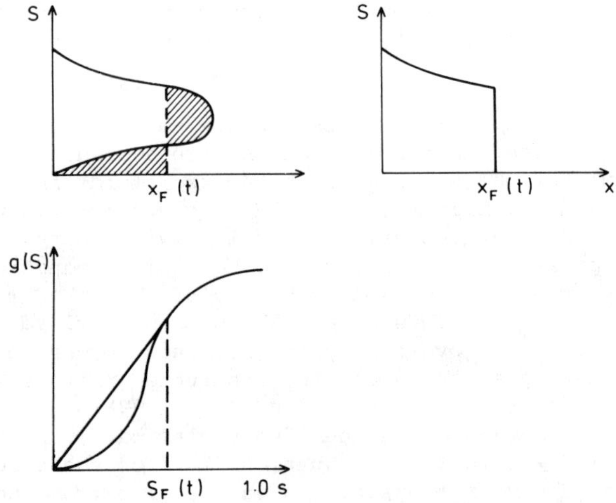

Figure 5 The Buckley-Leverett solution and Welge construction

4. NUMERICAL ASPECTS

4.1 Techniques arising from finite difference methods

The overwhelming majority of simulations of oil reservoir behaviour are made using well established techniques rooted in finite difference methods. A survey of these from an elementary engineering viewpoint is made by Crichlow (1977) and from a more mathematical side by Peaceman (1977). I do not intend to discuss these in detail here except to draw attention to the important rôle they have had in the past in the development of numerical solvers for partial differential equations. The work of Sheldon (1959) on successive over-relaxation methods, the Alternating Direction Implicit methods of both Peaceman-Rachford (1955) and Douglas-Rachford (1956) forms and the Strongly Implicit Method of Stone (1968), all have close ties with oil reservoir simulations. More recently these problems have helped to stimulate the rise of approximate matrix factorization methods (Dupont et al (1968)) and their combination with conjugate-gradient methods (Vinsome (1976)).

In their simplest forms the finite difference methods give matrices with sparsity patterns of just the kind one would expect with a small number of non-zero entries lying on regular off-diagonal bands. However the presence of more subtle phenomena, such as oil wells which are multiply completed and allow flow between non-neighbouring grid blocks, can produce awkward extra non-zeros which have to be treated carefully.

4.2 Variational methods

As was noted in section 3 it is in many ways natural to use variational methods for the reservoir simulation equations. Further, these methods have a long and interesting history in the numerical solution of both elliptic (Mikhlin (1964), Aubin (1967)) and parabolic (Price et al (1968), Douglas and Dupont (1970)) equations. Despite this, the relative ease with which engineers and other users of numerical methods can grasp the basis of finite difference methods and write programs to use them has given difference methods overwhelming popularity. There has been a recent growth of interest in variational methods and this is reflected in the papers presented at recent reservoir simulation symposia SPE (1976 and 1979). The earlier advocates of the methods (for example, Price et al (1968), Douglas and Dupont (1970)) claimed more accurate answers than could be obtained by comparable computing effort using difference methods. There was also a bonus in that answers came in the form of smooth or at least piecewise smooth functions well suited to the kind of graphical and numerical processing of results needed in simulation studies. More recently, with the rise of interest in finite element methods for flow problems (FEFP (1980) is the most recent illustration), other advantages have been claimed. These include its ability to handle complex geometries with curved boundaries, the possibility of local refinement or coarsening of the element mesh near regions of special character such as wells or water aquifers, and the use of special, perhaps singular, interpolation functions on some elements to take out difficult behaviour such as pressure singularities at wells.

The steps in the numerical procedure are by now well known and can be summarised briefly as follows. In a weak formulation of the equations, such as (16), (17) both the unknown fields P, S and the test functions Φ, Ψ belong to suitable function spaces which may be the same, as in the widespread Galerkin method, or different as in some other weighted residual methods (Finlayson (1972)). We approximate the solution to the equations by requiring that both fields and test functions lie in a finite dimensional subspace spanned by linearly independent basis functions, say ψ_i, $i = 1, .., N$. These are often constructed from finite element shape functions (Oden and Reddy (1976)) and so are typically, though not necessarily, piecewise low order polynomials which although not orthogonal have a localisation property which implies that each ψ_i has a non-zero inner product with only a few of the others. In equations (16) and (17) then, P and S would be unknown linear combinations of ψ_i, if the same basis functions are used for both, with coefficients say p_i and s_i. Note that this substitution has to be made, in some way, even inside the functions T_O, T_W and P'_c. The equations should be satisfied for all test functions in the subspace so, apart from some technical adjustments to cater for boundary conditions, it is equivalent to replace ψ and Φ by each Ψ_i in turn. When the known basis functions are substituted into (16), (17) in this way, the space integrals can be performed and this is usually done numerically (leading to another area of interest in its own right). The result is a set of nonlinear first order differential equations in $\frac{ds}{dt}i$, s_i, p_i which have to be solved to find the approximate S and P fields at any space point x and time t. Because of the localisation property of the basis functions these equations usually have reasonable sparsity.

The remaining steps for a solution are quite clear but at this point methods proliferate with almost every worker in the field adopting his own particular combination. The time integration is almost always done by finite difference methods with a general tendency to introduce as much explicitness as possible. In real reservoir simulations it often turns out that timestep size is limited not by the form of the equations but by changes in boundary conditions or source terms as oil-well behaviour is varied according to some external plan. For problems where this does not happen we, at Harwell, have used Gear's method with great success. The time discretised equations are nonlinear algebraic equations. Again a great many methods are used to linearise these with, for variational approaches, the Newton method or one of its close relatives perhaps being the most popular. Douglas and Dupont (1970) have proposed a predictor-corrector method in which the time discretisation also linearises the equations. Finally, the linear equations themselves have to be solved. Contrary to the finite difference approach, the variational approach has tended to use direct solution methods, although we should make it clear that the problems are usually two-dimensional in space. In particular the frontal method (Irons (1970), Hood (1976)), discussed elsewhere in these proceedings, has proved very popular when finite elements have been used and, in our experience, is hard to beat for general two-dimensional problems. For three-dimensions it is too costly and other methods have to be found. Block SOR methods can be readily adapted (Voss and Pinder (1978)) and look to be practicable for many 3-D cases and also in some explicit schemes where the very well behaved capacity matrix, the matrix multiplying the vector of $\frac{ds}{dt}i$, has to be inverted many times. There seems to be no consensus yet on the most appropriate iterative schemes for 3-D reservoir problems and this is clearly an area needing, and receiving, more research.

Having computed an approximate solution it is desirable to have some information on its relation to the true solution of the problem in hand. This is another large area of study. For "comfortably" parabolic problems there seems to be a good collection of approximation estimates; some idea of these can be obtained from Douglas and Dupont (1970) or Oden and Reddy (1976). For the hyperbolic case $P_c' \equiv 0$, or in practice whenever the diffusion coefficient is small compared with the ratio of the square of mesh spacing h to the time step, things are much more complicated. Such equations may have shock solutions, as discussed in section 3, and at the shocks gradients scale like inverse powers of h. The truncation error terms in Taylor expansions of the nonlinear discrete equations may not go to zero but remain finite even in the limit $h \to 0$. Thus the discrete equations may not be representing the hyperbolic equation at all but an equation with added higher order diffusion and dispersion terms (Jackson and Rae (1979)). Its shock solutions, which have been approximated in the calculation, may well have a speed (and therefore height) different from the physical shock and exhibit extraneous ripples. For small timestep and enough dissipation in the discretised equation it can be seen that the solution will approximate the true shock but exhibit severe spreading (Jackson and Rae (1979)), perhaps over 5-10 mesh spacings. As the time step is increased the time truncation error terms can lessen this spreading and in 1-dimensional cases even remove it for a particular ratio of time step to mesh size. It is interesting that many applications of variational methods to these hyperbolic problems have found, sometimes unexpectedly, just these effects (Dalen (1977), Langsrud (1976), Mercer and Faust (1977), Wilson and Casinder (1978), Douglas et al (1979)). The remedies proposed can also be understood in this framework. Reintroduction of dissipation terms either directly through P_c' or indirectly through upstream differences has the effects noted above, while lumping the capacity matrix changes the truncation terms in a way which suppresses ripples and lowers the accuracy of the discretisation. Since some of these fixups have become a traditional part of finite difference simulation programs the difficulties are seldom so apparent for them, but exactly the same points apply. It appears that for general reservoir problems with negligible capillary pressure, a naive application of variational methods (or finite difference methods) requires a fine space mesh for good enough resolution and a consequent high cost of computing. Current research is seeking clever ways round this.

REFERENCES

Ames, W.F. (1965 and 1972) Nonlinear Partial Differential Equations in Engineering, Vols. I and II, Academic Press.

Aubin J.P. (1967) Behaviour of the error of the approximate solutions of boundary value problems for linear elliptic operators by Galerkin's and finite difference methods, Ann. Scuola Norm. Sup. Pisa 21, pp. 599-637.

Bear, J. (1972) Dynamics of Fluids in Porous Media, Elsevier.

Beran, M.J. (1968) Statistical Continuum Theories, Wiley.

Buckley, S.E. and Leverett, M.C. (1942) Mechanism of fluid displacement in sands, Trans. AIME, 146, pp. 107-116.

Collins R.E. (1961) Flow of Fluids through Porous Materials, Rheinhold.

Courant, R. and Hilbert, D. (1962) Methods of Mathematical Physics, Vol. II.

Crichlow, H.B. (1977) Modern Reservoir Engineering, a Simulation Approach, Prentice-Hall.

Dafermos, C.M. (1974) Quasilinear hyperbolic systems that result from conservation laws, In Nonlinear Waves, (S. Leibovich and A.R. Seebass Eds), Cornell University Press.

Dake, L.P. (1978) Fundamentals of Reservoir Engineering, Elsevier.

Dalen. V, (1977) Immiscible flow by finite elements, in Finite Elements in Water Resources, (W.G. Gray, G.F. Pinder and C.A. Brebbia Eds) Pentech Press, pp. 3.69-3.90

Darcy, H.P.G. (1856) Les Fontaines Publiques de la Ville de Dijon, Victor Dalamont.

Douglas, J., Darlow, B.L., Wheeler, M.and Kendall, R.P. (1979) Self-adaptive Galerkin methods for one-dimensional two-phase immiscible flow, SPE (1979), pp. 65-72.

Douglas, J. and Dupont, T. (1970) Galerkin methods for parabolic equations, SIAM. J. Numer. Anal., 7, pp. 575-626.

Douglas, J. and Rachford, H.H. (1956) On the numerical solution of heat conduction problems in two and three space variables, Trans. AMS, 82, pp. 421-439.

Dupont, T., Kendall, R.P. and Rachford, H.H. (1968) An approximate factorisation procedure for solving self-adjoint elliptic difference equations, SIAM. J. Numer. Anal., 5, pp. 559-573.

Duxbury, P.D. and Rae, J. (1979) Flow of equally-mobile immiscible fluids in porous media, Report AERE-R.9522, Harwell, UKAEA, England.

FEFP (1980) Proceedings of the Third International Conference on Finite Elements in Flow Problems, Banff, Canada.

Finlayson, B.A. (1972) The Methods of Weighted Residuals and Variational Principles, Academic Press.

Friedman, A. (1964) Partial Differential Equations of Parabolic Type, Prentice-Hall.

Girault, V. and Raviart, P.A. (1979) Finite Element Approximation of the Navier-Stokes Equations, Springer-Verlag.

Hood P. (1976) Frontal solution program for unsymmetric matrices, Int. J. Num. Meth. Engng, 10, pp. 379-399.

Hopf, E. (1969) On the right weak solution of the Cauchy problem for a quasilinear equation of first order, J. Math. and Mech., 19, pp. 483-487.

Irons B.M. (1970) A frontal solution program, Int. J. Num. Meth. Engng., 2, pp. 5-32.

Jackson, C.P. and Rae, J. (1979) The effects of truncation error in numerical modelling of flood fronts in oil reservoirs, Report TP.813, Theoretical Physics Div., AERE Harwell, England.

Ladyzhenskaya, O.A. and Uraltseva, N.N. (1968) Linear and Quasilinear Elliptic Equations, Academic Press.

Langsrud, O. (1976) Simulation of two-phase flow by finite element methods, In SPE (1976), pp. 107-122.

Lax, P.D. (1957) Hyperbolic systems of conservation laws II, Comm. Pure and Appl. Math. 10, pp. 627-646.

Lax, P.D. (1971) Shock waves and entropy, In Contributions to Non-Linear Functional Analysis, (E.H. Zarantonello, ed.), Academic Press, pp. 603-634.

Mercer, J.W. and Faust, C.R. (1977) The application of finite-element techniques to immiscible flow in porous media, In Finite Elements in Water Resources, (Gray, W.G., Pinder G.F., and Brebbia, C.A. (eds), Pentech Press, pp. 1.21-1.57.

Mikhlin, S.G. (1964) Variational Methods in Mathematical Physics, Pergamon Press.

Morrey, C.B. (1966) Multiple Integrals in the Calculus of Variations, Springer-Verlag.

Oden, J.T. and Reddy, J.N. (1976) An Introduction to the Mathematical Theory of Finite Elements, Wiley.

Oleinik, O.A. (1963) Uniqueness and stability of the generalised solution of the Cauchy problem for a quasilinear equation, Trans. AMS., Ser. 2, 33, pp. 285-290.

Oleinik, O.A. (1964) Boundary value problems for linear elliptic and parabolic equations with discontinuous coefficients, Trans. AMS., Ser.2, 42, pp. 175-194.

Peaceman, D.W. (1977) Fundamentals of Numerical Reservoir Simulation, Elsevier.

Peaceman, D.W. and Rachford, H.H. (1955) The numerical solution of parabolic and elliptic differential equations, SIAM J. 3, pp. 28-41.

Price, H.S., Cavendish, J.C. and Varga, R.S. (1968) Numerical methods of higher-order accuracy for convection-diffusion equations, SPE J. 8, pp. 293-303.

Quinn, B.K. (1971) Solutions with shocks: An example of an L contractive semigroup, Comm. Pure and Appl. Maths., 24, pp. 125-132.

Robinson, P.C., Rae, J. and Jackson, C.P., (1980) A nonlinear test case for the finite element method in two-phase flow, Appl. Math. Modelling, 4, pp. 212-216.

Scheidegger, A. (1960) The Physics of Flow through Porous Media, Macmillan.

Sheldon, J.W. (1959) On the spectral norms of several iterative processes, J. ACM., 6, pp. 494-505.

SPE (1976) Society of Petroleum Engineers, 4th Symposium on Numerical Simulation of Reservoirs, Los Angeles.

SPE (1979) Society of Petroleum Engineers, 5th Symposium on Numerical Simulation of Reservoirs, Denver.

Stone, H.L. (1968) Iterative solution of implicit approximations of multi-dimensional partial differential equations, SIAM

J. Numer. Anal., 5, pp. 530-559.

Temam, R. (1979) Navier-Stokes Equations, North Holland.

Vinsome, P.K.W. (1976) Orthomin, An iterative method for solving sparse sets of simultaneous linear equations, Paper SPE. 5729, 4th SPE Symposium on Numerical Simulation of Reservoir Performance, Los Angeles, pp. 149-159.

Voss, C.I. and Pinder G.F. (1978) Block iterative finite element preprocessed scheme for simulation of large non-linear problems, Int. J. Num. Meth. Engng, 12, pp. 1543-1558.

Welge, H.J. (1952) A simplified method for computing oil recovery by gas or water drive, Petrol. Trans. AIME 195, pp.91-98.

Wilson, D.C. and Casinder, P.C. (1978) The use of finite element models in simulating north sea waterfloods, Paper EUR.86, Proc. EUROPEC Conference, London, pp. 207-218.

Zemansky, M.W. (1957) Heat and Thermodynamics, McGraw-Hill.

DISCUSSION

PROFESSOR A. JENNINGS (Belfast). Are your calculations principally on two or three dimensional problems?

RAE. We can do many calculations in 2-dimensions, for example where the reservoir is very thin or for cross-section calculations. However, for the full blown problem, we need to solve a three dimensional model.

MR. M.H. GOLDWATER (British Gas). At British Gas we use two-dimensional methods to simulate a three-dimensional problem by averaging through thicknesses in a vertical direction and using modified permeabilities (pseudo functions).

UNKNOWN. What solution technique do you currently use for the linear systems arising from the 3-dimensional discretization?

RAE. To date we only use finite differences to discretize three-dimensional problems. Although some years ago SOR was the preferred method, then SIP, preconditioned conjugate gradients is the current favourite. A hard look is also being given to methods suitable for vector machines.

Note added in proof: Since preparing this lecture the author has learned that a proof has been given by G. Chavent for an existence theorem of the type mentioned in section 3.2(G. Chavent (1977) About the identification and modelling of miscible or immiscible displacements in porous media. Lecture Notes in Control and Information Sciences 1, Springer.)

RAPID ELLIPTIC SOLVERS

E. Detyna

(University of Reading)

1. INTRODUCTION

The aim of this paper is to provide a review of the Rapid Elliptic Solvers (RES) for the solution of elliptic partial differential equations (e.p.d.e.) in one, two or three dimensions. (We shall use the expression "e.p.d.e." for one dimensional equations even though it is not correct.) The review is neither complete nor comprehensive but, hopefully, indicates the developments in this area in the last 15 years.

First of all the term "rapid" in RES requires a definition. With the aid of the finite element method (FEM) or finite difference method (FDM), the e.p.d.e. can be approximated by a set of linear equations $A\underline{x} = \underline{b}$, where A is an N × N matrix (usually very sparse) and $\underline{x}$ and $\underline{b}$ are N-vectors of unknowns and right hand sides respectively. To classify as a RES the algorithm must be able to provide the solution $\underline{x} = A^{-1}\underline{b}$ in α N floating point operations* (f.p.o.) <u>and</u> use no more than N to 2N memory storage locations, where α is a constant (or slowly varying function of N) of the order of 10. This last requirement excludes various sparse matrix techniques which often result in algorithms with $\alpha \sim N^{\frac{1}{2}}$, (George (1973), George and Liu (1978), Duff and Stewart (1979)).

It is clear that with such stringent conditions imposed on the performance of the algorithms not all e.p.d.e.'s on arbitrary domains can be solved with RES. In particular, there is no RES capable of solving a general non-separable e.p.d.e. in 2 or 3 dimensions. What is more the author feels that no such algorithm is possible.

The most general e.p.d.e. for which RES exists, at least on

---

* the term f.p.o. means here a single floating point operation of the type +, −, *, /.

special domains, is the general separable e.p.d.e.:

$$\sum_{\alpha=1}^{d} (a_\alpha \partial_\alpha^2 + b_\alpha \partial_\alpha + c_\alpha)u = f, \tag{1.1}$$

where a_α, b_α and c_α are arbitrary functions of one variable only ($a_1(x^1)$, $a_2(x^2)$ and so on) $\partial_\alpha \equiv \partial/\partial x^\alpha$ and d is the number of dimensions. The a_α's have to be all of the same sign for equation (1.1) to be elliptic. For certain boundary conditions (b.c.) the function f on the RHS may have to satisfy compatibility conditions. The equation (1.1) is to be satisfied on domain D with appropriate b.c. on ∂D (Dirichlet, Neumann, periodic or mixture of the three).

The equation (1.1) can be symmetrized by transforming out the terms $b_\alpha \partial_\alpha u$ with $u' = ug$, where

$$g = \prod_{\alpha=1}^{d} \exp(\tfrac{1}{2}\int b_\alpha / a_\alpha dx^\alpha).$$

Therefore without loss of generality we shall assume $b_\alpha = 0$.

In the case of isotropic diffusion,

$$a_1(x^1) = a_2(x^2) = a_3(x^3) = \text{const},$$

and we have a separable Helmholtz equation:

$$[\nabla^2 + \sum_{\alpha=1}^{d} c_\alpha(x^\alpha)]u = f. \tag{1.2}$$

Finally, when all the coefficients are constant we have a constant coefficient Helmholtz equation

$$(\nabla^2 + c)u = f \tag{1.3}$$

and if c = 0 it becomes the Poisson equation. In the following we shall discuss methods of solving these equations depending on the dimension d of the domain. We shall see that for d = 1 there is a number of RES's for the most general equation (1.1) while for d = 2 and 3 some of these methods are no longer RES even for the simplest Poisson equation.

2. ONE DIMENSIONAL EQUATIONS

The equation (1.1) in one dimension may be reduced to

$$(d^2/dx^2 + c(x))\, u(x) = f(x) \tag{2.1}$$

on a domain $0 \leq x \leq 1$ with b.c. at $x = 0$ and $x = 1$. Discretizing it on a uniform mesh with h^2 accuracy gives

$$U_{i-1} + C_i U_i + U_{i+1} = F_i, \tag{2.2}$$

where $C_i = -2 + h^2 c(ih)$, $F_i = h^2 f(ih)$, $U_i = U(ih)$, $h = 1/n$ and $i = 1, 2, \ldots, n-1$. For $i = 0$ and $i = n$ the equation (2.2) has to be supplemented by the b.c. We shall consider homogeneous Dirichlet b.c. $U_0 = U_n = 0$ unless otherwise stated.

The equation (2.2) is the simplest approximation to (2.1), we could use higher order formulae which would result in, say, a 5-diagonal matrix in equation (2.2). This does not increase the work for some algorithms as will be indicated later.

We shall now review methods of solving equation (2.2) bearing in mind possible applicability to higher dimensions.

2.1 Fast Fourier transforms (FFT)

The functions $\sin(\pi ik/n)$ are the eigenvectors of the coefficient matrix of equation (2.2) provided that the coefficient C_i is a constant: $C_i = C$. Taking the FFT of equation (2.2) gives $\hat{U}_k = G_k \hat{F}_k$, where ^ denotes the Fourier transform and

$G_k = (C + 2 \cos \pi k/n)^{-1}$ is the influence function (an inverse of the eigenvalue of the coefficient matrix). We note that the higher order schemes would merely change this function. Since it is normally precalculated and stored, its particular shape will not change the operation count for the algorithm.

Originally the FFT algorithm existed only for n being a power of 2 (Cooley and Tukey (1965)) but later (Singleton (1969)) the FFT was extended to any value of $n = 2^p\, 3^q\, 5^r \ldots$ although it is fastest when n is a product of small prime numbers. The operation count of the FFT is $\alpha\, n \log_2 n$, where α in the latest methods is about 2-2.5. Hence the algorithm has the operation count of $1 + 5 \log_2 n$ f.p.o. per mesh point. It is stable for all values of C provided that the influence function G_k is not singular.

2.2 LU decomposition

If we write equation (2.2) in the matrix form $A\underline{u} = F$, the matrix A can be decomposed into A = LU, where

$$L = \begin{bmatrix} \lambda_1^{-1} & & & \\ 1 & \lambda_2^{-1} & & \\ & \ddots & \ddots & \\ & & 1 & \lambda_n^{-1} \end{bmatrix} \qquad U = \begin{bmatrix} 1 & \lambda_1 & & \\ & 1 & \lambda_2 & \\ & & \ddots & \ddots \\ & & & 1 \end{bmatrix} \tag{2.3}$$

where $\lambda_1 = 1/C_1$, $\lambda_i = 1/(C_i - \lambda_{i-1})$ for i = 2, 3, ..., n. The solution is then obtained in two stages

1) $\underline{v} = L^{-1}\underline{F}$ that is: ($v_1 = F_1\lambda_1$, $v_i = (F_i - v_{i-1})\lambda_i$

for i = 2, 3, ... n)

and then

2) $\underline{u} = U^{-1}\ \underline{v}$ that is: ($u_n = v_n$, $u_i = v_i - \lambda_i\ u_{i+1}$ for

i = n-1, n-2, ..., 1).

The operation count is 4n plus 2n for the precalculation of the λ's regardless of whether C_i is constant or not.

There is one exception to this (O'Leary and Widlund (1979)) - if we retain the Dirichlet b.c. at x = 1 but impose mixed b.c. at x = 0, namely $\mu\ u_0 + u_1 = 0$, where

$$\mu = \frac{C}{2} + \left(\frac{C^2}{4} - 1\right)^{\frac{1}{2}}$$

and $C = C_i$ is a constant, then the matrix A takes the form

$$A = \begin{bmatrix} \mu & 1 & & & & \\ 1 & C & 1 & & & \\ & 1 & C & 1 & & \\ & & \cdot & \cdot & \cdot & \\ & & & \cdot & \cdot & \cdot \\ & & & & 1 & C \end{bmatrix}$$

which has a particularly simple LU decomposition (2.3) with all $\lambda_i = \mu^{-1}$, hence they do not have to be calculated. We note that for the Poisson equation $C = -2$, hence $\mu = -1$ and the b.c. at $x = 0$ is simply the Neumanncondition, but for the Helmholtz equation, $C \neq -2$, the mixed b.c. is less obvious. The memory required in this case is one location per mesh point whereas in the general case additional storage is needed for the λ's.

2.3 Cyclic reduction (CR)

We shall consider equation (2.2) with $C_i = C$ although the CR algorithm for variable C_i is only slightly more complicated. By taking an appropriate linear combination of 3 consecutive equations (i-1, i and i+1) we get

$$U_{i-2} + C^{(1)}U_i + U_{i+2} = F_i^{(1)} \equiv F_{i-1} - C\,F_i + F_{i+1}, \qquad (2.4)$$

where $C^{(1)} = 2 - C^2$. Thus only U_i's at every second point are connected. This process can be repeated ℓ-1 times for $n = 2^{\ell}$ resulting in a single equation, connecting boundary values U_0 and U_n with the centre point $U_{n/2}$, which can be solved. Then using the reduced equations in the reverse order the remaining U's can be calculated. We note that for Poisson's equation, $C = -2$, hence $C^{(\ell)} = -2$ for all levels ℓ. On the other hand, the most unstable harmonic will grow by a factor of 4 at each level of reduction in $F^{(\ell)}$. Therefore for $C = -2$ the algorithm is unstable with error ϵ growing as $\epsilon \sim 4^{\ell} = n^2$. For $C \neq -2$, the fastest harmonic grows during the reduction process as $\prod_{p=0}^{\ell-1} (2 + |C^{(p)}|)$, whereas the solution is found by dividing it by

$C^{(\ell)}$. Since for $C << -2$ the constants $C^{(\ell)} \approx C^{2^\ell}$ the algorithm is stable but for $|C + 2| << 1$ the error growth may be appreciable and needs evaluating especially for large n.

The CR method has been extended (Sweet (1974)) to cope with every value of n.

The operation count is 6n with additional storage of $\log_2 n$ for the $C^{(\ell)}$'s.

In the variable coefficient case the operation count grows to 18n with additional storage of 2n.

2.4 Marching algorithms

Marching methods are based on the fact that if the solution is known at two consecutive points then equation (2.2) may be used as a recursive formula

$$U_{i+1} = F_i - C_i U_i - U_{i-1}. \tag{2.5}$$

Assuming a value for U_1 with given U_0 we can calculate U_2^*, U_3^*, ..., U_n^*. The calculated U_n^* differs from the given boundary value U_n but this difference can be used to find the correct solution U_1 and then the march is repeated to find the corrected solution. In order to study the stability let us solve equation (2.5) for $C_i = C$ and $F_i = 0$:

$$U_i = A\mu_+^{\,i} + B\mu_-^{\,i}, \tag{2.6}$$

where $\mu_\pm = -\frac{C}{2} \pm \sqrt{\frac{C^2}{4} - 1}$ and A and B are constants. For $C < 0$, μ_+ is the larger root, hence the error grows as $\mu_+^{\,i}$. After marching over p points we loose D decimal places of accuracy where $D = p \log_{10} \mu_+$. With the exception of the Poisson equation, where $C = -2$ and $\mu_+ = \mu_- = 1$ and the march is stable, the procedure for $C \neq -2$ is exponentially unstable. One strategy (Lorenz (1976)) is to do partial marches of length $p = D/\log_{10}\mu_+$, where D is the number of digits we are prepared to lose, over n/p separate regions. The resulting n/p equations are then solved to get correct starting values. This could be an efficient method provided that μ_+ is not too large.

The operation count is $6n + O(D^{-1})$.

2.5 *Conclusions*

It seems that in the one dimensional case all the methods are rapid. The various properties of these methods are summarised in Table I and it is clear that LU decomposition is by far the best method.

3. TWO-DIMENSIONAL EQUATIONS

The RES for two-dimensional equations (1.1) exist only for regular domains either in a rectangle ($x^1 = x$, $x^2 = y$) or a circle ($x^1 = r$, $x^2 = \phi$). The arbitrarily shaped boundary can be solved by embedding the region of solution in a rectangular domain and applying RES to the regular domain. The required b.c. are imposed in the interior of the domain by a correction technique known as the Capacitance Matrix method. This method requires the numerical inversion of an $m \times m$ matrix C, where m is the number of internal boundary points (Hockney (1968), Proskurowski and Widlund (1976)). Therefore, for all but almost regular boundaries (small m), it cannot be classified as RES.

Since equations described on a circle and a rectangle are similar we shall continue this section by considering rectangular b.c. only.

Discretizing equation (1.1) on a rectangular $n \times n$ mesh one obtains the following interaction molecule at the mesh point (i, j)

$$\left\{\begin{matrix} & p_j & \\ q_i & r_i + s_j & q_i \\ & p_j & \end{matrix}\right\} U_{ij} = F_{ij}. \tag{3.1}$$

The above is the shorthand for the equation.

$$q_i(U_{i-1,j} + U_{i+1,j}) + p_j(U_{i,j-1} + U_{i,j+1}) + (r_i + s_j)U_{ij} = F_{ij},$$

where $q_i = a_1(ih)$, $p_j = a_2(jh)$, $r_i = c_1(ih)h^2 - 2q_i$,

$$s_j = c_2(jh)h^2 - 2p_j, \; F_{ij} = h^2 f(ih, jh); \; i,j = 1, 2, \ldots, n-1.$$

We have chosen equal spacing on the square domain since the extensions to the rectangle or different spacings in x- and y-directions are trivial.

TABLE I

One dimensional problems

Method	Equation	Op. Count in f.p.o.	Additional Storage	Stability
FFT**	Const. Coeff. Helmholtz	$n(1 + 5 \log_2 n)$	n	stable
LU decomp.	General	$4n + 2n$*	n	stable
CR	Poisson	6n	-	polynomially unstable
	Helmholtz			
	a) const.	6n	$\log_2 n$	stable
	b) variable	18n	2n	stable
March	Poisson	6n	-	stable
	Helmholtz	$6n + O(D^{-1})$*	$O(D^{-1})$*	exponentially unstable

* see text

** can be used for higher order discretization methods with no additional cost

3.1 Fourier transform

Fourier analysis can be applied in, say, the x direction only if $a_1(x)$ and $c_1(x)$ are constants which implies $q_i = 1$, $r_i = -2$. In that case the Fourier transform in variable i results in a set of n one-dimensional equations for each harmonic k:

$$p_j\hat{U}_{k,j-1} + (s_j-2)\hat{U}_{k,j} + p_j\hat{U}_{k,j+1} = \hat{F}_{k,j} \tag{3.2}$$

which can be solved by any method of the previous section. Assuming the operation count for solving (3.2) for each harmonic k to be 6n the total is $(5n \log_2 n + 6n)n$. Since this algorithm is a "product" of two one-dimensional algorithms all the remarks of the previous section apply.

3.2 LU decomposition

Unlike the one-dimensional case, where a tridiagonal matrix A factorized into bidiagonals L and U, the five diagonal matrix A describing equation (3.1) does not factorize into tridiagonal L and U. This can only be done approximately (Stone (1968)) LU - A + E, where E represents the two additional diagonals of "fill-in". Hence the equation $A\underline{u} = \underline{F}$ can be solved iteratively $LU\underline{u}^{(\ell+1)} = F + E\underline{u}^{(\ell)}$. The efficiency of this algorithm depends on choosing an LU decomposition of this kind which minimizes E. This is not a RES algorithm and since it does not depend strongly on the structure of matrix A it is best used for general e.p. d.e.'s. Details of this method can be found in Jacobs (1980).

3.3 Cyclic reductions

3.3.1 Global cyclic reduction

The equation (3.1) may be written in quasi-one-dimensional form

$$Q_i\underline{v}_{i-1} + C_i\underline{v}_i + Q_i\underline{v}_{i+1} = \underline{g}_i \tag{3.3}$$

where Q_i and C_i are matrices $Q_i = Iq_i$, $C_i = r_iI + T$, T is tri-diagonal $[p_j, s_j, p_j]$ and $\underline{v}_i$ and $\underline{g}_i$ are n-vectors made up of U_{ij} and F_{ij} where $\{\underline{v}_i\}_j = U_{ij}$ and $\{\underline{g}_i\}_j = F_{ij}$. Let us first study a constant coefficient Helmholtz equation ($Q_i = I$, $C_i = C$, constant diagonal). In this case, we can carry on as in section 2.3 cyclicly reducing equation (3.3). Of course, this time the

C's at higher levels $C^{(1)}$, $C^{(2)}$, ... are matrices, therefore the solution at level ℓ requires inverting matrix $C^{(\ell)}$ which is $(1 + 2^{\ell+1})$-diagonal. It was shown (Buzbee et al (1970)) that $C^{(\ell)}$ can be factorized into $2^{\ell+1}$ tridiagonal matrices and solved as for one-dimensional cases. Because the ratio of the maximum to minimum eigenvalues of the matrix $C^{(\ell)}$ grows exponentially with ℓ the algorithm is unstable. Buneman (1969) stabilized it by rephrasing the method of calculating cyclicly reduced right hand sides of equation (3.3). The resulting algorithm is stable with operation count $n^2(6 \log_2 n + \alpha)$ where $\alpha = 1$ or 3.5 depending on whether an additional memory of $\frac{1}{2} n^2$ is used or not.

The algorithm has been extended to cope with mesh sizes of arbitrary n (Sweet (1974)).

The general equation (3.3) with variable coefficients can also be solved using this method (Swarztrauber (1974)), but the factorization of the matrices $C^{(\ell)}$ is no longer a simple process and requires the numerical calculation of coefficients of the tridiagonal matrices. This can be done in the preprocessing stage and then the solution is found as in the constant coefficient case. The operation count for the solution phase alone is $(32 \log_2 n - 88)n^2$ while the preprocessing phase takes usually twice as long.

FACR algorithm

The equation (3.3) and the cyclicly reduced equations have the same form, therefore after r reductions the equations on the remaining $n/2^r$ lines can be Fourier analyzed (Hockney (1965)) as in section 3.1, provided that the coefficients are constant. We note that the amount of work required to invert the matrix $C^{(\ell)}$ doubles at each level ℓ, whereas the FFT is most efficient for a small number of points. Therefore one should be able to find the optimal number ℓ to design an algorithm with ℓ levels of cyclic reductions followed by FFT on the remaining lines, hence the name Fourier Analysis and Cyclic Reductions - FACR(ℓ). Its optimal operation count is $4n^2 \log_2(\log_2 n)$, that is to say 8-12 operations per mesh point. Temperton (1977) wrote one of the most efficient FACR(ℓ) algorithms but pointed out that the realistic operation count is 20-24 operations per mesh point. This algorithm is the most efficient combination of FFT, CR and LU decomposition methods for the constant coefficient two-dimensional problem.

3.3.2 Point cyclic reductions

The two-dimensional e.p.d.e. was discretized to obtain equation (3.1) using one h^2-accurate formula, but for the constant coefficient Helmholtz equation there is another h^2-accurate formula.

Let us write both interaction molecules in this case

$$\left\{\begin{matrix} & 1 & \\ 1 & C^{+} & 1 \\ & 1 & \end{matrix}\right\} \qquad \left\{\begin{matrix} 1 & & 1 \\ & C^{\times} & \\ 1 & & 1 \end{matrix}\right\} \tag{3.4}$$

where C^{+} and $C^{\times}$ are constants. In order to simplify notation let us introduce S^{+} and $S^{\times}$ operators

$$S^{+} = \left\{\begin{matrix} . & 1 & . \\ 1 & . & 1 \\ . & 1 & . \end{matrix}\right\} \text{ and } S^{\times} = \left\{\begin{matrix} 1 & . & 1 \\ . & . & . \\ 1 & . & 1 \end{matrix}\right\}.$$

We can now write the two types of discretized equations as

$$(S^{+} + C^{+})v_{ij} = F^{+}_{ij} \text{ and } (S^{\times} + C^{\times})v_{ij} = F^{\times}_{ij}, \tag{3.5}$$

where F^{+} and $F^{\times}$ are chosen in such a way as to minimise the difference between the two solutions of the equations (3.5). The two dimensional cyclic reductions are possible if we note (Detyna (1979)) that $(S^{+})^2 = S^{2+} + 2S^{\times} + 4$ and $(S^{\times})^2 = S^{2\times} + 2S^{2+} + 4$, where S^{2+} and $S^{2\times}$ stand for interaction molecules connecting points on twice as coarse a mesh. To put it another way, if we can sum the first of equations (3.5) centred at points (i-1,j), (i+1,j), (i,j-1) and (i,j+1) on the mesh and subtract the first equation multiplied by C^{+} and the second multiplied by 2, both at point (i,j), we will get:

$$(S^{2+} + C^{2+})v_{ij} = F^{2+}_{ij} \equiv (S^{+} - C^{+})F^{+}_{ij} - 2F^{\times}_{ij}, \tag{3.6a}$$

where $C^{2+} = 4 - 2C^{\times} - (C^{+})^2$. This equation is identical in structure to equation (3.5) but refers to the values of v_{ij} on twice as coarse a grid, hence it is an equation cyclicly reduced in both dimensions. We can cyclicly reduce the second equation in a similar manner, viz.

$$(S^{2^\times} + C^{2^\times})v_{ij} = F^{2^\times}_{ij}. \tag{3.6b}$$

In both cases the reduced densities $F^{2^+}_{ij}$ and $F^{2^\times}_{ij}$ - the right hand sides of equations (3.6) - are just linear combinations of the right hand sides of equations (3.5). Since the structure of the equations after point cyclic reduction remains unchanged we can repeat the process. After $\ell = \log_2 n - 1$ point cyclic reductions we obtain an equation on the coarsest mesh - it consists of the boundary points and the centre point only - hence it can be immediately solved. Using the reduced equations on progressively finer grids the solution can be found on all points of the mesh. The important feature of this algorithm is the need to have the two types of discretized equations (3.5) at some ¼n points of the mesh.

This is an "n^2-algorithm" and the exact operation count depends on the amount of work needed to calculate the right hand sides of F^+_{ij} and $F^\times_{ij}$. If we choose the standard "5-point" and "rotated 5-point" formulae, that is $F^+_{ij} = F_{ij}$ and $F^\times_{ij} = (1 + \frac{1}{4}S^+)F_{ij}$ respectively, the operation count is $9.5n^2$.

The stability of this algorithm is similar to that of the one-dimensional cyclic reduction method: the error growth is proportional to the square of the number of mesh points (polynomially unstable) for the Poisson equation, but for the Helmholtz equation, $C^+ < -4$, the algorithm is stabilized. The same applies to the error growth due to the difference in the two discretizations, equations (3.4).

3.4 Marching algorithms

The two dimensional marching method is identical in principle to the one described in 2.4, the difference being in that we now march over whole lines:

$$u_{i+1,j} = (F_{ij} - p_j(u_{i,j+1} + u_{i,j-1}) - (r_j + s_j)u_{ij})/q_j - u_{i-1,j}.$$

This also means that the instability is growing faster. For the

Poisson equation the error now grows as

$$(3 + \sqrt{8})^p \cong 5.8^p$$

for p lines marched. In practice, it means that we lose D digits of accuracy for p lines marched, where D = 0.77p. If we are prepared to lose D digits in the result we have to do 1.3n/D partial marches and solve the equations over 1.3n/D lines to get corrected starting values. Therefore the operation count is of the order of $(10 + \alpha/D)n^2$ and $(14 + \alpha/D)n^2$ for constant and variable coefficient cases respectively, where α depends on the method of solution for the correct starting values for partial marches. If we use FFT for the constant coefficient case then $\alpha \tilde{=} 10 + 4 \log_2 n$. In the variable coefficient case the solution has to be obtained iteratively, which means $\alpha \sim n$ in practice.

In any event this algorithm is only suitable for computers with a very long word length since only then can we afford a loss of a reasonably large number, D, of decimal digits of accuracy.

3.5 Conclusions

There is only one RES capable of solving general separable e.p.d.e.'s in two dimensions on a rectangle (Swarztrauber (1974)). On the other hand there are several algorithms for the constant coefficient Helmholtz equation. The best two seem to be FACR and PCR. Both of them are in practice "n^2-algorithms", the former being twice as slow as the latter but also far more accurate.

4. THREE DIMENSIONAL EQUATIONS

Discretizing equation (1.1) on a regular n × n × n grid with uniform spacing gives an equation which, written in "two dimensional" form, is

$$\left\{\begin{matrix} & p_j & \\ q_i, & r_i + s_j + t_\ell, & q_i \\ & p_j & \end{matrix}\right\} U_{ij\ell} + z_\ell (U_{i,j,\ell-1} + U_{i,j,\ell+1}) = F_{ij\ell}, \tag{4.1}$$

where the interaction molecule acts on the first pair of indices i and j only. The notation here is the same as in equation (3.1) and $z_\ell = a_3(\ell h)$, $t_\ell = c_3(\ell h)h^2 - 2z_\ell$.

TABLE II

Two dimensional problems

Method	Equation	Operation Count	Additional storage	Stability
FFT + LU	Const. coeff. in one variable	$(5 \log_2 n + 6)n^2$	n	stable
GCR	general separable	$(32 \log_2 n - 88)n^2$	$3 n^2$	stable
	const. coeff.	$(6 \log_2 n + 3.5)n^2$	-	stable
FACR	const. coeff.	$(20 - 24)n^2$	n	stable
PCR	const.coeff. Helmholtz	$9.5 n^2$	$\frac{1}{8} n^2$	stable
	Poisson	$9.5 n^2$	$\frac{1}{8} n^2$	unstable
March*	general	$(14 + \alpha/D)n^2$	-	controllable
	const. coeff.	$(10 + \alpha/D)n^2$	$O(D^{-1})$	instability

* see text

There does not seem to be a RES algorithm capable of solving this general equation. The only candidate is the marching algorithm: assume a trial solution on the two adjoining planes, use the equation (4.1) as a recurrence formula, find a solution on the boundary, compare it with the given boundary values and find a corrected starting solution and, finally, use the recurrence formula again to calculate a corrected solution. Unfortunately, the method is highly unstable - the error grows exponentially and it can be shown for Poisson's equation that we lose one decimal digit of accuracy per plane marched (the error grows as 9.9^p, where p is the number of planes). With such an error growth a method of partial marches, described for the two-dimensional case, becomes inefficient since, in order to keep reasonable accuracy, the length of a partial march could not be much longer than one or two steps. This would obliterate any advantages of the marching method.

In conclusion, the general equation (4.1) cannot be solved by RES at present and the solution should be sought either using iterative methods or general sparse matrix techniques.

On the other hand, if the coefficients are independent of one variable, say, ℓ (for example, $z_\ell = 1$, $t_\ell = -2$) then we can Fourier analyse equation (4.1) in that coordinate using FFT:

$$\left\{ q_i,\ r_i + \begin{matrix} p_j \\ s_j \\ p_j \end{matrix} + 2\cos \pi k/n - 2,\ q_i \right\} \hat{U}_{ijk} = \hat{F}_{ijk}, \qquad (4.2)$$

where $\hat{U}_{ijk}$ and $\hat{F}_{ijk}$ are Fourier transforms of $U_{ij\ell}$ and $F_{ij\ell}$ in the ℓ-th coordinate. The above equation for each harmonic k is now identical to the two-dimensional equation (3.1) and can be solved using any of the methods of the previous section. The solution $U_{ij\ell}$ can then be obtained by taking the inverse Fourier transform of $\hat{U}_{ijk}$ in the k-th variable. The operation count for this method is then $5\, n^3 \log_2 n + n\, O_2$, where O_2 is the operation count for the two dimensional method. The most efficient algorithms are, of course, obtained for the constant coefficient equation, which, after Fourier analysis in the ℓ-th direction, reads

$$\left\{ 1,\ -6 + \begin{matrix} 1 \\ \\ 1 \end{matrix} + 2\cos \pi k/n + ch^2,\ 1 \right\} \hat{U}_{ijk} = \hat{F}_{ijk}, \qquad (4.3)$$

where c is a (negative) constant for Helmholtz's equation and $c = 0$ for Poisson's equation. It should be noted that both Helmholtz and Poisson's three-dimensional equations separate, after Fourier analysis, into n two-dimensional Helmholtz's equations. This is because the centre of the interaction molecule, $-6 + 2 \cos \pi k/n + ch^2$, is less than -4 for $c \leq 0$. (The case $k = 0$ does not invalidate this argument since we never use the 0-th harmonic in the solution.) Therefore the PCR method used to solve equation (4.3) is stable for any $c \leq 0$.

If we use the two most efficient algorithms, FACR and PCR, to solve the equations (4.3), the resulting three-dimensional algorithms have the operation count $(20 + 5 \log_2 n)n^3$ and $(9.5 + 5 \log_2 n)n^3$ respectively.

5. CONCLUSIONS

In this very short review of RES algorithms only a few of the most important methods used for solving e.p.d.e.'s have been outlined. Their importance is far greater than it would seem at first. It is true they only solve simple "Poisson-like" equations on a rectangle with uniform meshes, but many other methods for solving general equations on arbitrary domains use RES iteratively and hence their efficiency depends on the performance of the simple RES. Just to mention one example: the whole rich field of molecular dynamics simulations was severely restricted until an efficient RES algorithm tailor-made for that purpose was developed, the so called P^3M method due to Hockney and Eastwood (Eastwood (1976)).

To conclude, it is instructive to compare the operation counts per mesh point for the best RES algorithms in different dimensions for the constant coefficient Helmholtz's equation.

We are comparing algorithms dealing with problems of different complexity so we shall account for that somehow. We propose to calculate the ratio of the operation count for the solution $\underline{u} = A^{-1}\underline{f}$, where A is the finite difference matrix approximating the e.p.d.e., and the operation count of the discretization, or in other words, the operation count of the product $A.\underline{u}$. The results are summarized in the following table:

number of dimensions	discretization f.p.o./mesh point	solution f.p.o./mesh point	ratio
1	3	6 (LU, CR)	2
2	5	9.5 (PCR)	1.9
		~ 20 (FACR)	4
3	7	~ 30 (FFT + PCR)	~ 4.3
		~ 40 (FFT + FACR)	~ 5.7

It is difficult to imagine a one-dimensional RES with lower operation count than the known ones hence we take the ratio of 2 as the limit. This limit has also been reached for the two-dimensional case but it would appear that it is still to be found for the three-dimensional equation.

One hopes that the future 3-D RES would have an operation count of the order of $15n^3$ or twice as good as the best existing ones. It will, most likely, be based on cyclic reductions.

REFERENCES

Buneman, O. (1969) A compact non-iterative Poisson-solver. Report 294 Stanford University Institute for Plasma Research, Stanford, California.

Buzbee, B.L., Golub, G.H. and Nielson, C.W. (1970) On direct methods for solving Poisson's equations, SIAM J. Numer. Anal. 7, pp. 627-656.

Cooley, J.W. and Tukey, J.W. (1965) An algorithm for the machine calculation of complex Fourier series, Math. Comp. 19, pp. 297-301.

Detyna, E. (1979) Point cyclic reductions for elliptic boundary value problems, I. The constant coefficient case, J. Comput. Phys. 33, pp. 204-216.

Duff, I.S. and Stewart, G.W. (ed) (1979) Sparse Matrix Proceedings (1978), SIAM Press.

Eastwood, J.W. (1976) In Computational Methods in Classical and Quantum Physics, (M.B. Hooper (ed)), Advance Publications Ltd., pp. 206-228.

George, A. (1973) Nested dissection of a regular finite element mesh, SIAM J. Numer. Anal. 10, pp. 345-363.

George, A. and Liu, J.W.H. (1978) An automatic nested dissection algorithm for irregular finite element problems, SIAM J. Numer. Anal. 15, pp. 1053-1069.

Hockney, R.W. (1965) A fast direct solution of Poisson's equation using Fourier analysis, J. ACM 12, pp. 95-113.

Hockney, R.W. (1968) Formation and stability of virtual electrodes in a cylinder, J. Appl. Phys. 39, pp. 4166-4170.

Jacobs, D.A.H. (1980) The exploitation of sparsity by iterative methods, these proceedings.

Lorenz, E.N. (1976) A rapid procedure for inverting del-square with certain computers, Monthly Weather Rev. 104, pp. 961-966.

O'Leary, D.P. and Widlund, O. (1979) Capacitance matrix methods for the Helmholtz equation on general three-dimensional regions, Math. Comp. 33, pp. 849-879.

Proskurowski, W. and Widlund, O. (1976) On the numerical solution of Helmholtz's equation by the capacitance matrix method, Math. Comp. 30, pp. 433-468.

Singleton, R.C. (1969) An algorithm for computing the mixed radix fast Fourier transform, IEEE Trans. Audio and Electroacoustics, AU-17, pp. 93-103.

Stone, H.L. (1968) Iterative solution of implicit approximations of multi-dimensional partial differential equations, SIAM J. Numer. Anal. 5, pp. 530-558.

Swarztrauber, P.N. (1974) A direct method for the discrete solution of separable elliptic equations, SIAM J. Numer. Anal. 11, pp. 1136-1150.

Sweet, R.A. (1974) A generalized cyclic-reduction algorithm, SIAM J. Numer. Anal. 11, pp. 506-520.

Temperton, C. (1977) On the FACR (ℓ) algorithm for the discrete Poisson equation, Report 14, European Centre for Medium Range Weather Forecasts, Bracknell.

DISCUSSION

DR. J.K. REID (Harwell). Some of the algorithms you have discussed are unstable. Shouldn't you remove them from the tables and further consideration?

DETYNA. Not really. Firstly, one should establish the type of instability: polynomial ($\epsilon \sim n^{\alpha}$), exponential ($\epsilon \sim a^{\alpha n}$) etc., where α and a are constants and n is the number of mesh points. Then one should calculate the actual growth in any given case - if the instability is, say, polynomial and the mesh is not too large, the error may be quite acceptable.

Secondly, there are algorithms (1-D cyclic reduction, PCR) which are unstable for Poisson's equation but stable for Helmholtz's equation - it is a warning that they should not be used without understanding their nature.

Thirdly, there are the marching algorithms. They are constructed on the assumption that although they are exponentially unstable we know in advance their error growth and we limit their use in such a way that the error does not exceed a certain a priori acceptable value. In other words, the partial march gives us controllable instability.

PROFESSOR J. McNAMEE (York, Ontario). How do multi-grid schemes compare with those you have discussed?

DETYNA. Since the multi-grid schemes are iterative by nature, it is impossible to obtain an exact operation count for comparison purposes. The underlying philosophy of these schemes is very similar to that of cyclic reduction methods. I believe that if RES can be applied it is preferable but, of course, multi-grid methods can be used effectively for more general equations for which no RES exists.

DR. A.M. ERISMAN (Boeing). We have successfully applied cyclic reductions to lumped models not derived from partial differential equations.

DETYNA. The cyclic reductions methods may be applied successfully to any equation A $\underline{x}$ = $\underline{b}$ provided that the sparse matrix A satisfies certain criteria. (For instance, any tridiagonal matrix A is a suitable case.)

DR. C. JESSHOPE (Reading). Is there a three-dimensional equivalent of the PCR method?

DETYNA. Unfortunately, no. Although there are three different, h^2-accurate, methods of discretization of the Poisson's equation and the appropriate difference operators are 'cyclicly reducible' - their squares are expressible in terms of the operators acting on twice as coarse a grid - the algorithm is not possible because the reverse process of solving the reduced equations cannot be carried out. This is because no matter in which order you try to solve the reduced equations, you always end up at certain mesh points with an equation with more than one unknown.

On the other hand one could produce a 3-D equivalent of the

global cyclic reduction - but it would be very cumbersome to implement the Buneman stabilization and the final operation count would not compare favourably with a simple algorithm based on the Fourier transform, as we described earlier.

FRONTAL METHODS FOR SOLVING FINITE-ELEMENT SYSTEMS OF LINEAR EQUATIONS

J.K. Reid

(AERE, Harwell)

ABSTRACT

Although used in computer programs before then, the frontal method was first explained in the published literature by Irons in 1970 and the method is hence firmly linked with his name. It depends on the observation that finite element assembly does not have to be completed before Gaussian elimination is commenced. Indeed any variable may be eliminated immediately after its last appearance in an assembly. This has led to a widely used and very successful method that is far more natural for element problems than any band or general sparse technique.

We describe in detail how this works for the usual case with a symmetric and positive definite matrix and then describe extensions for cases lacking these properties. We also show how the ideas can be integrated with those of substructuring and nested dissection and illustrate with results on a small number of realistic test problems.

1. INTRODUCTION

We consider the solution of sets of n linear equations

$$A\underline{x} = \underline{b} \tag{1.1}$$

that have arisen from finite element applications. The nxn matrix A is usually symmetric (that is $A^T=A$) and positive definite (that is $\underline{z}^T A\underline{z} > 0$ unless $\underline{z}=\underline{0}$) because an energy minimizing principle underlies the mathematical model. These properties are very helpful to both direct and iterative methods and in the main we assume that they hold, but we do not neglect more general cases.

The matrix A and the vector $\underline{b}$ always originate as sums

$$A = \sum_k B^{(k)}, \quad \underline{b} = \sum_k \underline{c}^{(k)} \tag{1.2}$$

of contributions from the finite elements. Each element k is associated with only a few variables and the matrix $B^{(k)}$ and the vector $\underline{c}^{(k)}$ have non-zeros only in those rows and columns that correspond to these few variables. Therefore $B^{(k)}$ and $\underline{c}^{(k)}$ may be stored in the computer in arrays of dimension the number of variables in element k, provided there is an associated index list specifying which variables are involved (that is to which row of the over-all matrix each row of the packed form corresponds).

The resulting over-all matrix A is certainly sparse and may be treated by any sparse matrix technique, for example as a band matrix if the variables are suitably ordered. In such a case the individual matrices $B^{(k)}$ are added (assembled) into a suitable data structure before the solution commences.

On the other hand the frontal method does not wait until assembly is complete before starting elimination operations. Because addition and subtraction may be reordered (apart from round-off effects, which are minor) the elimination operations

$$a_{ij} := a_{ij} - a_{i\ell}a_{\ell\ell}^{-1}a_{\ell j} \qquad (i,j > \ell) \tag{1.3}$$

$$b_i := b_i - a_{i\ell}a_{\ell\ell}^{-1}b_\ell \qquad (i > \ell) \tag{1.4}$$

may be completed before all the assembly operations

$$a_{ij} := a_{ij} + b_{ij}^{(k)} \tag{1.5}$$

$$b_i := b_i + c_i^{(k)}, \tag{1.6}$$

are complete. It is only necessary that $a_{i\ell}$, $a_{\ell\ell}$, $a_{\ell j}$, b_ℓ are fully assembled so that the correct quantities are subtracted in (1.3) and (1.4). In the frontal method the eliminations are performed as early as possible, that is the eliminations (1.3),

(1.4) for variable ℓ are performed as soon as its row and column are fully summed. After this, row and column ℓ play no further part in the elimination and may be written away to backing store (e.g. disk). At a typical intermediate stage we work only with those rows and columns that have been partially summed. These correspond to those variables that are associated with one or more assembled elements and with one or more unassembled elements. These variables lie on the border between the set of assembled elements and the rest, and as the assembly and elimination progresses they form a "front" that moves across the region. It is usual to hold the active rows and columns, corresponding to variables in the front, in a full matrix labelled with an index list of associated variables. All the actual assembly and elimination operations are performed within this full matrix, a varying amount of which is actually used as the front varies in size.

Notice that the elimination order is determined from the assembly ordering, apart from unimportant changes among variables that become fully assembled simultaneously. It is desirable to choose an ordering that results in the front-width remaining small. This seems far more natural than choosing an elimination order. Certainly the method is very successful and widely used. It is firmly linked with the name of Irons because he first explained the method in the literature (Irons (1970)), although it was in use in computer codes before 1970.

2. ORDERING FOR THE BASIC FRONTAL METHOD

A good ordering for the frontal method is one for which the front always stays modest in size. For instance for a cigar-shaped region, as illustrated in Figure 2.1, it is desirable to order the elements along its length for then the front will have its size determined by the number of variables across its width.

The frontal method is closely related to band and variable-band elimination of assembled matrices and here an established technique for automatic ordering is that of Cuthill and McKee (1969), see also Cuthill (1972). We can interpret this in an element setting as dividing the elements into "level sets" S_i, the first of which contains all elements involving a given starting variable. Later all elements which are neighbours of elements in S_j (technically we say that two elements are "neighbours" if they have one or more variables in common), but are not in any of the sets S_i, $i=1,2,\ldots,j$, are taken to be in set S_{j+1}. The element ordering finally taken is of those in set S_1 followed by those of set S_2, etc. For example the level sets

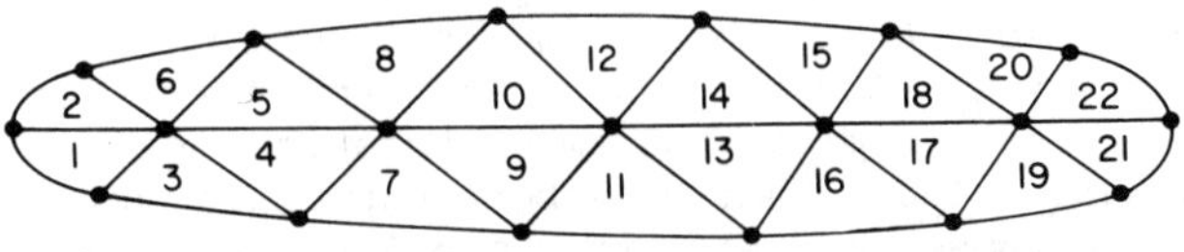

Figure 2.1 A good frontal ordering for a cigar-shaped region

for the region of Figure 2.1 might be {1,2},{3,4,5,6}, {7,8,9,10},... It is clear that if each level set has only a few elements then the resulting front size is likely to be modest. Therefore we aim for the level sets to be small and numerous. For a cigar-shaped region such as that of Figure 2.1, this requires starting at an end rather than the middle of a side, but a good choice is not always so apparent. Gibbs, Poole and Stockmeyer (1976) suggest that each variable with least degree (number of variables in all the elements containing the given variable, counting each variable just once) in the final level set S_k should be tried as a starting node. If one such node gives more than k level sets, then it is regarded as replacing the original starting node and the new level sets are examined, continuing until no further increase in the number of level sets is obtained. Gibbs et al reported that in all their test examples the number of level sets was actually maximized.

Rather surprisingly, George (1971) found that for variable-band elimination, reversing the order produced by the Cuthill-McKee algorithm often yields a worthwhile improvement, and a simple proof that reversing the order cannot increase the number of matrix coefficients within the variable-band profile has been given by Jennings (1977, pp. 162-163). The reason for the improvement is essentially that any variables associated with a single element or with a pair of adjacent elements in the same level set are eliminated early and therefore do not contribute to the bandwidth.

It is interesting that the advantages associated with reversing the Cuthill-McKee ordering for band matrices are automatically obtained in the frontal method since each variable is eliminated immediately it is fully summed. Thus a variable associated with a single element enters the front with the assembly of its element and immediately leaves it by elimination; a variable associated with a pair of elements enters the front with the first element and leaves it with the second.

The fact that the variables leave the front in a different order from that in which they enter is important to the efficiency of the method, but does lead to slight programming complications. An initial pass through the index lists associated with the matrices $B^{(k)}$ is needed to determine when each variable appears for the last time (i.e. when it can be eliminated). There is also some complication in organizing the actual assemblies and eliminations within the full frontal array. Normally the freshly entering rows and columns are placed at the end of the array and the eliminations are performed from the middle with the revised matrix closed up to fill the gaps. There is, of course, a need to maintain an index array to indicate which are the associated variables and in what order they lie.

3. THE CASE WHICH IS NOT SYMMETRIC AND POSITIVE DEFINITE

Wilkinson's backward error analysis (see for example Wilkinson (1965)) has established that Gaussian elimination is stable in the presence of floating-point roundoff provided that the matrix coefficients do not grow greatly in size. It is easily seen that growth may cause loss of accuracy for, if a large number is added to a small number using floating-point arithmetic, then some of the digits of the small number are lost.

In the important special case when A is symmetric and positive definite, Wilkinson (1961) has shown that there is no growth, in the sense that the largest intermediate coefficient does not exceed the largest original coefficient. It follows that in this case we can eliminate each variable as soon as it is fully assembled, as suggested so far in our description of the frontal method, without fear of numerical instability.

For the unsymmetric case Hood (1976) suggests delaying eliminations until the front reaches a critical size, then choosing as pivot the largest component in the submatrix of fully summed rows and columns. Rather more subtle pivotal strategies based on Hood's ideas have been used by Cliffe et al (1978) and Duff has used their code as a basis for subroutine MA32 which has recently been placed in the Harwell subroutine library (see Hopper (1980)). The principle change is that any component in the submatrix of fully summed rows and columns that satisfies the relative pivot criterion

$$|a_{ij}| \geq u \max_k |a_{kj}| \tag{3.1}$$

for some preset parameter in the range

$$0 < u \leq 1 \tag{3.2}$$

is immediately used as a pivot. This limits the growth in size of the maximum element to the factor $(1+u^{-1})$ at any one elimination step and permits eliminations to be performed when the front is much smaller than its maximum permitted size. Indeed it means that we can expect the front size to be only slightly bigger than it would have been without any pivoting. Note that because Hood's idea permits off-diagonal pivots to be chosen, separate index lists will be needed for the rows and columns (or at least for the fully assembled ones).

The symmetric non-definite case, which occurs for example when performing inverse iteration for the eigenvalue problem about a shift in the middle of the spectrum, may be treated as unsymmetric. Unfortunately this means that both the work and storage are approximately double that for the positive definite case. Bunch (1974) solves the problem for dense sets of equations by using block 2x2 pivots as well as ordinary pivots. The elimination operations are still represented by (1.3) and (1.4) if we regard the rows and columns as grouped into blocks of length 1 or 2 and the subscripts as labelling block rows and columns. For instance the pivot $a_{\ell\ell}$ may be an ordinary pivot or it may be a 2x2 block pivot. Bunch uses symmetric interchanges either to use a large diagonal element as a 1x1 pivot or to bring a large off-diagonal element into a 2x2 pivot as its off-diagonal element. Clearly a similar technique can be used in conjunction with the frontal method.

4. SUBSTRUCTURING

If a variable occurs in just one element, then its row and column is fully assembled within the single matrix $B^{(k)}$ and the associated elimination operations can be performed within $B^{(k)}$. This is known as "static condensation" and means that the true cost of adding internal variables to finite elements is very slight indeed. For instance the algebraic cost of working with 9-node biquadratic elements as shown in Figure 4.1 is virtually identical with that of working with the usual 8-node element (Figure 4.2).

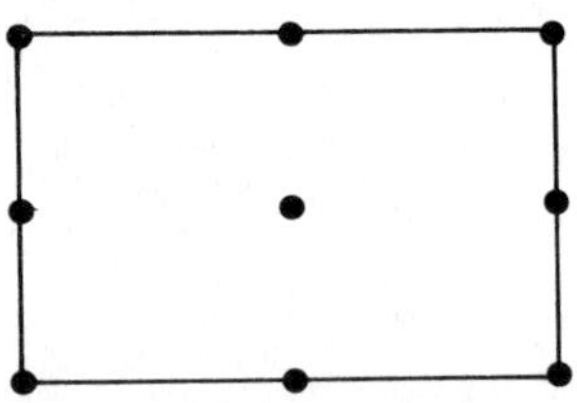

Figure 4.1 9-node biquadratic element

Figure 4.2 8-node element (biquadratic less term in x^2y^2)

This idea extends naturally to groups of elements. Any variables that are internal to a group may be eliminated by operations involving only those matrices $B^{(k)}$ that are associated with the group. There results a matrix whose rows and columns are associated with variables on the boundary of the group and which has a structure just like that of a large element parameterized by these variables. Indeed the term "generalized element" is often used (see Duff (1980) and George (1980)). A simple example is at an intermediate stage of frontal elimination: if ℓ elements have been assembled then the equation for the variables $\underline{x}^{(\ell)}$ not so far eliminated is

$$(F^{(\ell)} + \sum_{k>\ell} B^{(k)})\ \underline{x}^{(\ell)} = \underline{b}^{(\ell)} + \sum_{k>\ell} \underline{c}^{(k)} \tag{4.1}$$

where $F^{(\ell)}$ and $\underline{b}^{(\ell)}$ are the current frontal matrix and frontal right-hand side.

Where a problem has several natural substructures (for example the wings, fuselage and tailplane of an aircraft) it is natural to treat each of them independently, in just this way. The over-all problem then involves only the variables on the boundaries of the substructures but has the usual form (1.2) if the summation is interpreted as over the substructures and the matrices $B^{(k)}$ are those that result from elimination of internal variables from substructure k. Once the over-all problem has been solved, the internal variables may be calculated by back substitution by recalling the pivotal rows stored during elimination.

Several advantages accrue from substructuring in this way. The calculations involved with each substructure may be performed in separate computer runs or even on separate computer systems. If any change is made in one of the substructures then the eliminations associated with it and with the over-all structure must be recomputed, but not those associated with the other substructures. Really enormous problems may need several disk packs if treated as a whole, whereas data associated with each substructure and with the over-all interconnection may each fit on a pack. If several substructures are identical then the elimination operations may be performed just once on a "master" copy, though the forward and backward substitution operations must be performed for each since they are likely to be different on each substructure. Finally the total number of arithmetic operations may be reduced because the front size within the substructures is less than it would have been if the whole structure had been treated together.

The idea of substructuring may be nested, that is each substructure may itself be substructured and to any depth. For example seven levels of substructures were used by Araldsen (1972) to analyse the structure of an oil tanker, and he made great use of identical substructures. For further comments on the advantages of this approach see the reviews of Noor et al (1977) and Dodds and Lopez (1980).

A tree (see for example Figure 4.4, representing the propeller of Figure 4.3) is a suitable data structure to represent nested substructuring. Associated with each terminal node is a finite element and associated with each node is a substructure whose members are associated with its "sons" in the tree. It is convenient to associate with each terminal node the corresponding matrix $B^{(k)}$ and vector $\underline{c}^{(k)}$ together with the associated list of variables. Similarly other nodes may have associated lists of boundary variables for substructures and matrices and vectors $B^{(k)}, \underline{c}^{(k)}$ obtained by assembly of components and elimination of interior variables. The processing required for each node is just like the processing of the whole of a simple problem; provided the index lists and the quantities $B^{(k)}$ and $\underline{c}^{(k)}$ are available they can be treated as if we had ordinary elements. For example we might use the frontal method at each node.

No node of the tree may be processed before all its sons, but there is nevertheless considerable freedom in the ordering. Separate substructures may be handled by separate processors; for instance the four first level substructures of Figure 4.4 might be treated on four separate processors which pass their results to a fifth processor. On the other hand, if a single processor is in use there is some advantage in ordering according to a depth-first tree search. For instance for the case shown in Figure 4.4 we would process the substructures in the order 23,24,27,25,28,26,29. The advantage of a depth-first search is that intermediate results can be held in a stack. The stack contents for our example would be data associated with nodes (23),(23,24),(27),(27,25),(28),(28,26),(29).

So far we have assumed that there is a natural way of substructuring, usually because a complicated design is assembled from simpler sub-assemblies. It makes sense to exploit these natural substructures, but it can also be advantageous to introduce artificial ones. For example George (1973) showed that a two-dimensional problem consisting of a regular qxq grid of square elements could be solved in $O(q^3)$ operations and $O(q^2 \log q)$ storage locations instead of $O(q^4)$ and $O(q^3)$, respectively, for band or frontal methods. He did this by dividing the elements into four groups with a "+" shaped cut, then

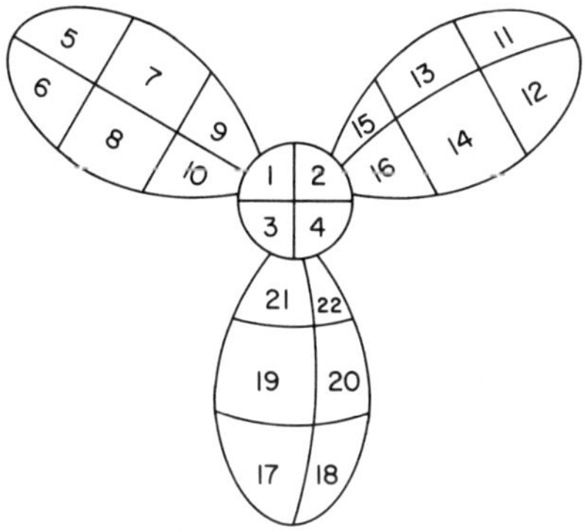

Figure 4.3 A propeller

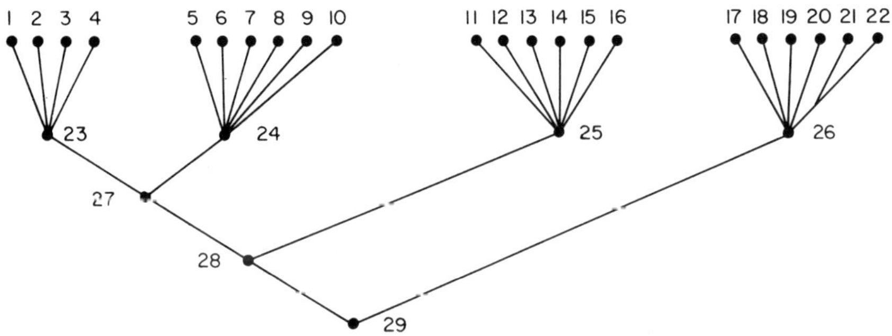

Figure 4.4 Tree representing propeller

dividing each group into four parts again with a "+" shaped cut and so on. Following a suggestion from Birkoff he called this "nested dissection" and this name has become well established. Actual gains for practical values of q, though worthwhile, are not enormous because the multiplying factors disfavour nested dissection. George's (1973) figures show, for instance, that if q=32 the number of operations is reduced to about 38% and the number of storage locations needed (excluding overheads) to about 52% of the corresponding figures for band elimination using page-wise ordering.

George and Liu (1978a) have a proposal for automating the dissection. They use the algorithm of Gibbs, Poole and Stockmeyer (1976) to find a pseudo-diameter and a set of level sets and take the middle level set as a separator. They then repeat on the halves, and so on. On the regular square problem they found results were very nearly as good as with the theoretical dissection of George (1973). They also make comparisons on a sequence of problems associated with using successively finer triangulations at the re-entrant corner of a L-shaped region. They compared this with a variable band solution based on a reverse Cuthill-McKee ordering and with treating the matrix as a general sparse matrix and using the Yale sparse matrix package (Eisenstat et al, 1977). The results showed surprisingly little difference in execution time, with the band algorithm performing more operations but making up with a faster speed per operation because of its simpler data structure. Nested dissection did gain on storage utilization, even allowing for its extra overheads, for example by about 25% over the band algorithm on the biggest problem, which had order 2233. Of course nested dissection is asymptotically superior and none of these problems was particularly large.

To obtain some indication of the behaviour on a bigger problem, the author has experimented with one kindly supplied by T.A. Manteuffel. Manteuffel himself (1980) describes this problem and shows a diagram. It is a cylindrical shell with some irregularities and is built from 784 20-node bricks to give a total of 17928 variables. Manteuffel estimates that a band technique would involve a semi-bandwidth of 1300, storage for $23x10^6$ non-zeros and $15x10^9$ arithmetic operations. The SPARSPAK code of George and Liu (1978b,1979) is unable to handle such a big problem, but we can make estimates from its analysis of the equivalent problem with one degree of freedom per node. This indicates that the reverse Cuthill-McKee algorithm would reduce the storage required by about the factor 2 and that nested dissection would further reduce it by about the factor 2.4.

A pivotal strategy that has proved remarkably successful for any sparse symmetric matrix is that of "minimum degree". At each stage of the elimination it chooses as pivot the diagonal element of a row with least non-zeros. Traditionally this has been implemented by holding all the non-zeros in a suitable list structure which expands with the fill-ins that result from the eliminations. Interestingly it can also be represented in terms of nested substructuring. We begin by computing the numbers of non-zeros there would be in the rows of the matrix if we assembled it. For each row this is just the total number of variables in the union of the elements that contain the variable that corresponds to the row, and it may be computed directly from the lists of variables in the elements. Elimination of the variable of minimum degree may be achieved by a local assembly of all the elements in which it is involved into a substructure. The chosen variable is, of course, interior to this substructure. Following its elimination any other internal variables have degrees one less and therefore must have minimum degree over-all. Thus having established the substructure we should eliminate all internal variables in the usual way. The resulting substructure, with its internal variables eliminated, can now be treated just like an ordinary element (see also Duff (1980)). The algorithm can thus be continued to build successively larger substructures. This process of substructure generation can be made efficient since only the variable lists need be manipulated (they are merged for assemblies and then internal variables are removed) and it is possible to update the degrees of the variables (only those adjacent to or in the assembled substructure change). The author has applied an experimental code of this kind, but with a different strategy (essentially a local minimum fill-in strategy) to four structural problems kindly supplied by A.J. Donovan of CEGB, and the results are summarized in Table 4.1. Notice that the time for analysis and symbolic factorization (finding a minimum degree ordering, rearranging the tree to correspond and computing the storage and operational cost of actual factorization) is, in all four cases, much less than the corresponding time for factorization (actual numerical factorization). This is because we work with index lists during analysis and the operations performed on them (merging and elimination of variables) are linear in their lengths, whereas during factorization we have to perform corresponding operations on full arrays of reals. The gain is particularly noticeable where the original elements are large (first and last cases). The time given is for Fortran code and includes all the overheads associated with depth-first search of the tree, access to disk storage and organization of intermediate results in a stack. Comparable code for in-core Gaussian elimination on a large full matrix executes at about 2 μsecs per multiplication. Comparison with minimum degree showed very similar fill-in and operation counts for cases 1,2 and 4 but in case 3 reduced the number of non-zeros in U to 69×10^3 and the number of multiplications to 1.1×10^6. We therefore propose to alter this code to minimum degree. We have also analyzed Manteuffel's matrix using the minimum degree algorithm and, in only 2 seconds of IBM 370/168 time, showed that the required storage would be for 4.8×10^6 reals and the number of operations would be 1.12×10^9. The storage gain over Manteuffel's estimate is very near to that predicted by SPARSPAK for automatic nested dissection and indeed SPARSPAK also tells us that on this problem minimum degree and nested dissection give similar performance. Notice that the minimum degree strategy has reduced Manteuffel's operation count by the factor 13 to make the computation a much more practicable proposition (he abandoned direct solution in favour of iteration using pre-conditioned conjugate gradients).

The success of the tree search technique for finite element problems naturally leads one to ask whether non-element problems might be handled efficiently by artificially converting them to

TABLE 4.1

Performance of experimental tree searching code on CEGB test problems

	3-dimensional cylinder with flange	2-dimensional reactor core section	Framework (essentially 2-dimensional)	Turbine blade
No. of variables	2919	3024	3306	2802
No. of elements	128	551	791	108
Max. element size	60	16	12	60
Thousands of non-zeros in factor U	357	116	100	236
Millions of mults to factorize	28	3	4	11
Time (IBM 370/168 secs) for Analyze	2	4	5	1
Time (IBM 370/168 secs) for Factorize	71	13	26	32
Time (IBM 370/168 secs) for Solve	3	1	1	2

this form. This can be done by regarding each diagonal non-zero as having come from a one-variable element and each off-diagonal non-zero as having come from a two-variable element, but having so many short index lists leads to slow execution of analyze. An alternative is the following algorithm.

Algorithm 4.1

Flag all non-zeros a_{ij} as unused;

for each row i *do*

  while row i contains a flagged non-zero *do*

  begin add an artificial element containing just variable i;

    unflag a_{ii};

    for each flagged non-zero a_{ij} in row i *do*

    if $a_{kj} \neq 0$ for all variables k of the current element

    then add variable j to the element and unflag

      a_{kj} for all variables k now in the element

  end;

We have used this algorithm on the Duff and Reid (1979) collection of sparse matrices, and four representative results are shown in Table 4.2. We do not include the time for algorithm 4.1 because our implementation is experimental. We show comparisons with the Harwell code MA31, written by Munksgaard (1980), which includes facilities for performing conventional minimum degree factorization. It appears that the tree searching technique can show worthwhile gains, though the advantages are not so dramatic as in the genuine element cases.

Finally we would like to remark that just as the simple frontal method can be extended to handle unsymmetric and symmetric non-definite cases in a numerically stable way, so can this tree-based generalization. Pivots are used only if they satisfy a suitable criterion, such as (3.1). As a result the matrices stored during factorization may be a little larger than in the corresponding positive-definite case, since eliminations must wait for suitable pivots to be available.

TABLE 4.2

Performance on some test problems

Order		147	1176	199	130
No. non-zeros		1298	9864	536	713
No. elements		173	916	338	462
No. non-zeros in factor U		2340	10348	1059	716
No. mults to factorize		22210	161366	4541	2674
Analyze time (IBM 370/168 secs)	MA31	0.42	1.8	0.10	0.09
	Tree	0.13	0.24	0.09	0.17

5. CONCLUSIONS

The frontal method is a remarkably successful technique that truly exploits the form of a finite element problem. We have described its conventional use as well as some recent encouraging results associated with substructuring. We have shown how the general purpose and highly successful pivotal strategy of minimum degree can be implemented efficiently in this way and that a not uncompetitive algorithm for non-element problems may be constructed by first artificially converting them to element form. We have also discussed extensions to non-definite and unsymmetric cases.

ACKNOWLEDGEMENT

I would like to thank I.S. Duff for reading a draft of this paper and making many helpful comments.

REFERENCES

Araldsen, P.O. (1972) An example of large-scale structural analysis. Comparison between finite element calculation and full scale measurement on oil tanker ESSO NORWAY, Seminar at Det Norske Veritas.

Bunch, J.R. (1974) Partial pivoting strategies for symmetric matrices, SIAM J. Numer. Anal., 11, pp. 521-528.

Cliffe, K.A., Jackson, C.P., Rae, J. and Winters, K.H. (1978) Finite element flow modelling using velocity and pressure variables, Report AERE-R.9202, Harwell, UKAEA.

Cuthill, E. (1972) Several strategies for reducing the bandwidth of matrices, In Sparse Matrices and their Applications, (D.J. Rose and R.A. Willoughby, eds), Plenum Press, pp. 157-166.

Cuthill, E. and McKee, J. (1969) Reducing the bandwidth of sparse symmetric matrices, Proc. 24th National Conf. ACM, Brandon Systems Press, N.J., pp. 157-172.

Dodds, R.H. and Lopez, L.A. (1980) Substructuring in linear and non-linear analysis, Int. J. Num. Meth. Engng., 15, pp. 583-597.

Duff, I.S. (1980) A sparse future, These proceedings.

Duff, I.S. and Reid, J.K. (1979) Performance evaluation of codes for sparse matrix problems, In Performance Evaluation of Numerical Software, (L.D. Fosdick, ed.), North-Holland, pp. 121-135.

Eisenstat, S.C., Gursky, M.C., Schultz, M.H. and Sherman, A.H. (1977) The Yale sparse matrix package, I. the symmetric codes, Report 112, Dept. Comput. Sci., Yale Univ.

George, J.A. (1971) Computer implementation of the finite element method, Report STAN-CS-71-208, Computer Science Dept., Stanford University, California.

George, J.A. (1973) Nested dissection of a regular finite element mesh, SIAM J. Numer. Anal., 10, pp. 345-363.

George, J.A. and Liu, J.W.H. (1978a) An automatic nested dissection algorithm for irregular finite element problems, SIAM J. Numer. Anal., 15, pp. 1053-1069.

George, J.A. and Liu, J.W.H. (1978b) User guide for SPARSPAK: Waterloo sparse linear equations package, Report CS-78-30, Computer Science Dept., University of Waterloo, Canada.

George, J.A. and Liu, J.W.H. (1979) The design of a user interface for a sparse matrix package, TOMS, 5, pp. 139-162.

George, A. (1980) Direct solution of sparse positive definite systems: some basic ideas and open problems, These proceedings.

Gibbs, N.E., Poole, W.G., Jnr. and Stockmeyer, P.K. (1976) An algorithm for reducing the bandwidth and profile of a sparse matrix, SIAM J. Numer. Anal., 13, pp. 236-250.

Hood, P. (1976) Frontal solution program for unsymmetric matrices, Int. J. Num. Meth. Engng., 10, pp. 379-400.

Hopper, M.J. (1980) Harwell Subroutine Library, a catalogue of subroutines, Report AERE-R.9185, Harwell, UKAEA.

Irons, B.M. (1970) A frontal solution program for finite element analysis, Int. J. Num. Meth. Engng., 2, pp. 5-32.

Jennings, A. (1977) Matrix Computation for Engineers and Scientists, Wiley.

Manteuffel, T.A. (1980) An incomplete factorization technique for positive definite linear systems, Math. Comp. 34, pp. 473-498.

Munksgaard, N. (1980) Solving sparse symmetric sets of linear equations by preconditioned conjugate gradients, TOMS, 6, pp. 206-219.

Noor, A.K., Kamel, H.A. and Fulton, R.E. (1977) Substructuring techniques - status and projections, Comput. and Struct., 8, pp. 621-632.

Wilkinson, J.H. (1961) Error analysis of direct methods of matrix inversion, J. ACM, 8, pp. 281-330.

Wilkinson, J.H. (1965) The Algebraic Eigenvalue Problem, Oxford University Press.

DISCUSSION

PROFESSOR A. JENNINGS (Belfast). I am concerned about the stability of frontal codes for unsymmetric or symmetric indefinite matrices.

REID. The stability properties will be just as good as for any relative-pivot strategy on a fully-assembled matrix. If, at any stage, there is no fully-assembled non-zero satisfying our numerical criterion we do more assemblies until one does.

DR. A.M. ERISMAN (Boeing). Is your element based minimum degree ordering the same as the clique strategy described by Duff on the first day of the conference?

REID. Yes

DR. I.S. DUFF (Harwell). I should like to point out another major change made to the code of Hood and Cliffe et al. Harwell subroutine MA32 will also accept input by equations as well as by elements and so can be used where finite difference discretizations have been employed. Arguments similar to those used by Reid show that it will always do at least as well as a variable band scheme with the same ordering of equations.

DR. J.G. MORGAN (Culham). Have you ever contemplated using another language than Fortran for your tree manipulation?

REID. Certainly my code is very complicated and no doubt a language like ALGOLW could make it simpler. However, I doubt it would be more efficient.

DR. A.M. ERISMAN (Boeing). Could you tell us how the factor time compares with the analyze time on the example of Manteuffel?

REID. I have not done a factorize run because I know how many operations are involved from the analyze run and know that it would be very expensive. Such a run is not justified for research purposes only.

PROFESSOR A. GEORGE (Waterloo). I would like to emphasise that the overhead problem is much worse when initial elements are small. This is evident from the results of Reid on non-element problems.

DIRECT SOLUTION OF SPARSE POSITIVE DEFINITE SYSTEMS: SOME BASIC IDEAS AND OPEN PROBLEMS

A. George

(Department of Computer Science, University of Waterloo, Waterloo, Ontario, Canada)

ABSTRACT

Numerous ideas have been developed over the past fifteen years for improving the efficiency of direct methods for solving large sparse systems of equations. The objective of this tutorial paper is to review some of these ideas and associated results. Some issues and unsolved problems which currently confront researchers in the field are also discussed.

1. INTRODUCTION

The solution of large sparse systems of linear algebraic equations has been a very active area of research for the past ten to fifteen years. Such research is quite broadly based; some examples of sub-areas are

1) design and complexity analysis of algorithms,
2) numerical roundoff error analysis,
3) data structure design,
4) design of software packages,
5) influence of non-standard computer architecture on sparse matrix computations, and the design of new architectures to support such computations.

Much of the research requires at least some expertise in a number of areas of mathematics and computer science, including graph theory, combinatorics, numerical analysis, algorithm analysis, design of efficient data structures, software engineering, and computer architecture.

I cannot hope to review all the important advances. Instead, my objective is to consider a restricted (but nevertheless large and important) subclass of problems, and to review some of the ideas which experience and/or analysis have shown to be effective. A second objective is to expose non-experts in the field to some issues and problems which concern researchers in the field.

2. THE PROBLEM AND GENERAL METHODOLOGY

The class of problems I consider are symmetric positive definite systems of linear equations

$$A\underline{x} = \underline{b}, \tag{2.1}$$

where A is N by N, large, and sparse. The usual direct method for solving (2.1) is Cholesky's method. The matrix A is factored into LL^T, where L is lower triangular, and then $\underline{x}$ is obtained by solving the two triangular systems $L\underline{y} = \underline{b}$ and $L^T\underline{x} = \underline{y}$.

The fact that makes solving sparse systems of equations interesting (and difficult) is that when A is factored, it suffers fill; the matrix L will usually have far more non-zeros than the lower triangular part of A. A judicious reordering of the problem can often reduce the amount of fill, and hence the computation and storage requirements, assuming of course that sparsity is exploited. For any N by N permutation matrix P, the matrix PAP^T is also symmetric and positive definite and Cholesky's method still applies, so we can instead solve the equivalent system

$$(PAP^T)(P\underline{x}) = P\underline{b}.$$

As an example, suppose the system (2.1) has the structure depicted in Figure 2.1, where * denotes a non-zero. Obviously A suffers severe fill; its Cholesky factor L is full. On the other hand, reversing the ordering of the variables and equations to obtain $\bar{A}\,\bar{\underline{x}} = \bar{\underline{b}}$, as shown in Figure 2.2, yields a problem which suffers no fill at all. This is an extreme example, but it does serve to illustrate the importance of the choice of P. For large N, the arithmetic required to solve the system in Figure 2.1 is $O(N^3)$, compared to O(N) for the reordered system in Figure 2.2.

The choice of an appropriate P is usually referred to as the ordering problem, and algorithms for finding P are called ordering algorithms. Fill is only one of several criteria which may be used to measure the effectiveness of an ordering algorithm, although it is perhaps the most common one.

Suppose an ordering for A has been found, and denote by L the Cholesky factor of $\bar{A} = PAP^T$, the reordered A. There is still much work left to do before the actual numerical computation can begin. We must first set up a data structure to store L, and if

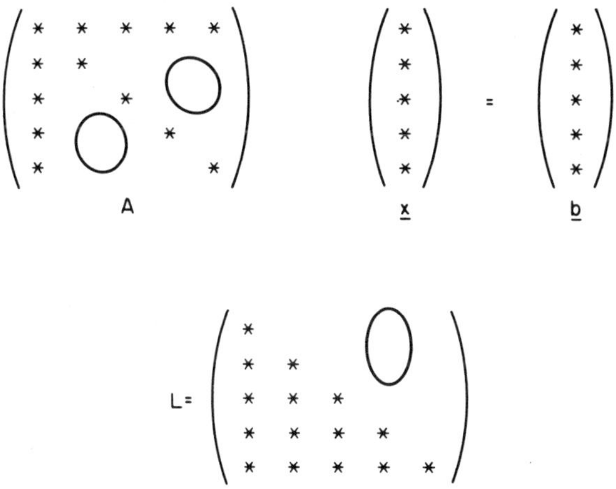

Figure 2.1 Structure of a sparse system $A\underline{x} = \underline{b}$, and the structure of the corresponding Cholesky factor L.

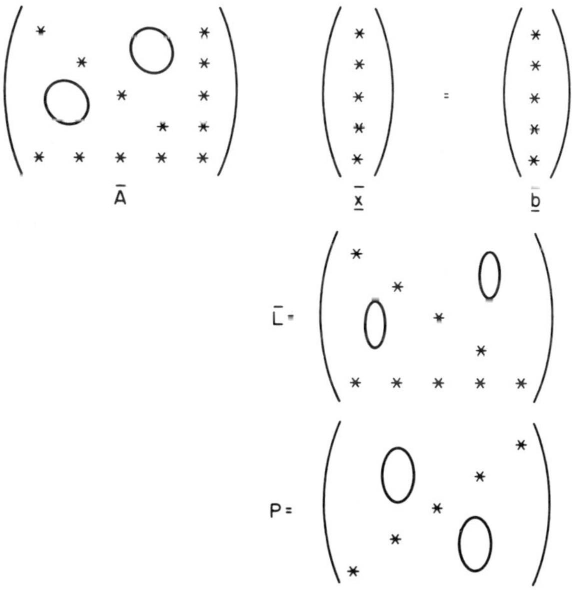

Figure 2.2 Structure of the system of Figure 2.1, after the ordering of the variables and equations has been reversed. The structure of the Cholesky factor $\bar{L}$ of $\bar{A}$ is also shown. Note that $\bar{A} = PAP^T$, $\bar{\underline{x}} = P\underline{x}$, and $\bar{\underline{b}} = P\underline{b}$.

we are going to exploit zeros, then we must determine the positions of some or all of the non-zeros in L, given the structure of $\bar{A}$. This task is easy for the A and $\bar{A}$ in Figures 2.1 and 2.2, but the example in Figures 2.3 and 2.4 illustrates that in general the problem is not trivial. Terms such as storage allocation, data structure set-up, and symbolic factorization are used to refer to this phase of the computation. Once this process has been completed, the actual numerical computation can be performed. Note that both the ordering and symbolic factorization steps depend only on the structure of A, and are independent of its numerical values, or those of L.

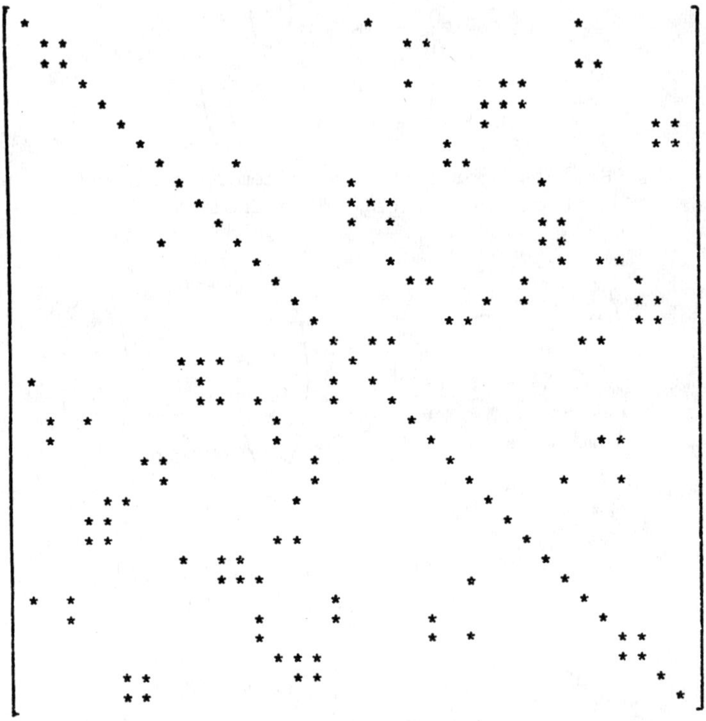

Figure 2.3 The structure of PAP^T, a symmetric permutation of a matrix A.

Thus, regardless of the actual ordering algorithms and data structures employed, the usual and generally accepted procedure for solving symmetric positive definite systems of equations involves four distinct phases:

Step 1 (Ordering) Find a "good" ordering (permutation P) for the given matrix A.

Step 2 (Storage Allocation) Determine the necessary information about the Cholesky factor L of PAP^T to set up the storage scheme for L.

Step 3 (Factorization) Factor the permuted matrix PAP^T into LL^T.

Step 4 (Triangular Solution) Solve $L\underline{y} = P\underline{b}$, $L^T\underline{z} = \underline{y}$, and set $\underline{x} = P^T\underline{z}$.

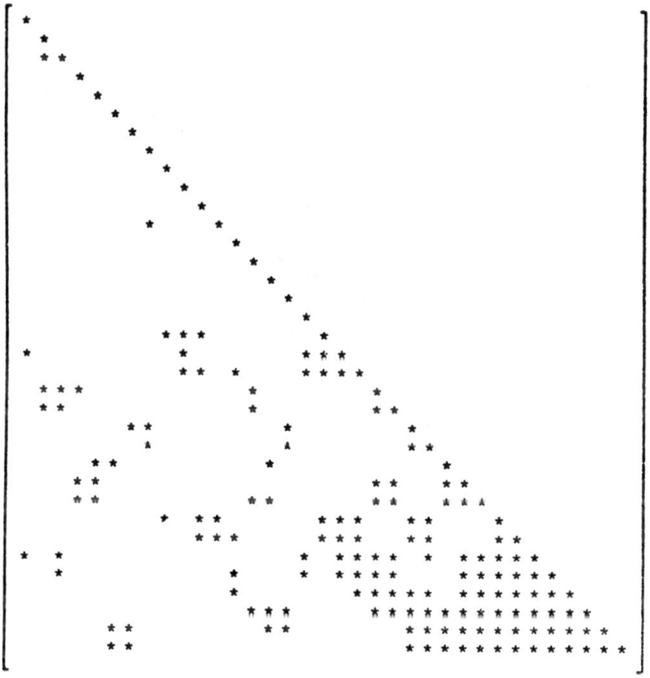

Figure 2.4 The structure of L, the Cholesky factor of PAP^T whose structure is shown in Figure 2.3.

With this perspective, we should note that the ordering and storage allocation problems for sparse positive definite systems are both different from and easier than those for general sparse A. For a general indefinite sparse matrix A, some form of <u>pivoting</u> (row and/or column interchanges) is necessary to ensure numerical stability. Thus, given A, one normally obtains a factorization of PA or PAQ, where P and Q are permutation matrices of the appropriate size. These permutations are determined <u>during the factorization</u> by a combination of (usually competing) numerical stability and sparsity considerations. Different matrices, even though they may have the same zero/non-zero pattern will normally yield different P and Q, and therefore have factors with different sparsity patterns. In other words, it is in general not possible to predict where fill will occur before the computation begins. Thus, we are obliged to use some form of dynamic storage scheme which allocates storage for fill as the computation proceeds. The three problems of

i) finding a suitable ordering,
ii) setting up the appropriate data structure, and
iii) the actual numerical computation cannot be isolated,

as they can for the symmetric positive definite case.

3. ORDERING ALGORITHMS

This section contains a description of some basic ordering ideas, along with indications of some important and apparently difficult open problems. Since the use of graph theory notation is quite common in the literature dealing with sparse matrix ordering algorithms, we begin by discussing the relationship of graphs and matrices.

For our purposes, a graph $G = (X,E)$ consists of a finite set X of nodes, together with a set E of edges, which are unordered pairs of vertices. An ordering of G is simply a labelling of its vertices from 1 to $|X|$, the number of elements in X. The ordered graph of the N by N symmetric matrix A, denoted by $G^A = (X^A, E^A)$, is one for which the N vertices of G^A are labelled from 1 to N, and $\{x_i, x_j\} \in E^A \Longleftrightarrow a_{ij} = a_{ji} \neq 0$, $i \neq j$. Here x_i is the node of X^A having label i. An example is provided in Figure 3.1.

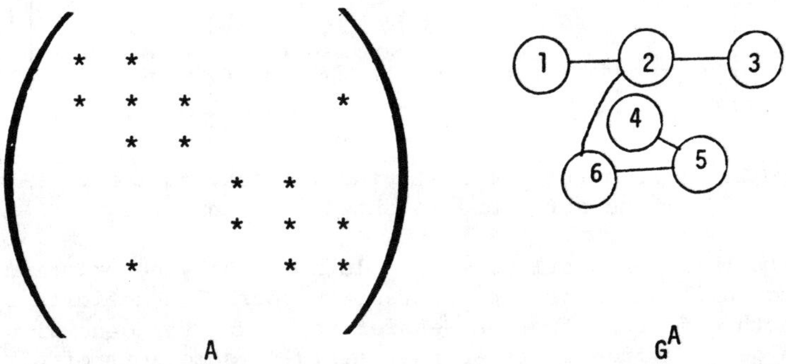

Figure 3.1 A matrix A and its labelled graph G^A, where * denotes a non-zero entry in A.

Note that the unlabelled graph of A represents the structure of A without suggesting a particular ordering. Finding a "good" permutation for A can be viewed as finding a good labelling for its graph. Figure 3.2 shows the structure of a matrix corresponding to the graph of Figure 3.1, but with a different labelling.

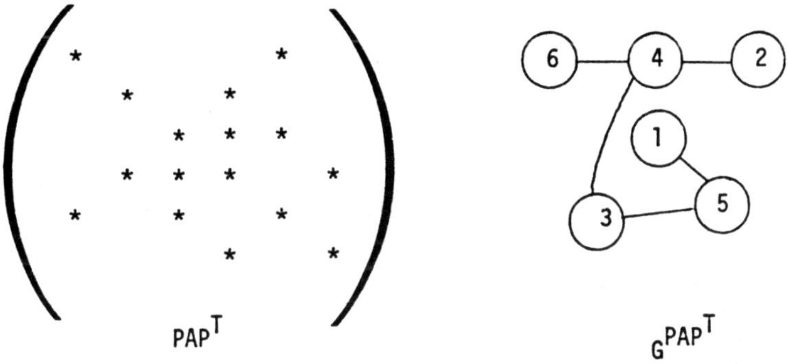

Figure 3.2 The permuted matrix of Figure 3.1 along with the correspondingly relabelled graph.

Although the notions and notations of graph theory are convenient and helpful in describing and developing sparse matrix algorithms, few if any classical results from graph theory have found application to the analysis of these algorithms. The graph theory results used in the sparse matrix field have largely originated there.

One of the simplest strategies is to choose P so that $\bar{A}=PAP^T$ has a small <u>bandwidth</u>, where the bandwidth of a matrix M is defined to be $m(M) = \max \{|i-j| \; |M_{ij} \neq 0\}$. The essential point of this strategy is that all fill is confined within the band, so that $m(L)=m(A)$. Implicit in this strategy is that zeros outside the band will be exploited, but those inside will be stored. Numerical software for solving systems of equations in band form is widely available and is quite efficient in the sense that the data structures used, and the programs which operate on them, are relatively simple and involve little overhead.

The problem of finding a P which minimizes $m(\bar{A})$ has received considerable attention, and several good heuristic algorithms for approximately solving the problem have been developed, the most effective of which are derivatives of an algorithm due to Cuthill and McKee (1969). Given a starting node r in G^A, the basic idea is to number the nodes in G^A in a level by level manner, starting with r, then those adjacent to r, then those adjacent to those just numbered, and so on. (The details are not important here.) An example of such an ordering for a graph is given in Figure 3.3. Note that in terms of the graph, the bandwidth of $\bar{A}$ is simply the maximum difference in labels of nodes which are adjacent.

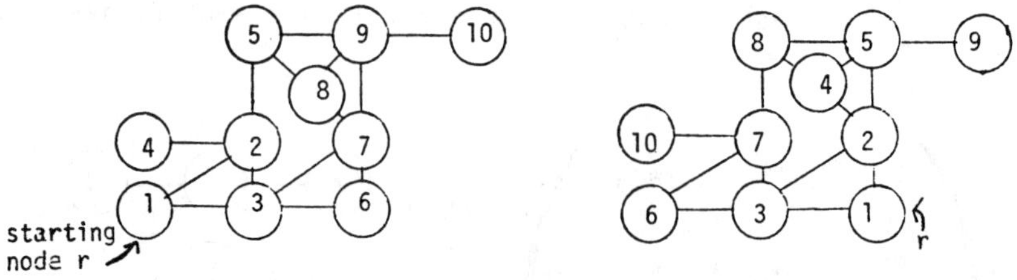

Figure 3.3 Two orderings of G^A provided by the Cuthill-McKee algorithm, with different starting nodes.

It should be intuitively clear that the effectiveness of the strategy depends crucially on the choice of starting node r, and the most costly part of the original algorithm was the process of finding a good r. A very important contribution to this aspect of the algorithm was provided by Gibbs et al (1976), who provided a very effective heuristic scheme for finding so-called pseudo-peripheral nodes in a graph (nodes of high eccentricity). Several useful refinements of it, along with software, is provided by Crane et al (1976) and George and Liu (1979).

These algorithms have not been proven to be good. That is, there are no known results which indicate how close they come to minimizing the bandwidth. However, they usually produce acceptable orderings for practical purposes. Moreover, the problem of actually minimizing $m(\bar{A})$ is known to be very hard - a so-called NP-complete problem (Papadimitriou (1976)).

It is well known among sparse matrix researchers that minimizing the bandwidth of $\bar{A}$ in general has very little connection with minimizing the amount of arithmetic required to solve a sparse system of equations, assuming that we are prepared to exploit all zeros. The same remark applies with respect to the fill that occurs during the factorization. The example in Figure 2.1 illustrates this point; if P is chosen to minimize the bandwidth, we get the structure of $\bar{A}$ and the corresponding L shown in Figure 3.4. For large N, solving the problem using this ordering involves $O(N^3)$ arithmetic operations and $O(N^2)$ storage, since a dense matrix of order N/2 is involved.

If minimizing the bandwidth is not closely related to minimizing arithmetic requirements and storage, why use such methods? There are at least two reasons: a) some people do not realize the above fact, and b) they realize it, but also recognize that there are other more practical issues involved. We can only hope that the class of people falling in a) is steadily decreasing! Point b) deserves some further explanation.

Assuming we exploit every zero, orderings that minimize arithmetic operations or fill tend to yield factors L having non-zeros scattered throughout their lower triangle, as illustrated in Figure 2.4. For some problems, such orderings can reduce arithmetic and fill by one or more orders of magnitude

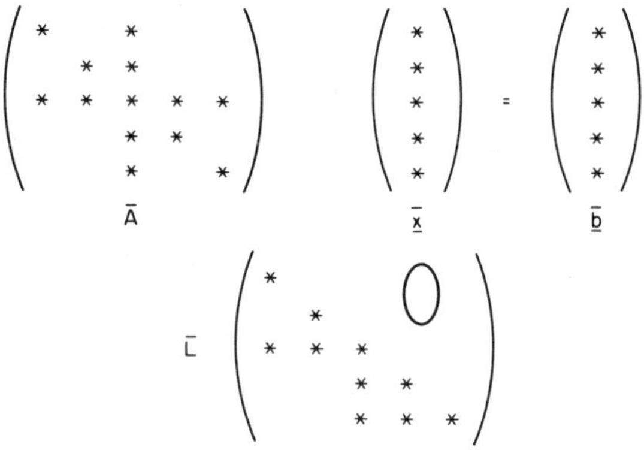

Figure 3.4 Minimum bandwidth ordering for the problem of Figure 2.1, along with the structure of the corresponding factor L.

over the corresponding levels for the minimum band ordering. However, the use of these near-optimal orderings has a price, and a comparison between the two approaches involves far more than simply counting arithmetic operations and/or the number of non-zeros in L. The more sophisticated orderings yield factors whose non-zeros are scattered, so one must use data structures that are more elaborate than those used for band ordered matrices. Storage used for sparse matrices consists of two components: primary storage, used for the actual numerical values, and overhead storage, used for pointers, subscripts, etc., which represent the structure of L and facilitate access to the non-zeros of L. The data structures required for the more sophisticated orderings usually have a much higher fraction of their storage devoted to overhead than do those used for band orderings, and the access to the non-zeros of L tends to be more expensive. Thus, the advantages enjoyed by the more sophisticated orderings may be far less significant in practice than a simple comparison of arithmetic operations or fill might suggest.

Despite these somewhat cautionary remarks, the more sophisticated ordering strategies often do pay off handsomely, and in the remainder of this section we discuss two such strategies, along with some results about them, and some interesting open problems.

The most commonly used algorithm for finding low-fill orderings is the so-called minimum degree algorithm,which is a symmetric variant of the Markowitz strategy(Markowitz(1957), George and Liu(1980b)). Basically, the idea is to interchange rows and columns at the beginning of each elimination step so that the pivot row and column have a minimum number of non-zeros. Experience has shown that this algorithm is very effective in finding low-fill orderings, for a wide class of problems. Thus, its efficient implementation has been an active area of investigation. In order to discuss some of these developments, we first provide an example in Figure 3.5 showing how the algorithm works, where $A=A_0$ and the A_i, $i=1,2,\ldots,8$ represent the structure of the part of the matrix remaining to be factored at the beginning of the i-th step. The †'s indicate the row-column pairs chosen as pivots at each step, and the ⊕ 's denote fill elements caused by the simulated elimination.

First note that ordering decisions depend on the structure of the partially factored matrix, which means that we need some representation of it in order to execute the algorithm. Also note that the choice of pivots is not strictly determined by the algorithm as stated, since ties occur. Although there is some evidence that tie-breaking can have a substantial effect on the quality of the resulting ordering, no inexpensive reliable tie breaking strategy has been developed, so ties are usually broken in some operationally convenient but otherwise arbitrary manner.

Another important point, not illustrated by our example but nevertheless usually true, is that some of the A_i may have far more non-zeros than does A_0. Thus, if we store an explicit representation of the partially factored matrix, then our storage requirement for the algorithm cannot be predicted, because it depends on the fill that occurs during the simulation of the factorization. This characteristic is particularly unpleasant if one wishes to include such an algorithm in a software package, since it is desirable to be able to tell the user about storage requirements before the software is executed.

Fortunately, efforts to overcome this undesirable aspect of the algorithm have been successful, and several fast $O(|A|)$-space implementations of the algorithm have been developed (see for example, George and Liu (1980b)) although space does not permit their description here. Here $|A|$ denotes the number of non-zeros in A. Experiments also suggest that for large classes of problems, good implementations run in $O(|A|)$ time. However, no useful execution bounds for <u>any</u> of these implementations have been found, except for essentially trivial problems. In other words, for good implementations the "space problem" has been solved, but the execution time, while empirically good, is theoretically unknown.

Another interesting open problem is the <u>quality</u> of the ordering produced by the minimum degree algorithm. How close to the <u>best</u> ordering, in the minimum fill sense, is that provided by the algorithm? As noted above, experience suggests that it usually does very well, but no one has been able to show that it is guaranteed to do so, even for fairly restricted classes of problems. There are some contrived examples showing that it can do badly, and some results for trivial problems showing that it produces the best ordering.

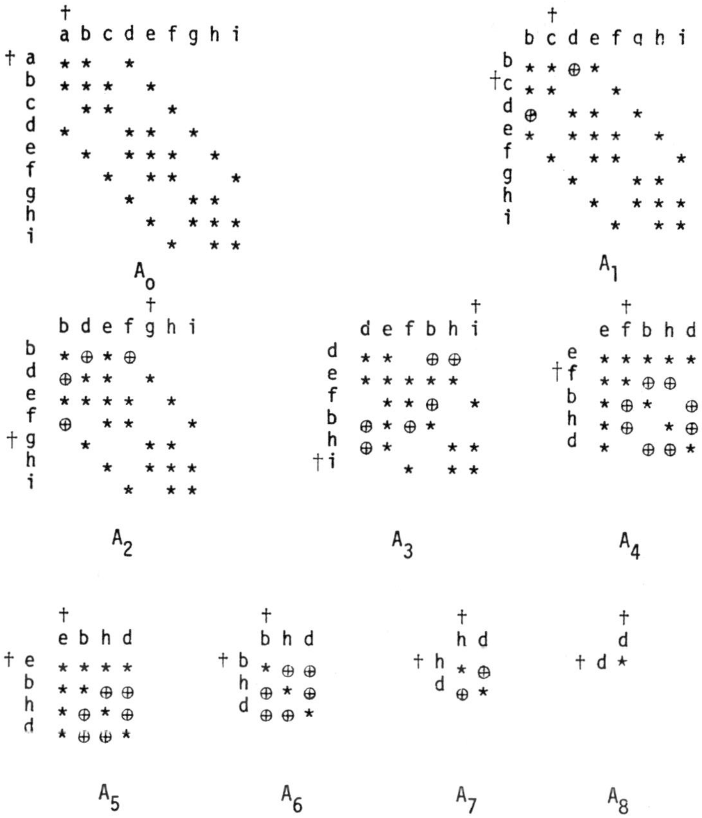

Figure 3.5 An example of the application of the minimum degree algorithm applied to a sparse matrix. The †'s indicate the row/column pairs chosen as pivots and the ⊕ 's denote fill elements caused by the (simulated) elimination.

There are several possible explanations for this rather unsatisfactory state of the theory. One is that tie-breaking does sometimes make a substantial difference, so the algorithm as usually described and implemented may not be well enough prescribed. Second, the difficulty of finding the best ordering, regardless of the algorithm used, is still unknown. (For unsymmetric A, it is known to be NP-complete (Rose and Tarjan (1978)), but the mechanism used in that proof does not seem to carry through for symmetric A.). Thus, we do not even know the difficulty of the problem that the minimum degree algorithm is trying to solve!

However, despite the absence of theorems about the quality of the orderings provided by the minimum degree algorithm, it continues to be heavily used in the field.

For certain matrix problems associated with planar meshes, there _is_ an algorithm available for finding asymptotically optimal orderings, and it is one of the few sparse matrix algorithms for which an adequate theory exists. This automatic nested dissection algorithm, due to Lipton et al (1979), is a generalization of the nested dissection algorithm described by George (1973). The algorithm executes in $O(|A| \log |A|)$ time and $O(|A|)$ space, and produces an ordering for which $|L|$ is $O(|A| \log |A|)$. All of these bounds are optimal.

Unfortunately, the algorithm is fairly complicated, and apparently a well-tested efficient implementation has not yet been developed. Another potential drawback is that the algorithm makes explicit use of planarity, so the problem class to which it applies is somewhat restricted. A heuristic automatic nested dissection algorithm which uses similar ideas and for which a good implementation does exist is described by George and Liu (1978b). It is not restricted to planar problems, but there are no results guaranteeing the quality of its output.

These algorithms work with the graph of the matrix A, and are essentially divide and conquer algorithms. The crucial observation is as follows. Suppose we can find a _separator set_ S of nodes of G which, if we remove them along their incident edges, disconnects the graph into two components C_1 and C_2, as shown in Figure 3.6. Then suppose we renumber the nodes of G so that those in C_1 and C_2 are numbered consecutively, followed by those in S, as depicted in Figure 3.6. The reordered matrix $\bar{A}$ will have the structure shown in (3.1).

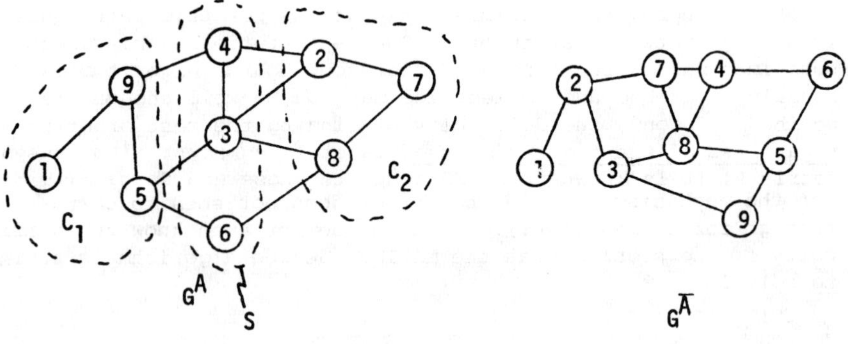

Figure 3.6 The graph of a matrix A, and a reordering induced by choosing a separator S.

$$\bar{A} = \begin{vmatrix} A_{11} & O & A_{13} \\ O & A_{22} & A_{23} \\ A_{13}^T & A_{23}^T & A_{33} \end{vmatrix} \tag{3.1}$$

The important point is that the zero matrix in $\bar{A}$ is preserved in its Cholesky factor L. This simple reordering device has created a large block of zeros in L.

We can obviously recursively apply the idea to the components C_1 and C_2, thereby inducing blocks of zeros in A_{11} and A_{22} which also would be preserved in L. This resulting ordering is called a nested dissection ordering.

4. SYMBOLIC FACTORIZATION (DATA STRUCTURE SET-UP)

Once an ordering has been found, the next task is to set up a data structure for the non-zero elements of L. For some orderings, such as band orderings, the storage schemes used are so simple that this phase of the computation is essentially trivial. Moreover, the structure of L that is needed in setting up the data structure can be read directly from that of the reordered $\bar{A}$. For example, all that is needed for one common storage scheme for band orderings is the column subscript of the first non-zero in each row (Jennings (1966)). We will say no more about these simple orderings here.

However, for more sophisticated orderings, such as those produced by the minimum degree algorithm, the non-zeros of L are scattered throughout the lower triangle, and in order to create an efficient data structure we must determine the positions of all the non-zeros in L. This involves some form of simulation of the factorization, and since the process does not need to involve the numerical values, it is often referred to as _symbolic_ factorization.

A detailed description of these algorithms is outside the scope of this paper, but it is possible to describe the important ideas in the algorithms, and the important associated results. One of the first significant complexity results in this area was to show that the structure of L can be determined in $O(|L|)$ time (Gustavson (1972), Rose et al (1975) and Sherman (1975)). One of the key observations in the development of the algorithm, and in its analysis, is illustrated in Figure 4.1. The crucial fact is that the structure of column i of L is simply the union of that of column i of A, along with those columns of L to the left of column i whose first non-zero below the diagonal is in position i[1]. Thus, in Figure 4.1, to determine the structure of column i, it is necessary only to examine column i of A, and columns k_2 and k_4 of L. Since columns k_2 and k_4 of L would not be examined when determining the structure of any other column, it is at least intuitively clear that the entire structure of L can be determined in $O(|L|)$ time.

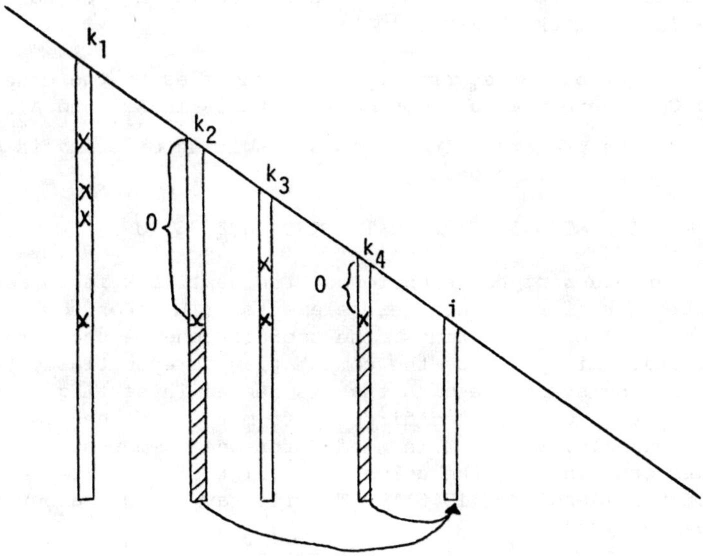

Figure 4.1 Illustration showing which columns of L determine the structure of column i.

As we mentioned in section 3, the storage required for L consists of two components, <u>primary</u> storage and <u>overhead</u> storage. Let R_L denote the overhead, consisting of pointers, subscripts, etc. needed to represent the structure of L and to access its elements. Thus, the total storage requirement for L is $|L| + |R_L|$.

Although it does not seem to be widely known, the state of the art for storing the triangular factor L is such that <u>usually</u>

---

1 It is left to the reader to convince himself/herself of the truth of this remark. As with many important results, the proof is relatively straightforward once the answer is known.

$|R_L|<|L|$, and often $|R_L| = o(|L|)$. In other words, less than one item of additional information is stored for each non-zero of L. This is achieved by using the fact that often sets of columns of L have similar structure; specifically, it is often true that columns j and k, j<k, have identical structure below position k. When this is true, we do not need to record the row subscripts of the non-zeros in column k if they are already recorded for column j.

Figure 4.2 contains an example showing how this fact is used to store a sparse L in an economical way. The diagonal elements of L are stored in an array DIAG. The off-diagonal non-zeros of L are contained column by column in the array LNZ, with the position of the beginning of the non-zeros of column i stored in XLNZ(i), i = 1,2,...,N. The row subscripts of the columns of L are stored in the array NZSUB in a similar manner, with the beginning position in NZSUB for the subscripts of column i given by XNZSUB(i). However, because of the observation in the previous paragraph, some subscripts in NZSUB serve "double duty" (or triple or ...) in those cases where the leading row subscripts of one column form a final subsequence of those of a previous column. This clever compressed storage scheme is due to Sherman (1975).

$$L = \begin{bmatrix} \ell_{11} & & & & & & \\ \ell_{21} & \ell_{22} & & & & & \\ & & \ell_{33} & & & & \\ \ell_{41} & \ell_{42} & & \ell_{44} & & & \\ & & \ell_{53} & \ell_{54} & \ell_{55} & & \\ & & \ell_{63} & & \ell_{65} & \ell_{66} & \\ & & & & \ell_{75} & \ell_{76} & \ell_{77} \end{bmatrix}$$

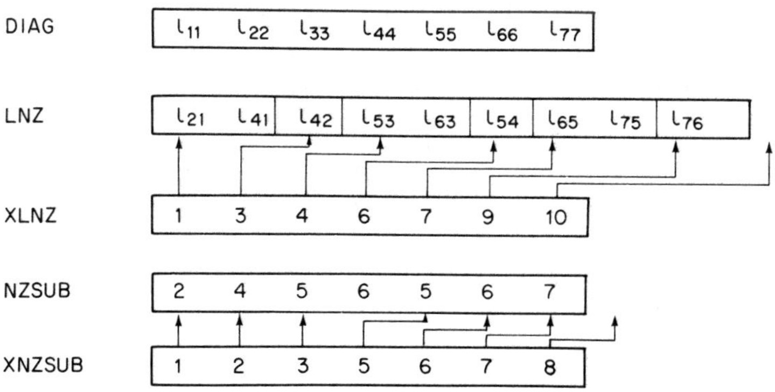

Figure 4.2 Compressed storage scheme

It has been argued that algorithms which compute the structure of L in $O(|L|)$ time are optimal. If the desired output of these algorithms is an explicit listing of the positions of the non-zeros in L, then their optimality is obvious. However, in practice, the objective of these algorithms is to construct a storage scheme (representation R_L) for L, and we have just seen that it can easily happen that $|R_L|<|L|$. See George and Liu (1980a) for an example where $|L| = O(|R_L|\log N)$. It seems reasonable to regard an algorithm for computing R_L as optimal only if its execution time is $O(|R_L|)$, regardless of $|L|$. In George and Liu (1980a) such an algorithm has been provided which generates a data structure similar to the one described above.

There is a very important practical problem associated with the computer implementation of these storage allocation algorithms. In general, we cannot predict how much storage will be required for R_L until after the algorithm has executed. That is, in Figure 4.2, we cannot predict the length of the array NZSUB. As we mentioned in section 3, such a characteristic is undesirable in mathematical software. Another disadvantage that some of these algorithms have is that they use their output during execution. That is, if insufficient space is available for R_L, it is not possible to continue executing the algorithm so that at least the number $|R_L|$ could be provided to the user. Fortunately, the $O(|R_L|)$ algorithm mentioned above does not use its output during execution, so $|R_L|$ can be provided even if there is not sufficient space for R_L.

Of course one could argue that we will need space for L and R_L anyway, if we are going to do the numerical computation, so the possibility of not having enough space for R_L during the data structure set-up phase of the computation is somewhat irrelevant. In some contexts this argument is valid, but not in others. Often the computation is segmented into phases as described in section 2, and each phase is executed as a separate job step which attempts to use the least storage possible. In any case, the user must still guess at $|R_L|$ or $|L| + |R_L|$, and he might guess low. If the algorithm simply aborts, without providing $|R_L|$, the execution has provided little useful information.

5. THROW-AWAY IDEAS

In this section we review some ideas which amount to simply discarding part of L during the factorization, and then either implicity or explicitly computing the discarded parts when they are needed later. We do not have space to explore all the uses of these ideas. We simply provide a few examples to illustrate the ideas.

Suppose we wish to solve $A\underline{x} = \underline{b}$, where A is ordered so that its bandwidth is m, and for convenience we suppose that for some integer k, $N = km$, and $k = O(\sqrt{N})$. We may regard A as a block tri-diagonal matrix as shown below, where the blocks are m by m.

$$A = \begin{pmatrix} A_1 & B_1 & & & \\ B_1^T & A_2 & B_2 & & \\ & B_2^T & A_3 & \ddots & \\ & & \ddots & \ddots & B_{k-1} \\ & & & B_{k-1}^T & A_k \end{pmatrix} \tag{5.1}$$

If we factor A into LL^T, where L is lower triangular, using Cholesky's method, we may again regard L as partitioned accordingly, as shown in (5.2).

$$L = \begin{pmatrix} L_1 & & & & \\ W_1^T & L_2 & & & \\ & W_2^T & L_3 & & \\ & & \ddots & \ddots & \\ & & & W_{k-1}^T & L_k \end{pmatrix} \tag{5.2}$$

Here the blocks are defined by

$$\bar{A}_1 = A_1 = L_1 L_1^T, \tag{5.3}$$

$$W_i = L_i^{-1} B_i, \; i = 1,2,\ldots,k-1, \tag{5.4}$$

and

$$\bar{A}_i = L_i L_i^T = A_i - B_{i-1}^T A_{i-1}^{-1} B_{i-1}, \tag{5.5}$$

$$= A_i - W_{i-1}^T W_{i-1}, \; i = 2,3,\ldots,k.$$

Band-orientated storage schemes usually treat the W_i and L_i as full lower triangular matrices (Note that the B_i are lower triangular). Thus, to store L we require $km^2 + O(N)$ storage locations.

From (5.5), it is clear that as far as the factorization is concerned, we do not need W_{i-1} after $\bar{A}_i$ is computed, so we simply discard it rather than save it. Instead, we retain only B_{i-1}; when we need to use W_{i-1}, we simply multiply by B_{i-1} and then solve the appropriate triangular system whose coefficient matrix is L_{i-1}. For example, to compute $\underline{z} = W_{i-1}\underline{y}$, we compute $\bar{y} = B_{i-1}\underline{y}$ and then solve $L_{i-1}\underline{z} = \underline{\bar{y}}$. Thus, we only store L_i, $i=1,2,\ldots,k$ and B_i, $i = 1,2,\ldots,k-1$, rather than L_i, $i=1,2,\ldots,k$ and W_i, $i = 1,2,\ldots,k-1$. This will often reduce storage requirements because in many situations the W_i are far less sparse than the B_i. For example, the total number of non-zeros in all the B_i may only be $O(N)$. If this is true, the storage requirement using this latter strategy is $\frac{1}{2}km^2 + O(N)$, rather than $km^2+O(N)$.

In connection with the computation (5.5), in some situations it may require fewer arithmetic operations to compute $\bar{A}_i$ as

$$A_i - B_{i-1}^T (L_{i-1}^{-T} (L_{i-1}^{-1} B_{i-1})) \text{ rather than as } A_i - (B_{i-1}^T L_{i-1}^{-T})(L_{i-1}^{-1} B_{i-1}),$$

which is implied by (5.5). (Consider the case when W_i is full lower triangular but B_{i-1} and L_{i-1} are very sparse.) Since we intend to discard W_i anyway, this "asymmetric" version of the computation usually requires less temporary storage than the first. That is, we can compute $\bar{A}_i$ one column at a time, discarding the column of $B^T_{i-1}A^{-1}_{i-1}B_{i-1}$ as soon as it has been used. On the other hand, the calculation of $\bar{A}_i$ as implied by (5.5) requires temporary storage for the whole matrix W_{i-1}.

It should be clear that the key to the usefulness of these ideas is the very simple form of the off-diagonal blocks of L. Ordering and partitioning algorithms which do not necessarily produce band orderings, but for which the off-diagonal blocks all have this simple form, have been developed (George and Liu (1978a)). In some contexts the computation and/or storage required is an order of magnitude less than that needed by the conventional band schemes.

We now review a slightly different throw-away idea. From (5.5) it is clear that as far as the factorization and forward substitution are concerned, once $\bar{A}_i$ and $\underline{y}_i$ are computed†, we could discard both W_{i-1} and L_i. Before discarding L_k we could solve $L^T_k\underline{x}_k = \underline{y}_k$, but we could proceed no further since the preceding blocks are no longer available. We now "start over", repeating the factorization only up to L_{k-1}, from which we can compute $\underline{x}_{k-1}$ and so on. The algorithm in "pseudo-Algol" is given below, and Figure 5.1 contains a schematic which illustrates which blocks and parts of vectors are involved in the computation during two different stages of the algorithm.

For $\ell = k, k-1, \ldots 1$ do the following:

A) Factor A_1, and if $\ell = k$, solve $L_1\underline{y}_1 = \underline{b}_1$

B) for $i = 2,3,\ldots,\ell$ do the following

B.1) Solve $L_{i-1}W_{i-1} = B_{i-1}$ and discard L_{i-1}.

B.2) Compute and factor $\bar{A}_i$ into L_i

B.3) If $\ell = k$, solve $L_i\underline{y}_i = \underline{b}_i - W^T_{i-1}\underline{y}_{i-1}$

B.4) If $i \neq \ell$, discard W_{i-1}.

C) Solve $L^T_\ell\underline{x}_\ell = \underline{y}_\ell$ and discard L_ℓ.

D) If $\ell \neq 1$, replace $\underline{y}_{\ell-1}$ by $\underline{y}_{\ell-1} - W_{\ell-1}\underline{x}_\ell$ and discard $W_{\ell-1}$.

---

†The vector $\underline{y}$ is assumed to be partitioned corresponding to A, and $\underline{y}_i$ denotes the i-th partition of $\underline{y}$.

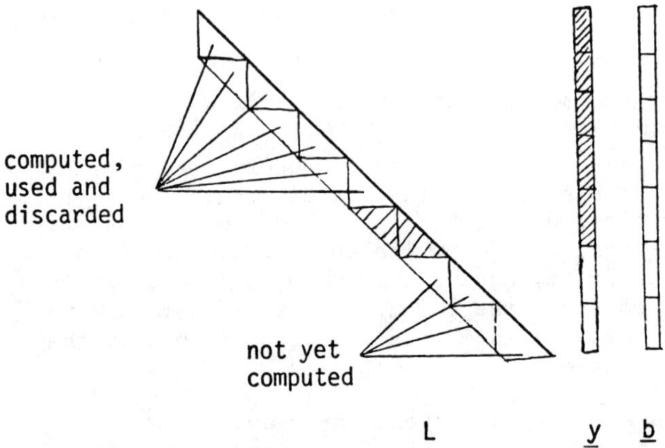

a) Situation when $\ell = k$ and $i = 5$. The first 5 parts of $\underline{y}$ have been computed.

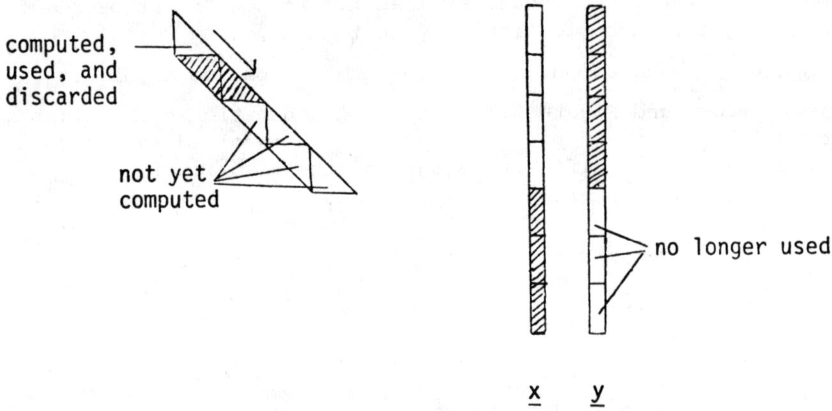

b) Situation when $\ell = 4$ and $i = 2$. The last 3 parts of $\underline{x}$ have been computed

Figure 5.1 Diagram showing which submatrices and vectors are involved in the computation, at two different stages of the algorithm. The variables i and ℓ refer to those in the pseudo-Algol description given above.

The point here is that the storage required is low, since only A, a few vectors of length N, and two or three blocks of L are required at any one time. For example, if we assume $|A|$ is O(N), in our example above the total storage required for the computation is O(N), compared to $O(N^{1.5})$ if the L_i are full lower triangular matrices, as they often are for band problems. Of course we pay a price in terms of computation, which will typically rise from $O(N^2)$ to $O(N^{2.5})$. However, in at least some contexts this tradeoff may be quite beneficial, particularly in a mini- or micro-computer environment where processor time may be inexpensive but storage is often severely limited.

In some contexts the tradeoff is even more attractive. For example, in Eisenstat et al (1976) it is shown that these ideas can be used in conjunction with nested dissection orderings of n by n grid problems ($N = n^2$) so that the storage for the solution is reduced from O(N logN) to O(N), but the optimal arithmetic operation count of $O(N^{3/2})$ is maintained. The operation count is roughly doubled, but the storage required drops by a factor of 1/log N.

Our brief discussion here of these ideas does not do them justice. For a detailed description of the numerous applications of these ideas, the reader is urged to consult Eisenstat et al (1976), Eisenstat et al (1978), George and Liu (1978a), or Sherman (1975) and the references contained therein.

6. CONCLUDING REMARKS

In this short article we have discussed a few of the many important ideas and problems related to the solution of sparse linear systems. Some of these problems are unique to symmetric positive definite systems, but similar or identical problems arise in general problems as well.

The problem of finding an ordering for which L has a minimal or nearly minimal number of non-zeros remains an interesting open theoretical problem. However, because the minimum degree algorithm can be implemented so well, and the orderings produced are almost always satisfactory, the absence of theoretical underpinnings appears to be a relatively unimportant practical issue. Some researchers speculate that the complexity of actually finding a best ordering is so difficult that we will continue to use heuristics anyway.

The storage allocation problem discussed in section 4 is in relatively good shape. The storage schemes that have been developed, and the symbolic factorization algorithms for constructing them, are efficient in practice, and have provable execution time bounds.

It seems fair to say that the "jury is still out" with regard to the throw-away ideas discussed in section 5. In theory, they are very effective, and it is relatively straightforward to develop new and novel methods which represent improvements over existing methods. However, implementing these methods on a computer is a challenge, and the actual savings realized in practice usually fall far short of those predicted by the theory. Whether or not net savings are realized is very much problem dependent. Another important point is that the estimates for storage and operation counts usually have large low order terms, so that methods which are asymptotically inferior to others may remain competitive or superior in practice for very large values of N.

REFERENCES

Crane, H.L., Gibbs, N.E., Poole Jr., W.G. and Stockmeyer, P.K. (1976) Algorithm 508: Matrix bandwidth minimization and profile reduction, TOMS 2, pp. 375-377.

Cuthill, E. and McKee, J. (1969) Reducing the bandwidth of sparse symmetric matrices, Proc. 24th Nat. Conf. Assoc. Comput. Mach., Brandon Systems Press, pp. 157-172.

Cuthill, E. (1972) Several strategies for reducing the bandwidth of matrices, in Sparse Matrices and their Applications (D.J. Rose and R.A. Willoughby eds.), Plenum Press, pp. 157-166.

Duff, I.S., Erisman, A.M. and Reid, J.K. (1976) On George's nested dissection method SIAM J. Numer. Anal. 13, pp. 686-695.

Duff, I.S. and Reid, J.K. (1974) A comparison of sparsity orderings for obtaining a pivotal sequence in Gaussian elimination, J. Inst. Math. Applics. 14, pp. 281-291.

Eisenstat, S.C., Schultz, M.H. and Sherman, A.H. (1976) Applications of an element model for Gaussian elimination, in Sparse Matrix Computations, (J.R. Bunch and D.J. Rose eds.), Academic Press, pp. 85-96.

Eisenstat, S.C., Schultz, M.H. and Sherman, A.H. (1978) Software for sparse Gaussian elimination with limited core storage, in Sparse Matrix Proceedings 1978, (I.S. Duff and G.W. Stewart eds.), SIAM Publications pp. 135-153.

Gibbs, N.E., Poole Jr., W.G. and Stockmeyer, P.K. (1976) An algorithm for reducing the bandwidth and profile of a sparse matrix, SIAM J. Numer. Anal. 13, pp. 236-250.

George, A. (1973) Nested dissection of a regular finite element mesh, SIAM J. Numer. Anal. 10, pp. 345-363.

George, A. (1974) On block elimination for sparse linear systems, SIAM J. Numer. Anal. 11, pp. 585-603.

George, A. and Liu, J.W.H. (1978a) Algorithms for matrix partitioning and the numerical solution of finite element systems, SIAM J. Numer. Anal. 15, pp. 297-327.

George, A. and Liu, J.W.H. (1978b) An automatic nested dissection algorithm for irregular finite element problems, SIAM J. Numer. Anal. 15, pp. 1053-1069.

George, A. and Liu, J.W.H. (1979) An implementation of a pseudo-peripheral node finder, TOMS 5, pp. 284-295.

George, A. and Liu J.W.H. (1980a) An optimal algorithm for symbolic factorization of symmetric matrices, SIAM J. Comput. 9, pp. 583-593.

George, A. and Liu J.W.H. (1980b) A fast implementation of the minimum degree algorithm using quotient graphs, TOMS 6, pp. 337-358.

Gustavson, F.G. (1972) Some basic techniques for solving sparse systems of equations, in Sparse Matrices and their Applications, (D.J. Rose and R.A. Willoughby eds.) Plenum Press, pp. 41-52.

Jennings, A. (1966) A compact storage scheme for the solution of simultaneous equations, Comput. J. 9, pp. 281-285.

Lipton, R.J., Rose, D.J. and Tarjan, R.E. (1979) Generalized nested dissection, SIAM J. Numer. Anal. 16, pp. 346-358.

Liu, J.W.H. and Sherman A.H. (1975) Comparative analysis of the Cuthill-McKee and reverse Cuthill-McKee ordering algorithms for sparse matrices, SIAM J. Numer. Anal. 13, pp. 198-213.

Markowitz, H.M. (1957) The elimination form of the inverse and its application to linear programming, Management Science 3, pp. 255-269.

Papadimitriou, Ch. H. (1976) The NP-completeness of the bandwidth minimization problem, Computing 16, pp. 263-270.

Rose, D.J., Tarjan, R.E. and Lueker, G.S. (1975) Algorithmic aspects of vertex elimination on graphs, SIAM J. Comput. 5, pp. 266-283.

Rose, D.J. and Tarjan, R.E. (1978) Algorithmic aspects of vertex elimination on directed graphs, SIAM J. Appl. Math. 7, pp. 176-197.

Sherman, A.H. (1975) On the efficient solution of sparse systems of linear and non-linear equations, Rept. No. 46 (Doctoral Thesis) Dept. of Computer Science, Yale University.

DISCUSSION

PROFESSOR J.M. McNAMEE (York, Ontario). Would it be fair to say that minimum degree is the best overall ordering strategy?

GEORGE. A year or so ago I would have argued that, for finite element types of problems, automated nested dissection was the best algorithm. However, the recent fast low-storage minimum degree algorithms make the situation less clear.

DR. A.M. ERISMAN (Boeing). It has frequently been observed that the tie-breaking stragey used with a minimum degree algorithm has a significant influence on the quality of the ordering produced. Do you have any comments on this?

GEORGE. I don't think tie-breaking strategies have very much practical importance. I think their importance is largely theoretical in as much as a particular choice of tie-breaking strategy might allow us to prove optimality results about the algorithm.

DR. J.K. REID (Harwell). Could you comment more on the likely asymptotic behaviour of the execution times of the fast ordering algorithms on practical problems?

GEORGE. For finite element problems, if the cliques that are encountered during the elimination process are "round", then the execution time of a good implementation of the minimum degree ordering algorithm can be shown to be linear in the number of variables. Liu and I have found this linearity to hold in practice for the ordering schemes references in my paper.

CONTRIBUTED PAPERS

The following papers have been submitted by the presenters of poster sessions.

THE AQ ALGORITHM

R.E. Borland

(Division of Numerical Analysis and Computer Science, National Physical Laboratory, Middlesex)

The AQ algorithm is a direct method for the solution of linear equations. It merits particular consideration when the system of equations is large and sparse and the matrix of coefficients, A, can be reordered so that it is near to lower Hessenberg form. Figure 1 illustrates the type of application for which one should consider the AQ algorithm.

The incursion, h(i), of the ith row of the matrix is defined as the minimum value of h for which

$$A_{ij} = 0 \quad \text{for all } j > i + h(i). \qquad (1)$$

The matrix is reordered to minimise h throughout the matrix ensuring that

$$h(i+1) \geq h(i) - 1.$$

It is a little difficult to be precise about the exact quantity which we seek to minimise in the reordering. As a first approximation the objective function is somewhere between the average value of h and the average value of h^2 but extremely low values of h are particularly useful even if these occur only intermittently.

This particular reordering is not as arbitrary as it might seem. It is fundamental for any direct method of solution of equations. Each equation can be used to eliminate one unknown but at each stage in the process one is left with a number of unknowns which must be handled in some form. The number of these unknowns is the incursion. If the incursion cannot be kept relatively small it is not clear, to my knowledge, how the sparsity can be fully exploited with a direct method and iterative methods seem to offer the main hope of exploiting the structure as in partial differential equations in three or more dimensions. In two dimensional partial differential equations for a square

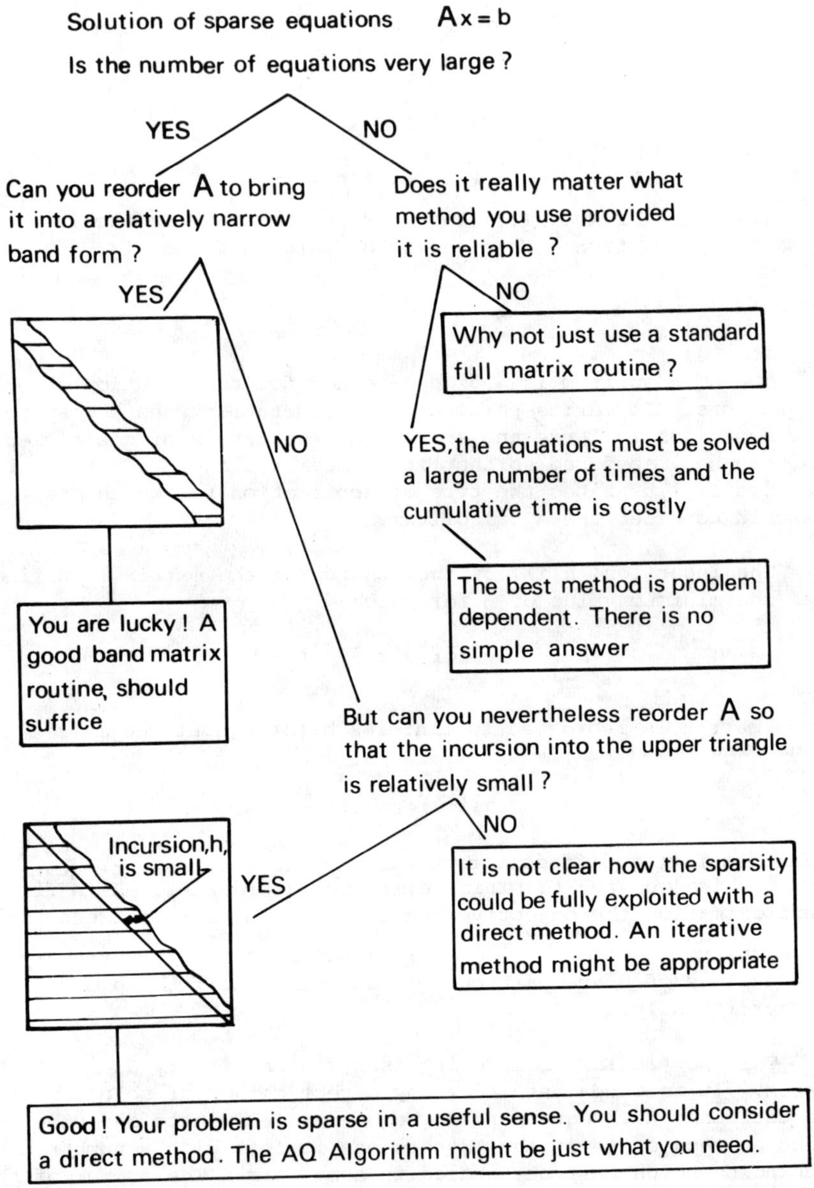

Figure 1 The decision tree

region we have a borderline case where direct methods can exploit the sparsity to some extent, but not fully, and are competitive with iterative methods.

It is not possible in this short paper to explain clearly all aspects of the AQ algorithm but we sketch below the main features. A full account of the AQ algorithm is given by Borland (1979).

The AQ algorithm is based on the orthogonal factorisation method for the solution of equations. The equations

$$A\underline{x} = \underline{b} \tag{2}$$

become

$$LQ\underline{x} = \underline{b} \tag{3}$$

using the factorisation of A into the form

$$A = LQ \tag{4}$$

where L is lower triangular and Q is orthogonal. Using the transformation $\underline{y} = Qx$, equation (3) becomes

$$L\underline{y} = \underline{b} \tag{5}$$

which can be solved by forward substitution. The solution $\underline{x}$ is then given by

$$\underline{x} = Q^T\underline{y}$$

where the ^{T} denotes the transpose.

However in the AQ algorithm the lower triangular matrix, L, is neither calculated nor used except for a small part near the diagonal. Instead we represent L by the form

$$L = AQ^T$$

obtained by postmultiplying equation (4) by Q^T.

It is because L is used in this form that the name "AQ" has been used. The transpose of an orthogonal matrix is itself orthogonal so it was not felt necessary to include the word "transposed" in the name.

Although in general Q is dense it has a sparse representation when the incursion, h, is small for all rows of A. It can then be specified by a set of small matrices of order h x h. In these circumstances $Q^T\underline{y}$ can be efficiently calculated. However, in the forward substitution process $Q^T\underline{y}$ must be calculated repeatedly as successive elements of $\underline{y}$ are computed. The work required to do this is minimised in two ways:

(1) A small part of L near the diagonal is calculated and used in the conventional way. If

$$\underline{y} = \underline{y}_1 + \underline{y}_2$$

Then

$$L\underline{y} = L\underline{y}_1 + L\underline{y}_2$$

$$= AQ^T\underline{y}_1 + L\underline{y}_2 .$$

In this context $\underline{y}_2$ contains a set of the most recently calculated elements and $\underline{y}_1$ the previously calculated elements. At intervals $Q^T\underline{y}$ is updated and the part of L that is used consists of a set of triangular portions near the diagonal.

(2) Instead of calculating $Q^T\underline{y}$ directly it is represented in a segmented form, part of which is a directly calculated contribution to the solution $\underline{x}$ and part of which is an indirect contribution. The segments are joined at intervals in such a way as to minimise the total work required. A similar segmentation is used in the factorisation process. For a full account of this see Borland (1979).

The work required for the factorisation and solution stages can be estimated after the matrix, A, has been reordered. If we take the case of N equations with a constant incursion, h, and a total number of non-zero elements, τ, then the number of multiplications required is:-

Factorisation stage	$h^2N \log_2(N/h) + 8h^2N + h\tau$
Solution for one R.H.S.	$(3h/2)N \log_2(N/h) + \tau$

A rough rule of thumb is that the number of multiplications required is

Factorisation	$18Nh^2$
Solution	$15Nh$

This is based on the assumption that τ is small compared to 8Nh and that a value of 10 is not too far wrong for $\log_2(N/h)$ over a wide range of practically occurring large sparse matrices.

COMMENTS

The method has the stability inherent in the traditional orthogonalisation method described above (equations 2 to 5). Consequently the size of pivots is irrelevant. It does not suffer from growing fill-in as $N \to \infty$ and for this reason should be more efficient than sparse LU decomposition for some problems when N is extremely large. The AQ algorithm should be ideally suited to network distribution problems (electricity? gas? water? communications?) in which many large conurbations with dense networks are connected by a relatively small number of main distribution lines. In such cases the incursion intermittently reduces to low values which the AQ algorithm can exploit.

Initial zeros in the lower triangular part of A can also be exploited, though a variant of the general algorithm is necessary to achieve this. For example the AQ algorithm could be useful for variable band problems particularly where lack of symmetry and pivotal problems cause difficulties with other methods.

REFERENCES

Borland, R.E. (1979), The AQ algorithm for the solution of a sparse system of linear equations, Report DNACS 17/79, National Physical Laboratory, Teddington, England.

A BRANCHING ENVELOPE REDUCING ALGORITHM FOR FINITE ELEMENT MESHES

N.G. Brown and R. Wait

(University of Liverpool)

1. INTRODUCTION

The problem considered in this note is that of solving the system of linear equations

$$A\underline{x} = \underline{b}$$

where A is a sparse N×N positive definite matrix arising in the application of finite element methods. The method of solution is standard; A is factored into the product LL^T, where L is lower triangular, and the triangular systems

$$L\underline{y} = \underline{b}$$

$$L^T\underline{x} = \underline{y}$$

are solved to obtain $\underline{x}$.

When the matrix A is factored it usually suffers fill-in; that is $L + L^T$ is usually more dense than A. Since PAP^T is positive definite for any N×N permutation matrix P, the above method can be used to solve the equivalent system

$$(PAP^T)(P\underline{x}) = P\underline{b}.$$

It is well known that a judicious choice of P can often drastically reduce fill-in and/or arithmetic requirements.

If only the envelope of the reordered matrix is to be stored for the factorization/solution of the system then an envelope reducing algorithm should be used to provide the new ordering. A well known algorithm which provides such an ordering is the reverse Cuthill-McKee (R.C.M.) scheme (for example, Cuthill and McKee (1969), Liu and Sherman (1975)). Although R.C.M. performs well on many finite element problems there are situations which it does not deal with satisfactorily.

The algorithm given in this note reproduces the R.C.M. ordering if the finite element grid does not contain corner nodes, singular nodes, holes or branches. If the grid does have some (or all) of these features the algorithm provides a significant improvement in terms of the amount of storage required for the envelope. In section 2 some simple examples are given which provide the motivation for the ordering algorithm discussed in section 3. Section 4 contains some numerical experiments with the algorithm and some concluding remarks are given in section 5.

2. MOTIVATION

It will not be possible to give a detailed discussion of the ideas behind the proposed algorithm but the two simple examples given here should explain why corner-nodes and singular-nodes are treated as special cases.

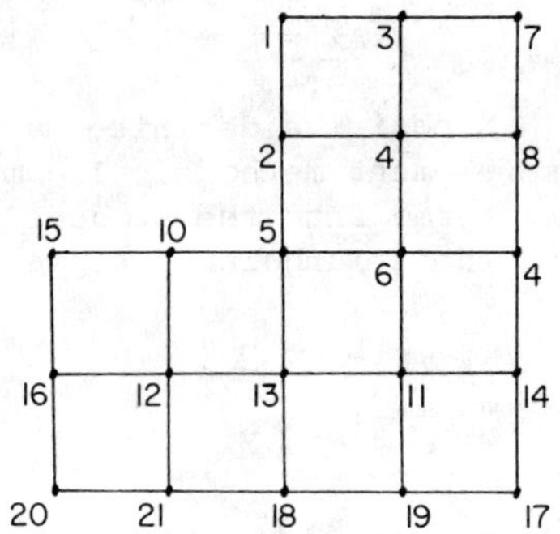

(a) Cuthill - McKee Ordering

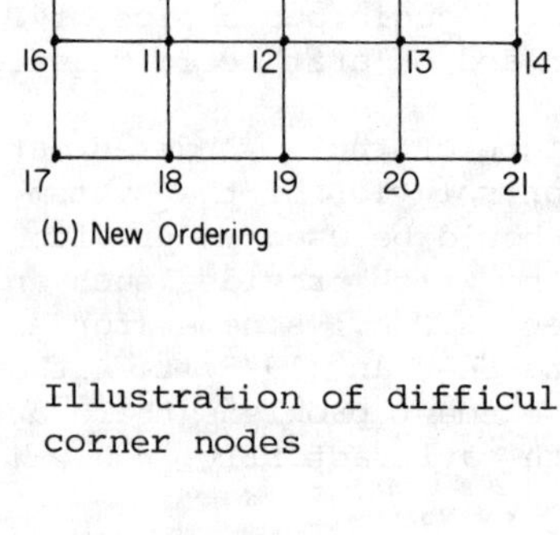

(b) New Ordering

Figure 1 Illustration of difficulties with corner nodes

A node is said to be singular if the unordered set of nodes associated with it in the next level set contains more than one vertex node. It should be noted that corner-nodes are particular examples of singular-nodes. Although this definition of singular nodes will automatically include the first vertex at each level, this will often be an edge node which must be checked anyway to see if it is a corner node. The edge nodes which are corner nodes are processed almost without penalty.

The term oscillation is used to indicate that the order in which the sets of nodes are taken, on a level, has more than one discontinuity.

In the simple example given here it is possible to overcome the difficulties encountered with the Cuthill-McKee algorithm by using a particular tie-breaking strategy. In slightly more complex situations, for example if Figure 1 was composed of 8-node rectangular elements, such a strategy would not be sufficient.

The problem associated with corner-nodes is shown in Figure 1(a) which shows an ordering obtained by the Cuthill-McKee algorithm. When ordering the neighbours of node 5 the nodes ordered 11, 12, 13 can be taken in any order, this causes an oscillation in the ordering of the rest of the grid. The ordering shown in Figure 1(b) is obtained by the new algorithm and is clearly superior.

The grid shown in Figure 2 is an example containing two singular-nodes (4, 6). The Cuthill-McKee algorithm again runs into difficulties, Figure 2(a), this time owing to the freedom allowed when ordering nodes 8, 9, 10, 11 which again leads to oscillations. No oscillations occur in the ordering given by the new algorithm, Figure 2(b).

That oscillations have an adverse effect on the storage requirements can be seen from the results given in section 4. Holes and branches are dealt with by tearing and examples showing the need for such a strategy can be found in George and Liu (1978).

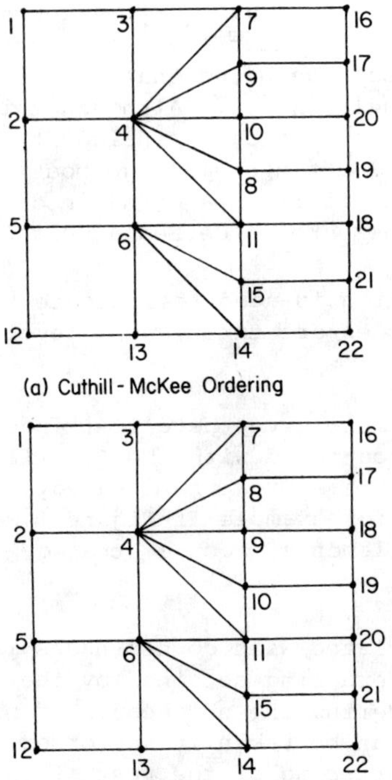

(a) Cuthill-McKee Ordering

(b) New Ordering

Figure 2 Illustration of effect of singular nodes

3. THE ORDERING ALGORITHM

The R.C.M. algorithm can be stated as follows.

R.C.M.

Find a suitable starting node x (for example, a pseudo-peripheral node).

$X \leftarrow \{x\}$
$S \leftarrow \{x\}$
While $|X| > 0$ do

begin

$$\left.\begin{array}{l} \text{For each } x_i \in X, \text{ do} \\ \\ \qquad Y_i \leftarrow \{y \in \text{adj}(x_i) \setminus (S \cup (\bigcup_{k=1}^{i-1} Y_k))\} \end{array}\right\} A$$

```
    For i = 1, ..., |X| do                          }
    Order members of Y_i by increasing degree       } B
    r ← |X|
    X ← {φ}

    For i = 1, ..., r do
        S ← S U Y_i
        X ← X U Y_i
end
```

Reverse the order of S.

<u>New Algorithm</u>

The proposed algorithm makes use of the following observations:

(i) Step B above is unnecessary if, as is usual after the first step at each level, Y_i contains only one vertex node.

(ii) If Y_i contains more than one vertex node step B is not, in general, sufficient to prevent oscillation in the ordering, as shown in section 2.

As a first step towards the new algorithm, step B above is replaced by: "If Y_i contains more than one vertex node then order Y_1 in a manner 'consistent' with the ordering accepted so far".

A detailed description of this step is not presented here but it should be noted that such a 'consistent' ordering will not cause oscillations. The test for branches and holes involves testing the Y's to see if they are connected. To facilitate this it is necessary to modify step A as follows:

$i \leftarrow 0$

For each $x_j \in X$ do

$$Z \leftarrow \{y \in \text{adj}(x_j) \setminus (S \cup (\bigcup_{k=1}^{i-1} Y_k))\}$$

If $Z \neq \{\phi\}$ then $i \leftarrow i+1$; $Y_i \leftarrow Z$.

This gives a set of non-empty Y's. Thus, given that no oscillations have occurred in the ordering so far, the test for the connected subset of Y's, containing at least a connected subset of Y_1, can be written as:

SWITCH ← 1

LIMIT ← i

For j = 1 to i - 1 while Y_j connected to Y_{j+1} do

SWITCH ← SWITCH +1.

If SWITCH ≠ LIMIT and Y_{SWITCH} connected to Y_{LIMIT} then

LIMIT ← LIMIT - 1.

For j = i to SWITCH + 2 by -1 while Y_j connected to Y_{j-1} do

LIMIT ← LIMIT - 1.

The set of Y's is connected if SWITCH = LIMIT.
If SWITCH ≠ LIMIT then the Y_k, k = SWITCH + 1, ..., LIMIT are not connected to the Y_j, j = 1, ..., SWITCH, LIMIT + 1, ..., i. The test for connectedness between Y_a and Y_b is complicated by the fact that either or both may be disconnected. If such a disconnection is found the set is modified to contain the connected subset connected to the previously accepted set of Y's. The remaining entries being placed in a new set Y_{i+1}. The entries in the non-connected Y's are further modified in that entries within each Y_k which have at least one neighbour which has not been ordered and is not in the current set of Y's, are removed and placed in a tearing set.

An outline of the steps involved in the proposed algorithm is now given.

N: an integer value giving the number of nodes to be ordered.
S: an array used to store the new ordering.
STAG, HSTAG: integer pointers associated with S.
STACK: a list of sets of integers.
STACKTAG: integer pointer associated with STACK.
X, Y_i, Z: sets of integers.

1. Find a suitable starting node x (for example, a pseudo-peripheral node).
2. X ← {x}
   S[STAG ← 1] ← x
   HSTAG ← N
   STACKTAG ← 1

3. If STAG = HSTAG go to 11
4. $i \leftarrow 0$
   For each $x_j \in X$ do

$$Z \leftarrow \{z \in adj(x_j) \setminus (S \cup (\bigcup_{k=1}^{i-1} Y_k))\}$$

   If $Z \neq \{\phi\}$ then

      $i \leftarrow i+1$;

      $Y_i \leftarrow Z$.

5. If $i = 0$ go to 10
6. Test the set of Y's for connectedness, forming a minimal tearing set Z if appropriate (as indicated above).

   If $Z \neq \{\phi\}$ then

      STACK[STACKTAG ← STACKTAG + 1] ← Z;
      For j to $|Z|$ do
         S[HSTAG] ← z_j;
         HSTAG ← HSTAG 1.
7. For j = 1 to i do
      If Y_j contains more than one vertex node then reorder the contents of Y_d in such a way as to prevent oscillations.
8. For j = 1 to i do
      For k = 1 to $|Y_j|$ do
         S[STAG ← STAG + 1] ← Y_{j_k} .
9. $X \leftarrow \{\phi\}$.
   For j = 1 to i do
     $X \leftarrow X \cup Y_j$.
     go to 3
10. X ← STACK[STACKTAG]
    STACKTAG ← STACKTAG - 1
    go to 4
11. Reverse the order of S[1 : STAG].
    The reordering is now complete with the torn nodes, if any, in S[STAG +1: $|S|$].

Although the above algorithm is more complicated than R.C.M. it should be appreciated that the more complex steps are only executed when problem nodes occur and, although a test for connectedness is required between the sets Y_i, the time spent checking for problem nodes is acceptable.

In the subsequent tables of numerical results this new

TABLE I

Statistics for R.C.M. and ALBERT on some test examples

TEST EXAMPLE	ORDER TIME		NO. OF NODES	PRIMARY STORAGE		OVERHEAD STORAGE	
	RCM	ALBERT	REMOVED BY ALBERT	RCM	ALBERT	RCM	ALBERT
a.1	0.53	2.0	0	25553	25553	1090	1092
a.2	1.20	10.9	0	23800	22870	962	964
b.1	0.64	2.3	0	26078	26077	1010	1012
b.2	3.40	12.5	0	28005	28065	1010	1012
c.1	0.70	3.1	48	25982	14320	1181	1402
c.2	1.70	5.7	84	19858	11146	946	1397
d.1	1.70	3.3	43	21682	14634	1378	1574
d.2	1.00	6.8	64	10993	7615	806	1144
e.1	0.45	2.1	26	22753	15444	937	1065
e.2	1.40	9.6	48	24797	15937	937	1219
f.1	0.70	3.3	48	28218	17442	1441	1658
f.2	1.40	6.1	84	22200	13290	1153	1597

algorithm is referred to by the acronym ALBERT - ALgorithm for Branching, Envelope Reduction and Tearing.

4. NUMERICAL EXPERIMENTS

A comprehensive set of results will not be presented here. The results in Tables I and II are to be compared with the results in George and Liu (1978), this reference should be consulted for the full description of the test problems.

TABLE II

Storage requirements for algorithms

TEST EXAMPLE	TOTAL STORAGE			
	RCM	RQT	RH-RQT	ALBERT
a.1	26643	20152	20152	26645
a.2	24762	19363	19363	23834
b.1	27088	21788	21788	27089
b.2	29015	22259	22259	29077
c.1	27163	15646	15646	15722
c.2	20804	13927	13927	12543
d.1	23060	17147	17147	16208
d.2	11799	10726	10726	8759
e.1	23690	17675	13769	16509
e.2	25734	19903	15889	17156
f.1	29659	23766	19195	19200
f.2	23353	20401	17197	14887

5. CONCLUDING REMARKS

The results presented here are only the preliminary findings of a continuing programme.

However, they do show that there is a significant improvement to be made in the storage requirements even in fairly simple problems. It has been found that with more complex grids the savings are even larger. The main advantage of the proposed algorithm over other tearing algorithms is that tearing only takes place when the graph, associated with the grid, branches thus the amount of extra work required during factorization is minimised.

REFERENCES

Cuthill, E. and McKee, J. (1969) Reducing the bandwidth of sparse symmetric matrices, Proc. 24th Nat. Conf. Assoc. Comput. Mach., Brandon Systems Press, pp. 157-172.

George, A. and Liu, J.W.H. (1978) Algorithms for matrix partitioning and the numerical solution of finite element systems, SIAM J. Numer. Anal. 15, pp. 297-327.

Liu, J.W.H. and Sherman, A.H. (1975) Comparative analysis of the Cuthill-McKee and the reverse Cuthill-McKee ordering algorithms for sparse matrices, SIAM J. Numer. Anal. 13, pp. 198-213.

SIMULTANEOUS ITERATION APPLIED TO THE SOLUTION OF FINITE ELEMENT MODELS IN SHIP VIBRATION

D. Catley

(The British Ship Research Association, Wallsend Research Station, Wallsend, Tyne and Wear, NE28 6UY)

1. BRIEF INTRODUCTION TO FINITE ELEMENT MODELS

The finite element method is one way in which a complex structure may be represented by a mathematical model. In a dynamic analysis of a structure, the model should represent the salient structural vibration for the frequency range of interest. In formulating a finite element model of the real structure, the infinite number of degrees of freedom, that is displacements and rotations, are approximated to by an assemblage of discrete parts (that is the individual finite elements), each with a finite total number of degrees of freedom at the grid points of the model. The elements are connected to each other at only a finite number of points (usually at the vertices of the elements) and these are the grid points of the model. The elements may be one, two or three-dimensional.

The idealisation, that is the description of the structure in terms of a mesh of finite elements, may be carried out in any number of different ways (see Figure 1) depending on such factors as the type of problem, the accuracy of the solution required and the time and money available. A coarse idealisation involving a small number of large elements would provide a comparatively rapid but approximate solution, whereas a finer idealisation, although using the same element types, would produce more accurate results, would take longer and would consequently cost more. Graded meshes can be adopted to some advantage by using small elements in regions where high detail is required and larger elements elsewhere.

Models in ship vibration require a representation of the mass distribution which, in addition to the mass of the structure, will include the non-structural mass of the lightship, the cargo and the added mass effect of the entrained water. One method of representing these masses is to lump them to suitable local grid points of the model. This results in a diagonal mass matrix requiring less storage than a distributed mass matrix which would

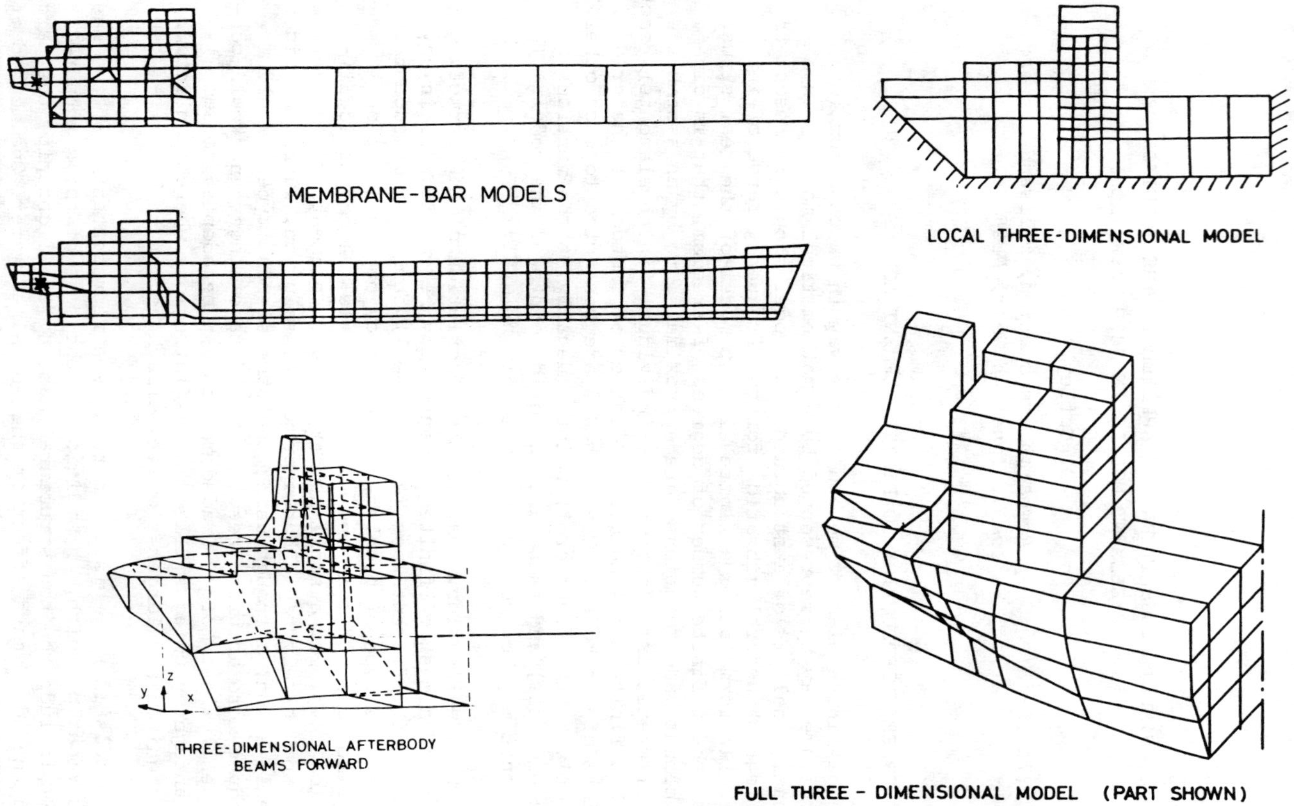

Figure 1 Examples of Finite Element Model Types

be of similar banded form to the stiffness matrix. The lumped mass model also requires less computational effort and a tendency to give lower frequencies. This tendency compensates for the degree of over stiffness in the finite element models used and produces a better correlation with measured frequencies.

A system of dynamic equations for the free vibration of the ship model is then defined using matrix techniques. The matrix eigenproblem is solved to obtain the free, undamped vibration modes and their corresponding natural frequencies. Using this natural mode model, one may then calculate the vibration response of the ship to a general system of forces at prescribed locations. In the calculation of response amplitudes it is necessary to take damping into account in order to obtain the correct effect of combining the calculated natural modes of the structure.

The BSRA system, VAS, has been written in such a way as to permit perturbations of this basic model to be considered in an efficient manner. As detailed below, the basis for the solution procedure is an adaptation of simultaneous iteration. In particular, this scheme is suitable for design studies. Using disc files, structural or inertial modifications are specified to form a revised model. The solution time for similar and subsequent analyses may thus be much less than the effort for the basic model.

2. SOLUTION OF THE EIGENVALUE PROBLEM

The equations of free vibration lead to the generalised eigenvalue problem:-

$$(M - \lambda K)\underline{q} = \underline{0}, \qquad (1)$$

where λ is an eigenvalue solution. If there are n degrees of freedom, there are n possible eigenvalues corresponding to the n natural frequencies, that is

$$\omega = \frac{1}{\sqrt{\lambda}}.$$

The n associated eigenvectors $\underline{q}$ are the corresponding vibration modes. In most dynamic analyses, only the lowest k frequencies are required. These correspond to the k most dominant eigenvalues of the system.

Simultaneous iteration takes advantage of both symmetry and bandform of the mass and stiffness matrices and enables eigenvectors of close or equal eigenvalues to be resolved. The problem is reduced to standard form by means of a Cholesky factorisation of the symmetric, positive definite matrix K into its upper and lower triangular factors, $K = LL^T$, to give the standard form:-

$$(A - \lambda I)\underline{q}_1 = \underline{0} \tag{2}$$

L may overwrite K in the computer storage and

$$\underline{q}_1 = L^T\underline{q}, \; A = L^{-1}M\,L^{-T}$$

This method continues with the definition of an orthogonal matrix of k + 3 starting vectors, U. The three additional vectors accelerate convergence of the higher modes. To ensure that a mode is not missed, a random vector is generated at each iteration. Apart from the random vector, the matrix U may be a matrix of zeros and ones; it may be randomly generated, or in a design study reasonable approximations to the true vectors may be available from some previous and similar analysis in which case convergence is usually obtained in a few iterations. These vectors are orthogonalised initially, and at each subsequent iteration, to maintain linear independence. The results of an interactive analysis are utilised to uncouple the vectors. The matrix A is never explicitly formed and thus there is no need to work with a full matrix. Convergence of the predominant vectors is rapid for large banded forms.

3. THE MOBILITY APPROACH FOR VIBRATION RESPONSE

The natural modes of free undamped vibration obtained above may be used as generalised freedoms to solve the problem of forced vibration using an estimate for the modal damping. The complete problem of defining and estimating the dynamic response of a general ship structure is extremely complex due in part to the complex excitation system. A technique which circumvents the requirements for an estimate of the exciting forces in determining the vibration characteristics of the ship is the mobility technique. This involves the calculation and/or measurement of the mobility, that is the response due to a unit force of varying frequency at a single point of excitation. When estimates of the actual force components are available, it is then a simple task to calculate the response of the ship at a general frequency of excitation. Many full-scale measurements of mobility have been carried out by BSRA and compared with calculations. An example is illustrated in Figures 2 and 3.

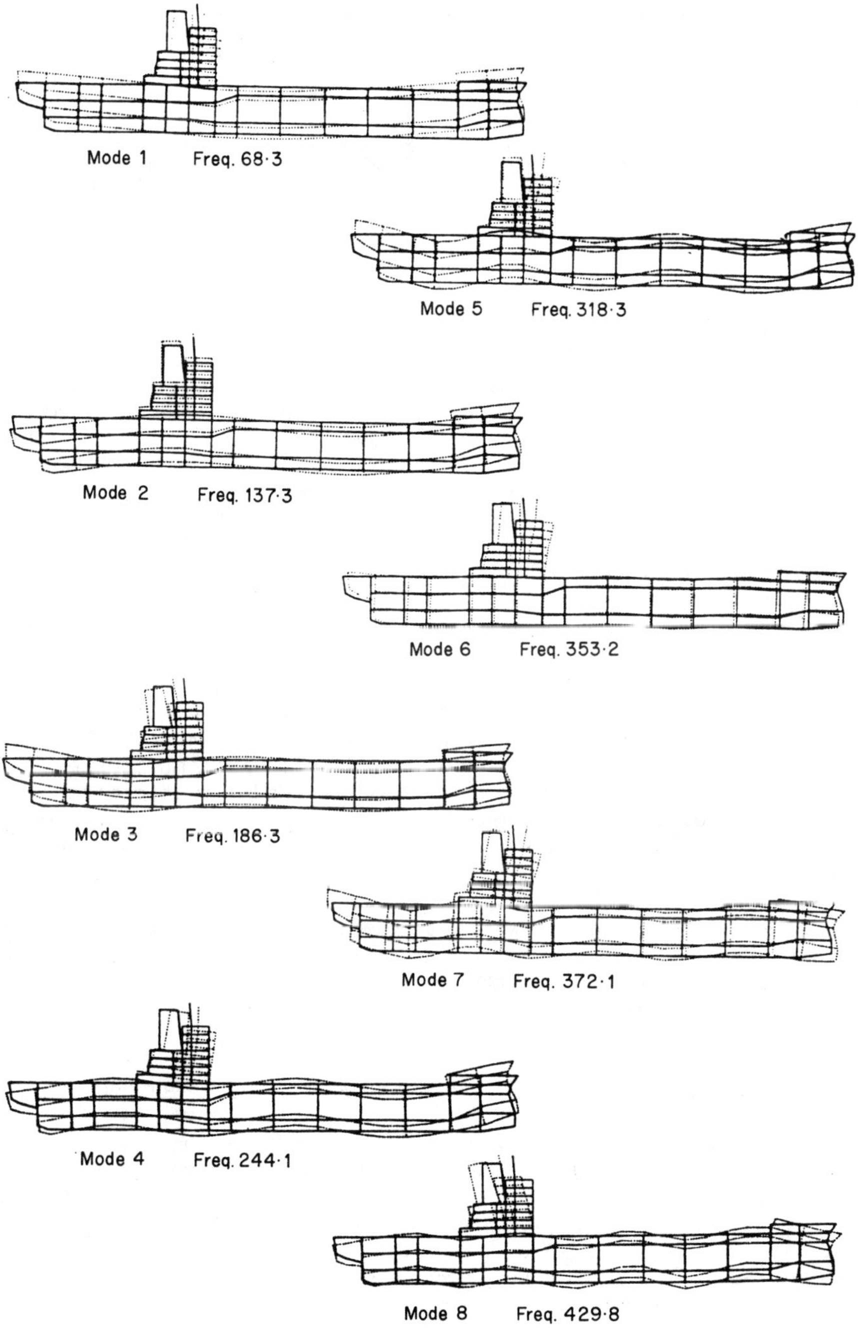

Figure 2 Calculated natural mode shapes. Frequencies in cycles/ min.

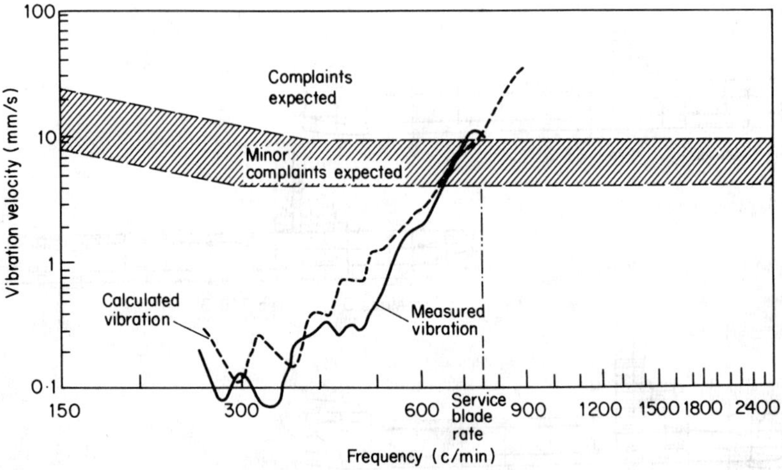

Figure 3 Calculated Vertical Propeller Excited Vibration

4. SUBSTRUCTING TECHNIQUES

As an alternative to the solution of a single large problem, there are two approaches which seek to reduce the problem size, whilst maintaining a degree of accuracy which is consistent with the larger model. The first is the so-called super-element approach which reduces a detailed definition of part of the structure (for example a deckhouse) to an equivalent smaller problem using selected master grid points and a static condensation procedure. This thus results in a presolution for each super-element and makes the computation more manageable. Results are sensitive to the size of the super-elements and the choice of master grid points. The second approach is the Component Mode Synthesis method in which complete branches of the structure are replaced by a suitable number of generalised freedoms of the substructure, referred to the boundary of the substructure with the rest of the model.

The second approach has been adopted and simultaneous iteration used to solve for the modes of each substructure. The adopted substructuring technique in VAS defines so-called attachment modes, involving systematic relaxation of each boundary freedom in turn, and substructure modes where the boundaries are fully clamped (Figure 4). Assembly of the model using an extended matrix scheme requires compatibility of displacements to be satisfied at the common boundaries and thence the solution of a much reduced eigenvalue problem.

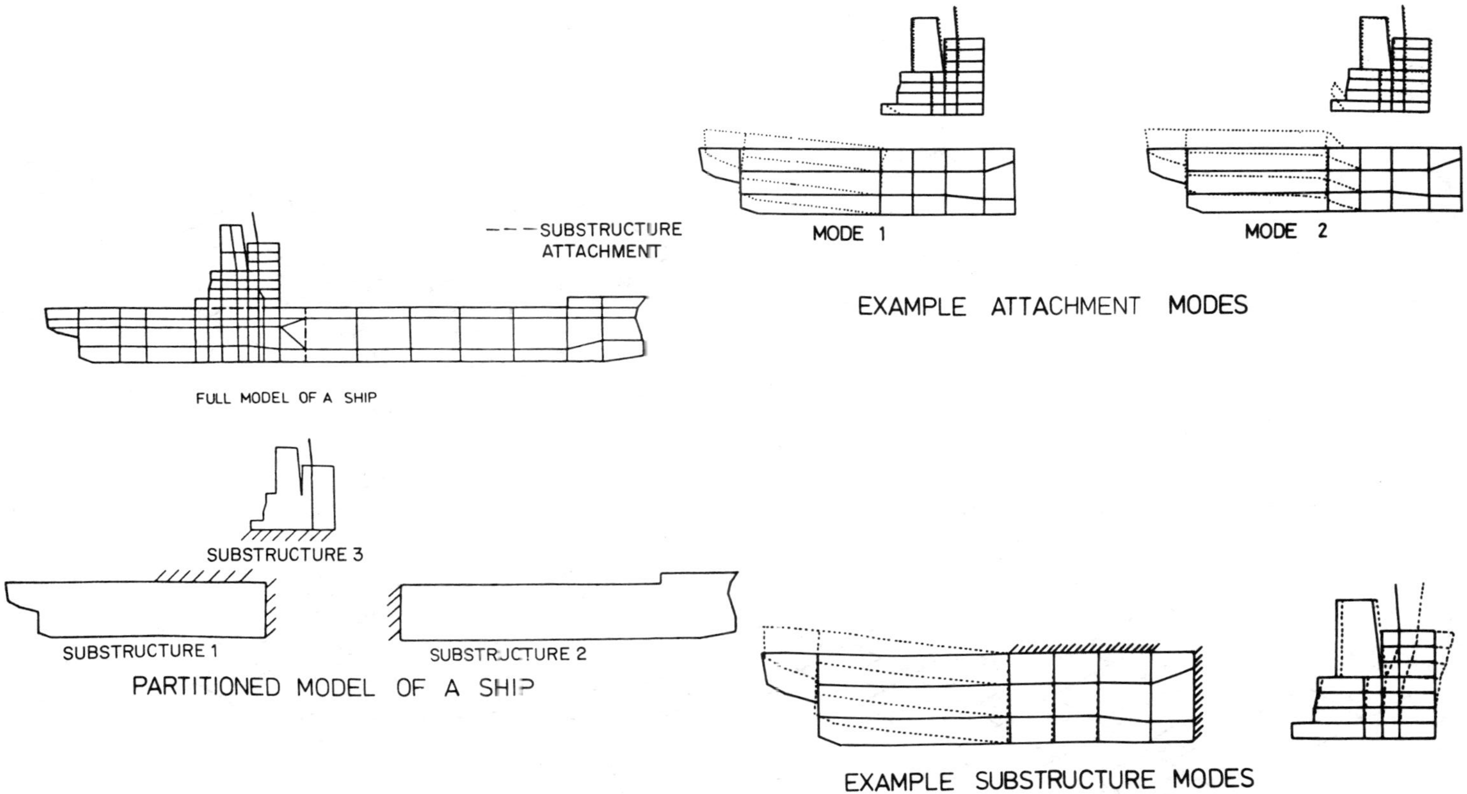

Figure 4 Substructuring of Membrane-Bar Model

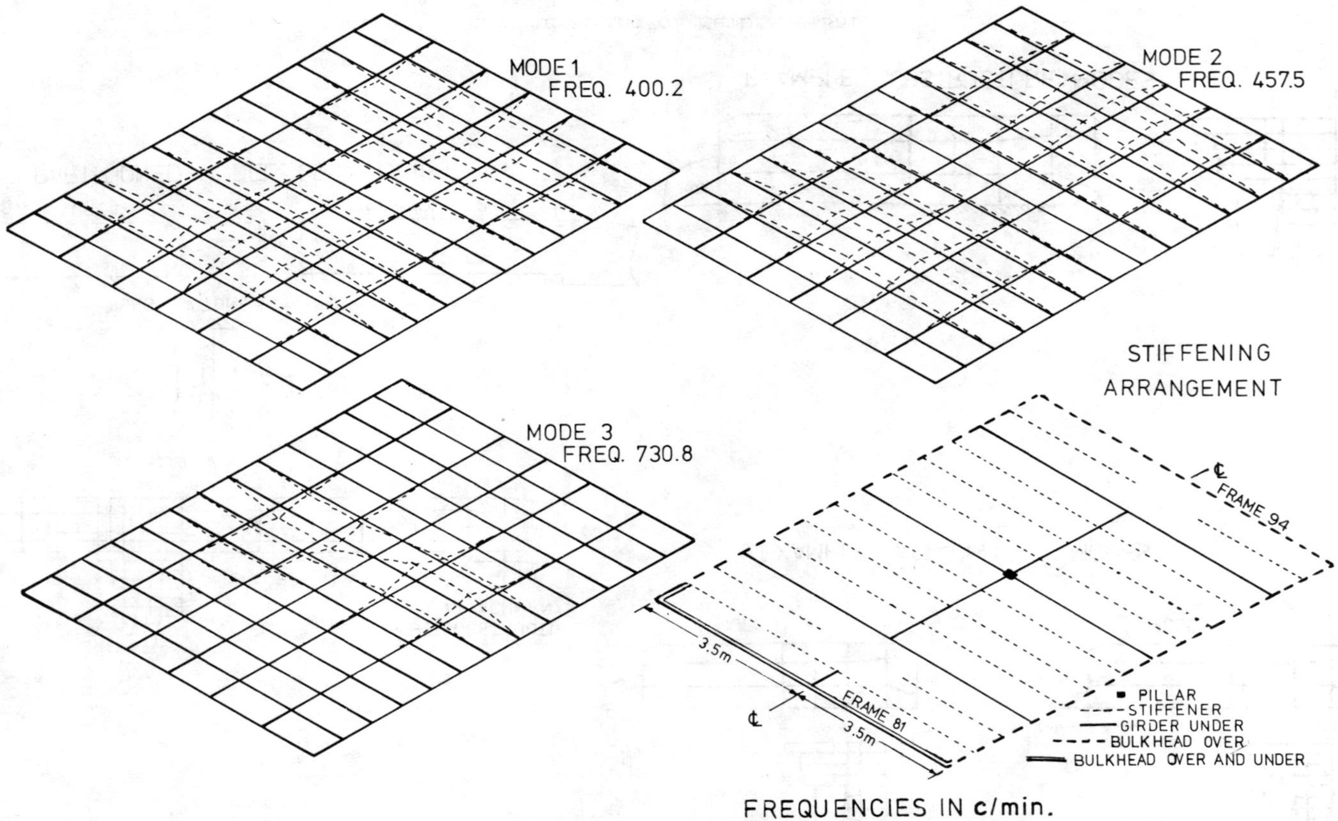

Figure 5 Examples of Deck Vibration Modes

5. LOCAL VIBRATIONS

Finite element models are also created for local vibrations, for example vibration modes of a local deck area are shown in Figure 5. Results depend on the assumptions made as to the boundary conditions; these are usually a combination of simply supported and/or fully clamped edges.

ACKNOWLEDGEMENT

The author expresses his thanks to the Council of the British Ship Research Association for permission to publish this paper.

REFERENCES

Catley, D and Norris, C. (1976) Theoretical prediction of the vertical dynamic response of ship structures using finite elements and correlation with ship mobility measurements. Presented at the 11th ONR Symposium, Department of Mechancial Engineering, University College London.

Norris, C and Catley, D. (1979) Applications of two-dimensional finite element model to ship vertical vibration and comparison with ship mobility measurements. Presented at RINA Vibration Symposium.

Jennings, A. and Orr, D.R.L. (1971) Applications of the simultanoous iteration method to undamped vibration problems, Int. J. Num. Meth. Engng, 3, pp. 13-24.

THE USE OF INTEGER PACKING TECHNIQUES AT THE RRZN

U. Harms

(Regionales Rechenzentrum f. Niedersachsen (RRZN) bei der Universität Hannover, Hannover, West Germany)

ABSTRACT

This paper describes the development of portable software for manipulating sparse matrices based on the Sparse Matrix Package of McNamee (1971) and the NAG subroutines FO3AJF, FO4APF and FO3AKF (NAG (1977) and Curtis and Reid (1971)) for solving sparse linear systems.

1. INTRODUCTION

RRZN is a scientific regional computer centre supplying the universities of Lower Saxony with high speed computing power (CDC Cyber 76-12, Cyber 73-16, Cyber 73-12), especially the University of Hannover. Because of several geodesy and finite element problems, there was a need for sparse matrix subroutines. After implementing the subroutines by McNamee (1971) we found some difficulties. Some of the resulting modifications including a failure parameter are described by Harms et al (1980).

The resulting package contains about 110 FORTRAN subroutines with the main components:

(1) modifications and extensions of the McNamee package (SPARSE),
(2) user interface for SPARSE and the NAG subroutines,
(3) modifications of the NAG subroutines to hold index arrays in packed form (needed because CDC integers occupy 60 bits)

and

(4) interface subroutines from SPARSE to NAG and vice versa.

A manual for the package and a detailed description of all the subroutines is written in German. A shorter manual in English is now in preparation.

2. STORAGE SCHEMES USED

The storage schemes for our packages are as follows:

SPARSE: row oriented

A = vector of the non-zeros (length nz)

M = array/vector of the index information (index matrix) containing the number of rows and number of columns (a half word each), the number of non-zeros (one full word), the number of non-zeros in row i, i=1,2,...,m (m half words), the column indices of the non-zeros (nz half words).

The total storage required is thus 2+(3nz+m)/2 words.

The order of the matrix A is thus limited to 9999, as the element that is to be put in the left half word is multiplied by 10000. This limitation can easily be changed by modifying only the subroutine that creates this compact index information and the function that unpacks this information. The factor 10000 is chosen for reasons of portability only.

NAG Mark 5: column oriented

A = vector of the non-zeros in column order with IA elements

IND = row indices of the non-zeros with IA elements

IND2 = auxiliary array for internal usage with IA elements

IW = auxiliary array for pivoting information and pointers for the first element in a column with (n+1)*13 elements

IA = the number of elements in the decomposed form. This must be estimated by the user. It will often be around 2 to 4 times nz. So the total storage required here is 3*IA+(n+1)*13 although, of course, we have allowed room for fill-in during Gaussian elimination.

3. MODIFICATIONS AND EXTENSIONS OF SPARSE

Our modifications and extensions to this package of McNamee (1971) are as follows:

- introduction of a failure parameter (as in NAG) for user controlled error processing
- tolerance factor eps. If an element is less than eps, the element will be treated as zero

- memory reduction for some subroutines, especially the product of a scalar and a matrix
- replacement of the product of $A*B^T$ by $A*B$
- changing of the sign of a matrix
- manipulation of the diagonal
- scalar product of a row of the matrix and a vector
- product of a matrix and a vector
- the result/index matrix is automatically written to tape for the arithmetic operations +,-,* between two matrices if the memory allocated is exceeded
- reading the above result matrix
- solving a linear system of equations using the SOR-iteration (successive overrelaxation)
- inverting a sparse matrix using SOR-iteration.

These subroutines enable a user to perform many different operations on sparse matrices.

4. USER INTERFACE FOR SPARSE AND NAG

We have written routines to perform the following operations:

- reading a row- or columnwise stored full matrix (with zeros) from tape/disc and changing it into the internal format of SPARSE or NAG
- transforming fully stored rows or columns in natural order to the internal format of SPARSE or NAG directly in central memory without using tape/disc
- filling up a row or column with zeros and writing it onto tape/disc
- filling up a row or column and giving it directly to the calling subroutine.

Thus the user can easily change between sparse-matrix techniques and normal matrix operations.

5. NAG SUBROUTINES WITH PACKING

We have provided three versions of packing for the NAG subroutines:

(a) only IND and IND2 packed for small systems (small N) and many non-zeros

(b) only IW is packed for big systems and few non-zeros

(c) IND, IND2 and IW are packed.

Internally some columns of the auxiliary index-array IW are equivalenced with parts of the real array of non-zeros in the NAG subroutines. In version (b) and (c) the dimension of IW is reduced (for example by a factor of 3 to 4), so it is impossible to use IW for the non-zeros. We therefore use a scratch file for the array A that is written blockwise onto tape and is automatically read from the file as required.

The user defines the number of integers to be packed into a word by calling subroutine DEF1 for IND and IND2 and calling subroutine DEF2 for IW, so that it is possible to have different packing sizes for the one- and two-dimensional arrays.

Integers are stored in packed form by calling the subroutines NAGIPK for IND and IND2 and by calling SET2 for IW. Corresponding routines for retrieval are the functions NAGIND and IAUS2. These routines contain machine-dependent features.

These subroutines and functions contain tests for

- dimensions within the given bounds,
- integer within the packing bounds; this means we test if it is possible to put the integer in the given number of bits,
- sign defined,
- array index within the given bounds of the unpacked array.

For speeding up these subroutines it is possible to omit these checks but in consequence it will be impossible to find errors in the compact storing scheme.

The versions are fully portable since the user only has to change (if necessary) the values related to word length in the subroutines DEF1, DEF2, NAGIPK and SET2, and the functions NAGIND and IAUS2. For reduction of computer time it is planned to use in-line code for packing and unpacking in innermost loops but that is contrary to portability. We give some timings for our routines in Table 1.

There is an increase in central processor time, but it is impossible to solve a linear system with N=5000 and 25000 non-zeros on our CDC Cyber 76-12 using LCM as there are more than 140,000 words to be declared by the user while only about 100,000 words are available. With a packing factor of three integers per 60 bit-word the 115,000 integers are reduced to about 40,000. So the packed routines only require 65,000 words in total. That

TABLE 1

Timing on the Cyber 76-12 using LCM (large core memory) in seconds of central processor time. The subroutines are as follows:

FO3AJF : calculates the determinant and factorization
FO4APF : solves the linear system
FO3AKF : calculates the determinant and factorization of a matrix with the same sparsity pattern as a matrix factorized by FO3AJF.

Subroutine	standard NAG	version (a)	version (b)	version (c)	Order	Non-zeros
FO3AJF FO4APF FO3AKF	0.185 0.004 0.023	0.697 0.008 0.074	0.476 0.012 0.047	1.086 0.017 0.099	100	500
FO3AJF FO4APF FO3AKF	0.5 0.007 0.053	2.300 0.015 0.223	1.151 0.023 0.101	3.200 0.032 0.280	200	1000

means that it is possible to solve these large problems that otherwise would remain unsolved.

6. TRANSFORMATION SUBROUTINES FROM SPARSE TO NAG AND VICE VERSA

The user can call some of our subroutines to change from SPARSE to NAG and its versions and vice versa. Some of the subroutines use tapes if the given space for the target matrix is too small. This enables the user to do arithmetic operations in SPARSE and to solve linear systems in NAG.

7. FUTURE PLANS

We are now modifying NAG Mark 7 to get a packed version like Mark 5. We would anticipate that we will inherit the gains of Mark 7 over Mark 5 (see Duff (1977)) but the different storage organizations of the two algorithms are giving us difficulties.

A subroutine for sparse matrix problems in linear optimization is under development.

Sparse matrix techniques can be used in applications in social sciences and statistics as well, but they are unknown by the prospective users. Therefore we intend to modify SPSS and shall implement sparse techniques for MANOVA and FACTOR so that the user is not aware of them. Such a transparent interface is necessary for successful use of the developed software.

8. CONCLUSIONS

The combination of the SPARSE package (McNamee (1971)) and the NAG subroutines gives the user an efficient set of subroutines for handling sparse matrices. The memory requirements of the NAG subroutines makes it necessary to use packed arrays to solve very large systems. A user interface and transformation subroutines between SPARSE and NAG have been implemented, but there is still a need for more comfortable and easy to use sparse subroutines.

REFERENCES

Curtis, A.R. and Reid, J.K. (1971) The solution of large sparse unsymmetric systems of linear equations, J. Inst. Maths. Applics., 8, pp. 344-353.

Duff, I.S. (1977) MA28 - a set of FORTRAN Subroutines for Sparse Unsymmetric Systems of Linear Equations, Report R.8730, AERE Harwell, HMSO, London.

Harms, U., Kollakowski, H. and Möller, G. (1980) Remark on Algorithm 408. A Sparse Matrix Package (Part 1), TOMS, 6, pp. 456-457.

McNamee, J.M. (1971) A Sparse Matrix Package, Part 1, Algorithm 408, CACM, 14, pp. 265-273.

NAG (1977) NAG User Manual, Mark 5.

NAG (1978) NAG User Manual, Mark 7.

THE USE OF RECORDS IN SPARSE MATRIX PROGRAMS

N. Houbak

(Technical University of Denmark, Numerical Institute, Lyngby, Denmark)

ABSTRACT

In this report we will outline some aspects of using the RECORD data structure for storing and operating on a sparse matrix, instead of the ARRAY data structure commonly used. We will try to compare an ALGOLW program that uses this structure with programs that use the ARRAY structure.

1. INTRODUCTION

In many computational problems, the numerical solution of one or more systems of linear equations

$$A \underline{x} = \underline{y}, \; A \in R^{NxN}, \; \underline{x}, \underline{y} \in R^{N} \tag{1}$$

is a major task. It often happens that the system is large and sparse; that is N (the order of the matrix) is large and only relatively few of the elements of A are non-zero. In this report we are concerned with the case when (1) is solved by Gaussian elimination with pivoting; that is a decomposition

$$P A Q = L U$$

is performed, where P and Q are permutation matrices describing the row and column interchanges, respectively, and L is a lower triangular matrix with "ones" on the main diagonal and U is upper triangular.

Developments during the last 10-15 years resulted in programs, most of them written in FORTRAN, which solve (1) assuming various properties of A: symmetry, non-symmetry, banded structure or general structure. Many of these programs are undeniably of high quality and have good portability. The data structures used are traditionally 1-dimensional arrays for general sparse matrices and 2-dimensional arrays for banded matrices, in part because FORTRAN does not support other data structures. In the same period of time compilers have been developed for languages such as PASCAL and ALGOLW which support the RECORD data structure. At our computing centre (NEUCC*) the ALGOLW compiler is available. In this report we will show that well-known sparse matrix techniques can be implemented in ALGOLW using the RECORD data structure, and we will compare the efficiency (both time and storage) of a program based on such a data structure to that of sparse matrix programs, written in both ALGOLW and FORTRAN, which use the traditional data structure.

2. DESCRIPTION OF THE RECORD DATA STRUCTURE

Fascinated by the elegant properties of linked lists, one could wonder if they could be implemented in a more efficient way than is normally done.

Instead of storing matrix elements and links in ARRAYs we can use a RECORD. What is a RECORD then? It may be thought of as a special 2-dimensional ARRAY with the following properties. First, there is no limit to the number of "rows" (only limited by the storage available) and new "rows" may be added dynamically as the need arises. Second, the "columns" may be simple variables of different types (INTEGER, REAL, LOGICAL ...). Third, each row is addressed not by an INTEGER index but by the simple variables named REFERENCE containing the storage address of the "row". Fourth, when the storage is used up, the compiler system performs a "garbage collection", that is, non-referenced "rows" are removed and the other "rows" are compressed. Because the REFERENCE variable is a simple variable it may itself be stored in a RECORD. Letting a REFERENCE in one "row" hold the address of the next "row", it is easy to form a linked list. The entry to the list is a REFERENCE holding the address of the first "row", and the list is terminated whenever an address to the next "row" is NULL (a special value).

The advantage of the RECORD data structure is that there is no need for indices to be stored as indirect addresses. Instead, the physical storage address of the element is stored. For example, if one wants the value of the <u>next</u> element in a row, only two very simple operations (each is a transfer from memory to register) have to be executed in the RECORD case, whereas in the ARRAY case the same two operations have to be done <u>plus</u> two conversions from an index to a physical storage address. Each of these conversions involves at least one integer multiplication and one integer addition. If the program is compiled in an "index-checking" mode, then some extra operations are generated

---

* Northern European University Computing Center, Technical University of Denmark

in both cases. However, these are much more time-consuming for the ARRAY-case (assuming ALGOLW on the IBM 370). According to Gram et al (1978), the following CPU-times (in μ-sec $\pm$ 10%) are used to access a value in a 1-dimensional ARRAY or in a "row" in a RECORD, provided that the index or the address is available:

	ARRAY	RECORD
INDEX-CHECK	~ 2.5	~ 0.16
NO INDEX-CHECK	~ 0.44	~ 0.16

These figures indicate that under equal conditions, a RECORD-oriented program will be somewhat faster than an ARRAY-oriented program.

Another useful feature in the RECORD structure is the dynamic allocation of storage. The ALGOLW compiler administers all the storage reserved for a job in such a way that, for example the storage used for an ARRAY in an inner block may be used for new "rows" in a RECORD when the inner block is left. There is no need for the user to calculate approximate upper bounds for the ARRAYs used. Only an estimate of the total storage used for the job is necessary.

These useful properties cannot be achieved without a price, and the price is some extra system storage. Actually, our compiler allocates with each "row" extra storage equivalent to an INTEGER. Here some information about the "row" is stored, and the locations are also used when the compiler system performs garbage-collections.

3. COMPARISON WITH OTHER CODES

In this section we will compare our code with other codes available at our computing centre. These codes are SSLEST (ALGOLW) Barker et al (1976) and SSLEST (FORTRAN) Thomsen and Zlatev (1978). Both of these use an ordered list storage scheme and have partly the same facilities. That makes it possible to make a comparison under almost equal conditions. Unfortunately, we are not able to compare our code with an ALGOLW program using linked lists with ARRAY data structures, so the influence of the data structure itself cannot be found. But we may be able to see if there are major differences between the two strategies (linked lists and ordered lists) if they are implemented with a data structure suited for the particular storage scheme.

The codes are tested on a class of matrices called F2. Four parameters determine the matrix, these are N, C, INDEX and ALPHA. The matrix has a main diagonal of "ones", an upper "zero"-band of width C, then a non-zero band of width INDEX-1 and 55 non-zeros in the upper righthand and in the lower lefthand corner. The values of the non-zeros and thus the condition number for the matrix are controlled by the parameter ALPHA. In this test we have fixed C(=13), INDEX (=3) and ALPHA (=8 (well conditioned)). Only N (the order of the matrix) is varied between 50 and 400. We will consider two tests - one with an index-checking code and one without an index-check. To ensure that the SSLEST programs only make a few garbage collections large ARRAYs were used (20000 locations for a maximum of 5000 non-zeros). Using smaller ARRAYS for the SSLEST programs will reduce their efficiency until the program stops due to insufficient space. Running our code in smaller "REGIONS" will have much less effect on the efficiency so long as there is sufficient space for the program not to stop with "STORAGE AREA OVERFLOW".

Figure 1 shows the CPU-time for one factorization for the 3 codes, the ALGOLW codes being in the index-checking mode and in Figure 2 the same results appear with the non-checking code. From the figures it is clear that the CPU-time increases almost linearly with N. That is to be expected, because the mean number of non-zeros per row is almost constant. It should be noted that our code in the index-checking mode only takes half the time of the ALGOLW-SSLEST code. Because the FORTRAN compiler used (FORTRAN G) does not generate index-checking code, and the fact that FORTRAN-SSLEST is a refined version of ALGOLW-SSLEST, the FORTRAN code is generally faster than the other codes. When the index-checking in ALGOLW is omitted, our code is as fast as the FORTRAN code for small systems. For high order systems the FORTRAN code is somewhat better. One should also notice the high reduction in computing time for the ALGOLW-SSLEST code, when the index-check is omitted. The reduction factor is almost equal to ½, assuming that half of the time in the index checking code is spent on checking indices.

4. CONCLUSION

We have shown, in this report, that it is possible to implement efficiently a linked list strategy using the ALGOLW RECORD structure. With this structure we have a simpler storage administration than if ARRAYs were used, we have not lost efficiency with respect to CPU-time compared with programs using ARRAYs and ordered lists, and we only have a small increase in storage requirements. The simplicity in storage management might be illustrated by the fact that our program occupies 18 kBytes with index-check and 13 kBytes with no index-check, including the iterative refinement procedure. The ALGOLW-SSLEST occupies 24 kBytes and 17 kBytes respectively.

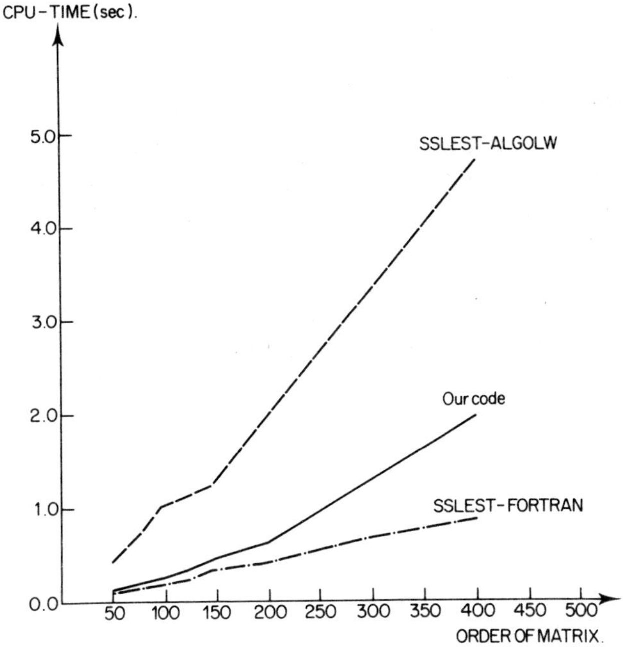

Figure 1 Index-checking codes

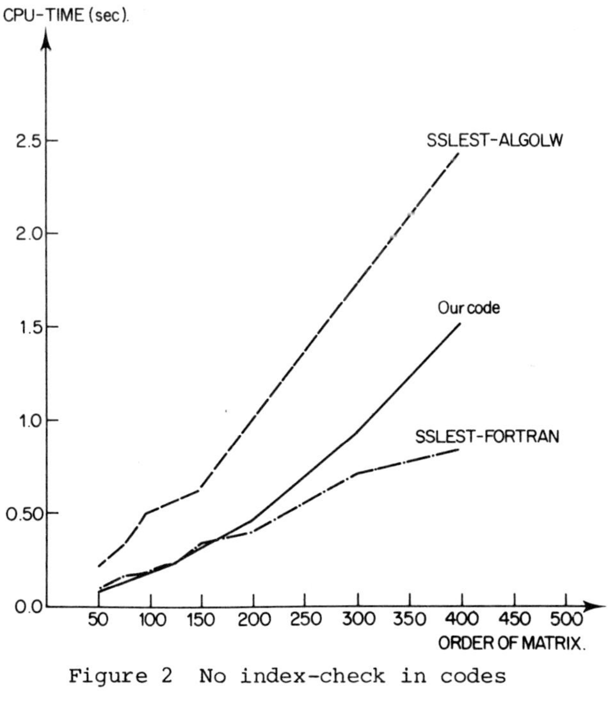

Figure 2 No index-check in codes

REFERENCES

Gram, C., Kristensen, J.T., Sharp, R.I., Wessel, A. (1978), ALGOLW Håndbog, Instituttet for Datateknik, DTH, Danmark.

Barker, V.A., Thomsen, P.G., Zlatev, Z. (1976), Logical procedure SSLEST - an ALGOLW procedure for solving sparse systems of linear equations, Report NI 76-13, Numerisk Institut, DTH, Danmark.

Thomsen, P.G., Zlatev, Z. (1978), SSLEST - a FORTRAN IV subroutine for solving sparse systems of linear equations, Report NI 78-01, Numerisk Institut, DTH, Danmark.

PRACTICAL LANCZOS ALGORITHMS FOR SOLVING STRUCTURAL ENGINEERING EIGENVALUE PROBLEMS

J.G. Lewis and R.G. Grimes

(Boeing Computer Services Co., Seattle, Washington, USA)

ABSTRACT

In this synopsis of our poster presentation we describe the implementation of a Lanczos eigenvalue calculation module, which is used to solve the generalized eigenvalue problems $K\underline{x} = \omega^2 M\underline{x}$ which arise in structural dynamics calculations. We present some numerical results which demonstrate the effectiveness of the algorithm. We conclude with remarks directed at further, more robust, implementations of the Lanczos algorithm.

1. IMPLEMENTATION

The eigenvalue module described in this paper was created to augment the dynamics capability of the popular structural engineering package GTSTRUDL. This package, which is marketed by Georgia Technological Institute (USA), has only limited capabilities for solving large sparse eigenvalue problems (the vendor recommends eigenanalysis only for fewer than 400 dynamic degrees of freedom). The Lanczos algorithm has become widely accepted in the numerical analysis community as a general method for computing eigenvalues of large, sparse symmetric matrices (see Parlett (1980) for a thorough discussion of the general algorithm). The implementation of the Lanczos algorithm described herein was intended as the first step in providing GTSTRUDL users with the capability of solving large dynamics problems.

The BCS Lanczos module was designed to run on CDC CYBER 175 series computers, without using secondary storage for linear equation solvers, using as little dynamic storage as possible, and yet allowing as large a problem as possible within these constraints. GTSTRUDL is primarily used for problems which do not have rigid body modes, so we assumed that K is positive definite. We further assumed that the typical user of the BCS Lanczos module would want to compute the lowest ten eigenvalues and eigenvectors. The Lanczos implementation was intended to solve this problem much more rapidly and more robustly than the software now available to the GTSTRUDL user.

The problem assumptions allow us to reduce the generalized eigenproblem to a standard eigenproblem after finding the Cholesky factors of $K = LL^T$. (Note that the assumption that K is nonsingular can be relaxed here to one that $K + \sigma M$ be nonsingular for some σ; it is, however, essential for the convergence of the Lanczos algorithm that we not use the standard reduction factoring M, even when M is diagonal.) We use a standard Lanczos recurrence on the reduced problem $L^{-1} M L^{-T} \underline{y} = \mu \underline{y}$. No reorthogonalization of any sort (global, local, selective) is performed.

The limitations on memory require careful implementation of the Cholesky factorization. SPARSPAK (George et al (1979)) is used to find a good ordering for factoring a matrix with the nonzero structure of $K - \nu M$. The code tries, in turn, the reverse Cuthill-McKee ordering, the minimum degree ordering and the automatic nested dissection ordering, and uses the first of these which results in a factorization which can be performed in available core. The condition number of K is estimated by extensions of the heuristics used in LINPACK (Dongarra et al (1979)). This provides a reliable way of detecting singularity (rigid body modes) and is also used in the stopping heuristics.

In the following, p is the number of eigenvalues requested. The Lanczos iteration is begun either with a random vector or a user-provided vector. The iteration is stopped every 2p steps to evaluate the eigenvalues of the tridiagonal matrix T produced by the algorithm. The convergence of the desired eigenvalues is evaluated by the heuristics of Cullum and Willoughby (1978), which were extended to allow for the reduction to standard form.

When the termination criteria are satisfied, the inertia of the matrix $K - \nu M$ is computed, where ν is between the p-th and p+1-st distinct eigenvalues of T. This is found by computing an LDL^T factorization with the data structures obtained from SPARSPAK. Stability of this factorization is monitored by using the error bounds from Erisman and Reid (1974). The results of the inertia calculation are used to direct the eigenvector computation which follows.

The eigenvectors of $K \underline{x} = \omega^2 M \underline{x}$ are computed by back-transforming from the eigenvectors of T, which can proceed in two different ways. If the matrix inertia was found with negligible error and indicates that all of the p lowest eigenvalues are singletons, a single eigenvector of T is found for each cluster, using a special tridiagonal inverse iteration. This is the normal case, and (almost) always occurs if the true eigenvalues are singletons. If the matrix inertia could not be found with small error, or if the inertia shows that there are multiple eigenvalues among the lowest p, an elaborate and baroque scheme is entered to obtain as much information from T as possible. All eigenvectors are obtained for each eigenvalue cluster of T, and a singular value decomposition is used to obtain the numerical rank of the resulting subspace. In the very worst case, a Rayleigh-Ritz projection for all of the accepted eigenvectors may be required.

2. EFFECTIVENESS OF THE ALGORITHM

The resulting eigenvalue computation is more than an order of magnitude faster than inverse iteration in NASTRAN and subspace iteration in SAP IV. It is several orders of magnitude faster than vector iteration in GTSTRUDL. It is capable of distinguishing very close eigenvalues and of finding multiple eigenvalues at the low end of the spectrum. The program runs out of memory somewhere between 1200 and 1500 dynamic degrees of freedom.

The capabilities of the eigenvalue module are demonstrated by the results in Tables 1 and 2, which are difficult eigenproblems for other current eigenvalue algorithms. The BCS Lanczos algorithm found these problems to be of no special difficulty.

The problem which results in Table 1 is a real application, a buckling analysis of half of a heated circular washer, with 1086 degrees of freedom. Ten eigenvalues and eigenvectors were computed in 21.49 seconds including overheads of reading and writing data to disk (CDC CYBER 750), all with absolute eigenvector residuals smaller than 10^{-9}. Computing twenty eigenvalues and eigenvectors took 26.16 seconds again including overheads, for similar accuracy. Details of the computational times are given in Table 1.

The problem which results in Table 2 is derived from an analysis of a real frame building, but with the masses altered drastically in order to create a particularly difficult spectrum. There are 1074 dynamic degrees of freedom. There are two interesting aspects to this problem. First, an envelope reordering of K would require 288,370 words of memory; a minimum degree reordering requires only 50,907 words. Second, the lowest ten distinct eigenvalues include two pairs of eigenvalues, with very close singleton neighbours. The total computation required 48.7 cpu seconds, of which only 13.47 seconds was

TABLE 1

BUCKLING ANALYSIS OF HEATED WASHER

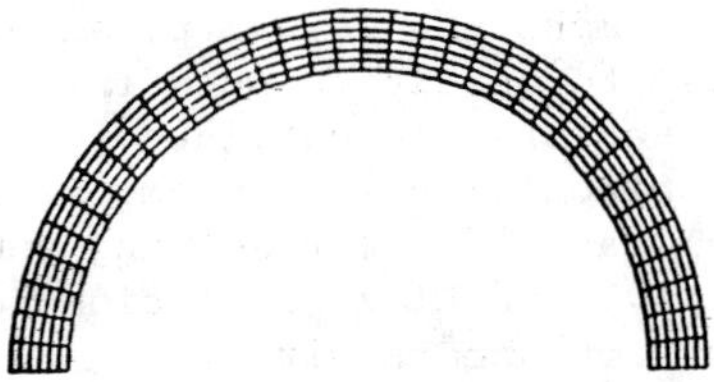

$\|K\|_1 = 5.2 * 10^7$ $\|M\|_1 = 4.1 * 10^4$ $K_1(K) = 5.2 * 10^5$

# of eigenvalues and vectors	10	20
# of iterations	100	120
Time for		
eigenvalues	9.97 seconds	11.80 seconds
stopping heuristics	0.42	0.38
eigenvectors	1.39	3.00
reordering K	0.20	0.20
factoring K	1.19	1.19
computing matrix inertia	1.22	1.22

The first ten eigenvalues are:

.07848	.07939	.07961	.08151	.08280
.08482	.08906	.08919	.09455	.09977

TABLE 2

FRAME BUILDING

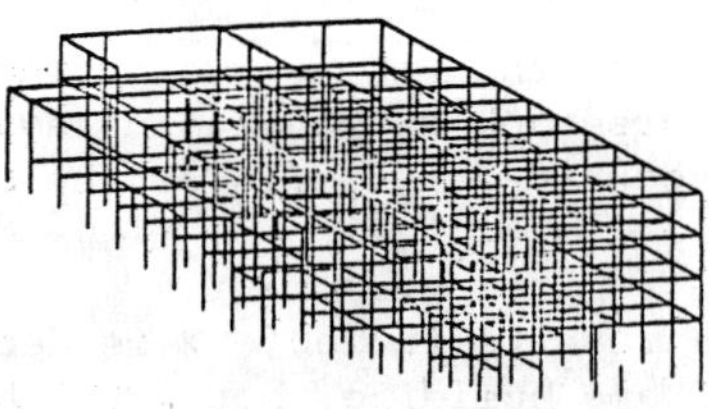

$\|K\|_1 = 8.96 * 10^{10}$ $\|M\|_1 = 1.44 * 10^6$ $K_1(K) = 1.26 * 10^7$

The lowest 10 distinct eigenvalues:

6.9007	18.1420		18.1423	=	18.1423
84.7862	84.7864	=	84.7864		85.537
91.049	93.445		130.93		131.49

used in the 140 Lanczos iterations. The matrix reordering required 18.15 seconds (exceptionally high for SPARSPAK), and the factorization of K required 2.27 seconds. Computing the eigenvectors took 8.23 seconds and exercised almost all of the checks for multiplicity.

3. INCREASING THE ROBUSTNESS OF LANCZOS EIGENSOLVERS

The BCS Lanczos algorithm was designed for a particular class of users, and works very well on the problems for which it was designed. However, this code can be inefficient if the number of eigenvalues desired is large. This is due primarily to the relatively rapid convergence of the very lowest eigenvalues and to the cost of evaluating the eigenvalues of T when the number of iterations is large. For the same reasons, eigenvalues well removed from the lowest eigenvalues cannot be found efficiently.

The tests for multiplicity work well for the expected cases where there are no multiple eigenvalues. The rank calculations work well when there are multiple eigenvalues, but they are expensive. Since the convergence of the eigenvalues of T gives no knowledge of multiplicities, the iteration will stop before finding the full multiplicities of multiple eigenvalues near the upper (interior) end of the desired set of eigenvalues. (This problem is viewed as being of only academic interest by some engineers.)

These limitations can be relieved by incorporating other variations on the Lanczos algorithm and on the reduction to standard form. Use of a block Lanczos algorithm (Cullum and Donath (1974), Lewis (1977), Scott (1979) or Underwood (1975)) will improve the capabilities for handling multiple eigenvalues and will also improve the convergence rates. Shifting (Ericsson and Ruhe (1979), Jennings and Agar (1979)) can be used to change the form of the reduced problem to allow for computation of interior eigenvalues. Both increase the storage required, and so were not used in the current code, but they could be combined with an out-of-core linear equation solver to relieve the current limitations on problem size.

Block Lanczos algorithms present no particular difficulty if implemented carefully. The use of shifts presents some real problems. We must factor an indefinite symmetric matrix; the usual sparsity considerations suggest that we do so without pivoting, which may lead to numerical instability in the factorization. Our experience with the Erisman-Reid error bounds indicates that these bounds may provide an entirely satisfactory check on the loss of numerical stability. In addition, the double factorization obviously requires more space, time and possibly I/O. The double factorization can be avoided by maintaining a set of M-orthogonal vectors, in a manner similar to that used in preconditioned conjugate gradient algorithms, but there are questions regarding the local loss of orthogonality for such an algorithm. Buckling analyses would require working with another, related, double factorization. The current scheme does not allow a general singular mass matrix. However, the important special case of a nonnegative diagonal mass matrix goes through cleanly.

REFERENCES

Cullum, J. and Donath, W.E. (1974) A block generalization of the symmetric s-step Lanczos algorithm, Technical report No. RC 4845, IBM T.J. Watson Research Center, Yorktown Heights, N.Y.

Cullum, J. and Willoughby, R.A. (1978) Lanczos and the computation in specified intervals of the eigenvalues of large, sparse real symmetric matrices, Technical report No. RC-7456, IBM T.J. Watson Research Center, Yorktown Heights, N.Y.

Dongarra, J.J., Bunch, J.R., Moler, C.B. and Stewart, G.W. (1979) Linpack Users' Guide, SIAM Press.

Ericsson, T. and Ruhe, A. (1979) The spectral transformation Lanczos method for the numerical solution of large sparse generalized symmetric eigenvalue problems, Report UMINF - 76.79, Dept. of Numerical Analysis, University of Umeå.

Erisman, A.M. and Reid, J.K. (1974) Monitoring the stability of the triangular factorization of a sparse matrix, Numer. Math., 22, pp. 183-186.

George, A., Liu, J.W.H. and Ng, E. (1979) User guide for SPARSPAK, Technical report, Dept. of Computer Science, University of Waterloo.

Jennings, A. and Agar, T. (1979) Progressive simultaneous inverse iteration for symmetric linearized eigenvalue problems, Technical report, Dept. of Civil Engineering, Queen's University, Belfast.

Lewis, J. (1977) Algorithms for sparse matrix eigenproblems, Technical report CS-77-595, Computer Science Department, Stanford University.

Parlett, B.N. (1980) The Symmetric Eigenvalue Problem, Prentice-Hall.

Scott, D. (1979) Block Lanczos software for symmetric eigenvalue problems, Technical Report CSD-48, Oak Ridge National Laboratory.

Underwood, R. (1975) An iterative block Lanczos method for the solution of large sparse symmetric eigenproblems, Technical report CS-75-496, Computer Science Department, Stanford University.

USING SPARSE MATRIX TECHNIQUES TO SOLVE A MODEL OF THE WORLD ECONOMY

D.B. Szyld

(Institute for Economic Analysis, New York University)

1. INTRODUCTION

In this paper we review some sparse matrix techniques which lead to a very efficient algorithm for the solution of a model of the world economy.

We give a short mathematical description of the model, from which we can study its underlying structure, and thereby choose the appropriate method to solve its equations, a suitable data structure, and an algorithm that can utilise high quality software.

2. STATEMENT OF THE PROBLEM

The Input-Output model of the world economy (IO model) developed at the Institute for Economic Analysis divides the world into r regions ($r \sim 15$). For each of the regions there is a set of linear algebraic equations of the form

$$A_i \underline{y}_i + S_i \underline{w} = \underline{0} \qquad (i = 1, \ldots, r). \tag{1}$$

In addition there are global equations

$$\sum_{i=1}^{r} G_i \underline{y}_i = \underline{0}. \tag{2}$$

A more detailed description of the IO model can be found in Duchin and Szyld (1979), Petri (1977), and Leontief, Carter and Petri (1977).

All the matrices involved are very sparse. For example

A_i could be 200 × 250 with 2500 non-zeros.

S_i could be 200 × 50 with 50 non-zeros.

G_i could be 50 × 250 with 100 non-zeros.

Each matrix A_i has more columns than rows and therefore some components of $\underline{y}_i$ have to be prescribed.

Denote by $\underline{z}_i$ the prescribed components of $\underline{y}_i$, and the unknowns by $\underline{x}_i$, and partition A_i and G_i accordingly; then we can rewrite (1) and (2) as

$$M_i\underline{x}_i + S_i\underline{w} = -B_i\underline{z}_i = \underline{b}_i \quad (i=1,\ldots,r). \tag{3}$$

$$\sum_{i=1}^{r} E_i\underline{x}_i = -\sum_{i=1}^{r} E'_i\underline{z}_i = \underline{b}_{r+1}. \tag{4}$$

In fact the structure of the matrices G_i guarantees that $\underline{b}_{r+1} = \underline{O}$, but this fact is not exploited in the implementation design.

The entire model is thus a linear system of equations with a nonsymmetric bordered block diagonal matrix of coefficients of the form:

$$\begin{pmatrix} M_1 & & & & S_1 \\ & M_2 & & & S_2 \\ & & \ddots & & \vdots \\ & & & M_r & S_r \\ E_1 & E_2 & \cdots & E_r & O \end{pmatrix} \begin{pmatrix} \underline{x}_1 \\ \underline{x}_2 \\ \vdots \\ \underline{x}_r \\ \underline{w} \end{pmatrix} = \begin{pmatrix} \underline{b}_1 \\ \underline{b}_2 \\ \vdots \\ \underline{b}_r \\ \underline{O} \end{pmatrix} \tag{5}$$

where the blank blocks in the matrix are zero blocks.

Our aim was to solve the system exploiting its block structure and the sparsity in each block, using state of the art techniques, to design a flexible algorithm to handle any r and M_i of different sizes, and to use a storage structure compatible with existing software and which is efficient for data management.

In contrast the Institute had a code that stored the given matrices in packed form and computed and stored all the inverses M_i^{-1}. The approximate computer time was 4 hours on a PDP-11. The (dense) inverses were saved for subsequent runs. When designing the solution algorithm we had in mind the size of the matrices, and that each matrix M_i is non singular, making Block

Gaussian Elimination suitable. Some research was performed on similar problems using iterative methods and we favoured a direct method approach and the use of available high quality software.

3. REVIEW OF BLOCK GAUSSIAN ELIMINATION

Block Gaussian Elimination goes back at least to Householder (1964) and has special advantages in sparse cases. Given a linear system

$$A\underline{x} = \underline{b},$$

we can partition the matrix as:

$$A = \begin{pmatrix} A_{11} & S \\ E & A_{22} \end{pmatrix}.$$

If we factor $A_{11} = L_1U_1$, then we have

$$A = \begin{pmatrix} L_1 & O \\ EU_1^{-1} & I \end{pmatrix} \begin{pmatrix} U_1 & L_1^{-1}S \\ O & C \end{pmatrix}$$

where C is the so-called Schur complement

$$C = A_{22} - EA_{11}^{-1}S = A_{22} - EU_1^{-1}L_1^{-1}S.$$

In fact, the off diagonal blocks EU_1^{-1} and $L_1^{-1}S$ do not have to be stored; their columns are computed one at a time, used and discarded.

We can now rewrite $A\underline{x} = \underline{b}$ as:

$$\begin{pmatrix} L_1 & O \\ EU_1^{-1} & I \end{pmatrix} \begin{pmatrix} U_1 & L_1^{-1}S \\ O & C \end{pmatrix} \begin{pmatrix} \underline{x}_1 \\ \underline{x}_2 \end{pmatrix} = \begin{pmatrix} \underline{b}_1 \\ \underline{b}_2 \end{pmatrix}$$

The solution $\underline{x}$ is obtained by solving first

$$\begin{pmatrix} L_1 & O \\ EU_1^{-1} & I \end{pmatrix} \begin{pmatrix} \underline{g}_1 \\ \underline{g}_2 \end{pmatrix} = \begin{pmatrix} \underline{b}_1 \\ \underline{b}_2 \end{pmatrix}$$

i.e.,

(1) obtained $\underline{g}_1$ from $L_1\underline{g}_1 = \underline{b}_1$

(2) obtain $\underline{e}$ from $U_1\underline{e} = \underline{g}_1$

(3) obtain $\underline{g}_2$ from $E\underline{e} + \underline{g}_2 = \underline{b}_2$

and then solving

$$\begin{pmatrix} U_1 & L_1^{-1}S \\ O & C \end{pmatrix} \begin{pmatrix} \underline{x}_1 \\ \underline{x}_2 \end{pmatrix} = \begin{pmatrix} \underline{g}_1 \\ \underline{g}_2 \end{pmatrix}$$

i.e.,

(1) solve $C\underline{x}_2 = \underline{g}_2$

(2) solve $L_1\underline{e}' = S\underline{x}_2$

(3) solve $U_1\underline{x}_1 + \underline{e}' = \underline{g}_1$.

Thus, the off diagonal blocks are not needed explicitly for the solution but only when computing the Schur complement C: denote any column of C by $\underline{c}$, the corresponding column of S by $\underline{s}$ and that of A_{22} by $\underline{a}$.

Solve $$A_{11}\underline{u} = L_1U_1\underline{u} = \underline{s}$$

and compute $$\underline{c} = \underline{a} - E\underline{u}$$

The step $$C\underline{x}_2 = \underline{g}_2$$

can be done by factoring $C = L_{22}U_{22}$, or by partitioning C and using Block Gaussian Elimination recursively. Further application of these ideas can be found in George (1980). In the first case we have,

$$A = \begin{pmatrix} L_1 & O \\ EU_1^{-1} & L_{22} \end{pmatrix} \begin{pmatrix} U_1 & L_1^{-1}S \\ O & U_{22} \end{pmatrix}$$

We have then recovered the LU factors of A. The advantage of our approach is that we only store the factors of A_{11} and C, and that fill-in occurs in these diagonal blocks.

Typically C is full, and Block Gaussian Elimination is advantageous specially when the order of C is small compared with that of A.

Note that if A is block diagonal, that is E = O and S = O, its LU factors are block diagonal, with its blocks being the factors of the diagonal blocks of A. Therefore, if A_{11} is block diagonal, L_1 is block diagonal and so is U_1.

We now consider our special problem (see section 2). Since A_{11} is block diagonal, its diagonal blocks being the matrices M_i, we can write the Schur complement

$$C = EA_{11}^{-1}S = \sum_{i=1}^{r} E_i M_i^{-1} S_i$$

We use the subroutines from MA20 (Duff (1977)) to solve the systems for each matrix M_i. This code looks for a block triangular form for each M_i, and factors its diagonal blocks. The Schur complement is then computed by accumulating it column by column as explained before. We use a square array to store C, since it is expected to be full, and factor it using suitable LINPACK subroutines (Dongarra et al (1979)). The complete solution algorithm is given by:

1. For i = 1,...,r

    1.1 Obtain M_i and $\underline{b}_i = B_i\underline{z}_i$

    2.1 Obtain the factors of M_i

2. For $i = 1,\ldots,r$

2.1 For each column of S_i and for $\underline{b}_i$,

2.1.1 Assign that column or vector to $\underline{v}$

2.1.2 Solve $M_i\underline{u} = \underline{v}$

2.1.3 Compute $E_i\underline{u}$ and subtract it from the corresponding column of E_{r+1} or $\underline{b}_{r+1}$.

(E_{r+1} and $\underline{b}_{r+1}$ are zero initially)

The resulting matrix is the Schur complement C and the resulting vector is called $\underline{f}$.

3. Factor C obtaining L_{22} and U_{22}

4. Solve $C\underline{w} = \underline{f}$

5. For $i = 1,\ldots r$

5.1 Compute $\hat{\underline{b}}_i = \underline{b}_i - S_i\underline{w}$

5.2 Solve $M_i\underline{x}_i = \hat{\underline{b}}_i$

4. DATA STRUCTURES

Each matrix A_i (the columns of which generate M_i and B_i) is stored in compressed columnwise form, as described by Gustavson (1978). We show how the non-zero elements are being handled, while the row indices and pointers are treated in a similar way.

A large array ABIG is used as storage for the non-zero entries of the matrices A_i, M_i and its factors. The location in ABIG of the first non-zero element of A_i is stored in the i-th position of a short array IDIVA.

We show in detail step 1 of the algorithm

For i = 1,...r

- Set IDIVA (i) (where IDIVA (1) = 1)
- Read the non-zeros of A_i into core.

ABIG	Non-zeros of A_i	

↑ IDIVA (i)

- Read a set of integers defining B_i, and read $\underline{z}_i$.
- Accumulate $\underline{b}_i = B_i\underline{z}_i$.
- The remaining columns of A_i correspond to M_i. Compress the non-zeros of A_i to obtain M_i.

ABIG	Non-zeros of M_i	Work space for factorization

↑ IDIVA (i)

- Obtain factorization of M_i (Analyze-Factor step of MA28)

ABIG	Non-zeros of factors of M_i	

↑ IDIVA (i) ↑ IDIVA (i+1)

5. EXPERIMENTAL CONCLUSIONS AND FURTHER REMARKS

In our runs, we have observed about 10% fill-in in each of the matrices M_i.

Runs on a IBM 370/168 required about 150 seconds CPU time for 4 decades, that is for 4 sets of linear systems, using double precision and 1 Megabyte of storage.

If 1 Megabyte of storage is not available, the factors of the matrices M_i could be stored out of core and be recalled when needed.

We shall now explore the feasibility of a "hybrid" method. Namely, the use of an iterative method for each diagonal block M_i, but we still compute the Schur complement and use Block Gaussian Elimination as described before.

For a typical matrix M_i the analyze and factor step takes between 300 and 455 milliseconds CPU time, while the solve step takes about 11 milliseconds CPU time. Because of the small amount of fill-in observed, the time for backsolving is comparable to the time of a matrix vector multiplication, and therefore comparable to the time of a single iteration in an iterative method with M_i as the matrix involved in the iteration. The time for the analyze and factor step is comparable with the time for about 27 to 42 iterations.

In our algorithm we have to solve about 50 systems of the form $M_i\underline{v} = \underline{f}$ for each matrix M_i (steps 2.1.2 and 5.2). The convergence properties of an iterative method for $M_i\underline{v} = \underline{f}$ have not been studied, but in any case an iterative method will be superior only if it converges in 2 iterations to an accurate solution. Thus, the "hybrid" approach of using an iterative method for each diagonal block is not competitive with our algorithm.

To study the feasibility of an iterative method involving the whole matrix we should bear in mind that computing the Schur complement requires as much time as the factorization of all the diagonal blocks M_i. Therefore, the time involved in our solution algorithm is comparable to 50 to 80 iterations using the entire matrix of the system (5). Yet it is not clear that there is a reliable iterative method for this particular problem, that is for a system with a matrix of the structure described.

REFERENCES

Dongarra, J.J., Bunch, J.R., Moler, C.B., and Stewart, G.W. (1979) LINPACK, User's Guide, SIAM Press.

Duchin, F. and Szyld, D.B. (1979) Application of sparse matrix techniques to inter-regional input-output analysis, Economics of Planning, Vol. 15 , in press.

Duff, I.S. (1977) MA28 - a Set of FORTRAN Subroutines for Sparse Unsymmetric Systems of Linear Equations, Report R. 8730, A.E.R.E. Harwell, HMSO, London.

George, A. (1980) Direct solution of sparse positive definite systems: some basic ideas and open problems, These proceedings.

Gustavson, F.G. (1978) Two fast algorithms for sparse matrices: multiplication and permuted transposition, TOMS 4, pp. 250-269.

Householder, A.S. (1964) The Theory of Matrices in Numerical Analysis, Blaidsell.

Leontief, W., Carter, A., and Petri, P. (1977) The Future of the World Economy, Oxford University Press.

Petri, P. (1977) Guide to the operation of the United Nations world model, Technical Report No. 3, Department of Economics and Brandeis Economics Research Center, Brandeis University.

SPARSE MATRICES - EFFICIENT DECOMPOSITION AND APPLICATIONS

Z. Zlatev and P.G. Thomsen

(Institute for Numerical Analysis, The Technical University of Denmark)

1. STATEMENT OF THE PROBLEM

Let $A \in \mathbb{R}^{n \times n}$ and $\underline{b} \in \mathbb{R}^{n \times 1}$ be given. Consider the problem: find an approximation $\bar{\underline{x}}$ to the solution of $A\underline{x}=\underline{b}$ for which $\|\bar{\underline{x}} - \underline{x}\| \le \bar{\varepsilon}$ where $\bar{\varepsilon}$ is some error tolerance prescribed in advance. The problem is often solved in the following way.

Step 1	DECOMPOSITION	$LU = PAQ + E$
Step 2	FIRST SOLUTION	$\underline{x}_1 = H\underline{b}$, $H = QU^{-1}L^{-1}P$
Step 3	IMPROVEMENT	$\underline{r}_i = \underline{b} - A\underline{x}_i$, $\underline{d}_i = H\underline{r}_i$, $\underline{x}_{i+1} = \underline{x}_i + \underline{d}_i$, $i=1(1)p-1$

Step 3 is optional. One can accept $\underline{x}_1$ (hoping that $\|\underline{x}_1 - \underline{x}\| \le \bar{\varepsilon}$). Then $\bar{\underline{x}} = \underline{x}_1$ is called "the direct solution" (DS). If Step 3 is carried out and if the process converges, then $\bar{\underline{x}} = \underline{x}_p$ ("the iteratively refined solution", IR) is normally more accurate than $\underline{x}_1$ and an estimation of $\|\bar{\underline{x}} - \underline{x}\|$ is computed by $\|\underline{d}_{p-1}\|$. Unfortunately, some extra storage and time are also required when Step 3 is applied. If a dense matrix technique (DMT) is used, then the extra storage is about 100%. If a sparse matrix technique (SMT) is used, then the extra storage depends on both the code and the computer. For the code Y12M on a UNIVAC 1100/82 the extra storage is only about 22% (Zlatev (1979)). The experiments show that relatively little extra time is needed by either DMT or SMT. Nevertheless, the use of extra storage and extra time with Step 3 should be avoided when this is possible. The main purpose of this paper is to show that sometimes the use of Step 3 can even lead to a reduction in storage and/or computing time (compared with DS).

2. THE COMBINATION SMT + LDT + IR

A large drop tolerance, LDT, should be added to the combination SMT + IR in order to improve the efficiency of Step 3. Each element of A which is smaller (in absolute value) than the drop tolerance T is considered as a zero element in further computation. T=0 or a very "small" T should be specified when DS is used.

Consider array A1(NN), where the non-zero elements of L and U have to be stored, and array A2(NZ), where the non-zero elements of A are kept. The second array is not needed for DS. NN should be larger than the number, NZ, of non-zero elements in A, but the optimal value of NN is as a rule not known in advance. Let $COUNT_1$ be the maximum number of elements kept in A1 when T=0 and DS is used; let $COUNT_2$ be the corresponding number when IR is used and T is large (note too that COUNT without any index will be used when there is no danger of misunderstanding). Assume that:

$$COUNT_1 - COUNT_2 > NZ. \tag{2.1}$$

In this situation (Zlatev et al (1980a)): (i) the storage requirements are smaller when Step 3 <u>is</u> applied, (ii) the same is true for the computing time, (iii) if the iterative process is convergent, then both a high degree of accuracy and a reliable error estimation will normally be obtained.

Of course, there is no guarantee that the condition (2.1) will be satisfied for the problem under consideration. If (2.1) is not satisfied, then some extra storage and/or some extra time may be needed. However, note that condition (2.1) is very conservative; therefore it may happen that the use of Step 3 gives better results even if (2.1) is not satisfied. It should be emphasized that for "expensive" problems (problems which produce <u>many</u> fill-ins) (2.1) is often satisfied and the use of the combination SMT+LDT+IR is extremely efficient.

3. EFFICIENCY OF THE COMBINATION: SMT + LDT + IR

Some experiments with matrices of classes D(n,c) and E(n,c) are given in Tables 1 and 2 to illustrate the efficiency of the combination: SMT + LDT + IR. Any matrix $A \in D(n,c)$ is of order n with non-zeros on the main diagonal, on three bands at a distance c above the diagonal and reappearing cyclicly under it, and in a 10 x 10 upper triangle in the upper right-hand corner of A. Any matrix $A \in E(n,c)$ is of order n whose non-zeros are: 4's on the main diagonal, -1's on the two side-diagonals and in two bands at a distance c from the main diagonal.

TABLE 1

An experiment with matrices D(n,84) on a UNIVAC 1100/82 with the FTN compiler (the numbers of iterations are given in brackets)

n	DS option of Y12M; $T=10^{-12}$			IR option of Y12M; $T=10^{-2}$		
	COUNT	TIME	ACCURACY	COUNT	TIME	ACCURACY
650	11826	6.18	9.31E-5	4986	2.36(5)	1.19E-7
850	14225	7.33	5.66E-5	6246	2.42(4)	0.0
900	15482	8.30	3.02E-5	6686	2.83(5)	1.04E-7
1000	16263	8.18	2.98E-4	7263	3.06(4)	1.49E-8

4. APPLICATIONS: LARGE BAND SYMMETRIC AND POSITIVE DEFINITE MATRICES (LBSPDM'S)

The combination SMT+LDT+IR is very efficient in the solution of some systems with LBSPDM's. Experiments with the code Y12M and the NAG version of a subroutine from Wilkinson and Reinsch (1971) (this subroutine is based on Cholesky's decomposition and exploits both the symmetry, the definiteness and the band structure of the matrix) have been carried out by Zlatev et al (1980b). The matrices used are very similar (but not the same) to the matrices obtained in the discretization of the Laplace operator on a square domain by the five-point rule. The results shown in Table 2 indicate that a code based on the combination SMT+LDT+IR and exploiting the symmetry, the band structure and the definiteness of the matrix may be competitive with the best codes based on band matrix techniques in the solution of similar problems (Y12M is a general code which does not exploit these special properties of the matrix). The maximum number (16) of iterations allowed has been used by Y12M in all examples. Under "R" the average number of non-zero elements per row, which are kept in array A1, is given.

5. APPLICATIONS: LINEAR LEAST SQUARES PROBLEMS (LLSP'S)

Consider the LLSP: $B\underline{y}=\underline{c}-\underline{r}$, $B^T\underline{r}=\underline{0}$ where $B\in\mathbb{R}^{m\times n}$, $\underline{c}\in\mathbb{R}^{m\times 1}$, $\underline{y}\in\mathbb{R}^{n\times 1}$, $m\geq n$. This problem is rewritten (see, for example, Björck (1967)) in the form $A\underline{x}=\underline{b}$ where

$$A = \begin{bmatrix} I & B \\ B^T & O \end{bmatrix} \quad \underline{x} = \begin{bmatrix} \underline{r} \\ \underline{y} \end{bmatrix} \quad \underline{b} = \begin{bmatrix} \underline{c} \\ \underline{0} \end{bmatrix} \quad B = \begin{bmatrix} B_1 \\ B_2 \end{bmatrix} \tag{5.1}$$

TABLE 2

An experiment with matrices E(n,c), $c^2=n$, on a UNIVAC 1100/82 with the FTN compiler.

n	BANDWIDTH	BAND SUBROUTINE		IR option of Y12M with $T=10^{-2}$			
		TIME	ACCURACY	COUNT	R	TIME	ACCURACY
900	31	5.17	1.21E-4	11680	12.9	8.15	1.76E-6
1024	33	6.59	1.45E-4	13406	13.1	9.41	3.57E-5
1225	36	9.25	1.88E-4	16123	13.2	11.23	6.22E-5
1296	37	10.28	2.03E-4	17001	13.1	11.86	1.34E-4
1600	41	15.35	2.76E-4	21137	13.2	14.76	3.06E-4
2304	49	30.66	3.32E-4	30623	13.3	21.45	4.16E-4

TABLE 3

Solving LLSP's by augmented matrices (m=250, n=125, NZ=2470, where B_1 and B_2^T in equation (5.1) are of the form D(n,c)) on an IBM 370/165.

	Tolerance	c=24	c=44	c=64	c=84	c=104
TIME	T=0.0	9.89	10.89	5.06	7.70	12.11
	$T=10^{-4}$	3.68	3.57	2.54	3.18	3.92
	$T=10^{-2}$	2.23	2.48	2.31	2.37	2.62
COUNT	T=0.0	13598	20623	14159	17936	21761
	$T=10^{-4}$	11357	11273	9001	10605	11483
	$T=10^{-2}$	6896	6809	5991	6609	6705

An improved version of SSLEST (the original code is described in Zlatev et al (1978)), which is still in preparation, has been used in the solution of such problems. The combination SMT+LDT+IR is extremely efficient for these problems (see Table 3 and Zlatev (1979), Zlatev et al (1980b)). A fairly general presentation of the use of the combination SMT+LDT+IR with direct methods other than Gaussian elimination is given by Zlatev (1980b).

6. APPLICATIONS: ORDINARY DIFFERENTIAL EQUATIONS (ODE'S)

The mathematical models of some problems arising in nuclear resonance theory contain large linear systems of ODE's: $\underline{y}' = A(t)\underline{y}+\underline{c}(t)$, $\underline{y}(a)=\underline{y}_0$, $t\in[a,b]$, $\underline{y}\in R^{n\times 1}$. These systems are moderately stiff and highly oscillatory. The use of the methods in Zlatev (1980c) leads to large sequences of systems of linear algebraic equations. The fact that many (usually several hundred) decompositions are needed in the numerical integration can be exploited to obtain an "optimal" drop tolerance T (beginning with a large T and reducing T each time when the iterative process is not convergent or some other difficulty is found). The efficiency of the combination: SMT+LDT+IR as well as the rule for obtaining the "optimal" T (this rule is called IRADT) is illustrated in Table 4, where: (i) n=255, (ii) NZ=7715 (i.e. the average number of non-zero elements per row is about 30), (iii) under "TIME" the computing time per integration step is given (an integration step consists of one decomposition, two back substitutions, two matrix computations and four matrix-vector multiplications when DS is used), (iv) under "COUNT" the largest number of non-zero elements kept in array A1 during the whole integration is given, (v) under "ITERATIONS" the average number of back substitutions per algebraic system is given, (vi) under "T" the drop-tolerance found at the end of the integration is given; the starting values of T are given in brackets (they are the same as the values at the end of the integration for the first two algorithms). For more details see Schaumburg et al (1980). The general case (non-linear systems of ODE's) is treated by Houbak and Thomsen (1979) and Thomsen (1979).

7. APPLICATIONS: PARABOLIC PARTIAL DIFFERENTIAL EQUATIONS (PPDE'S)

SMT was used in the solution of some PPDE's arising in thermo-dynamics (Zlatev and Thomsen (1977, 1979)). In this case long sequences of linear systems of algebraic equations with the same matrix are often obtained (see Example 3 in Zlatev and Thomsen (1979)). Therefore the use of DS with a "small" T is more efficient than SMT+LDT+IR. These problems show that the

TABLE 4

Solution of chemical problens described by linear systems of ODE's. Times on a UNIVAC 1100/82

ALGORITHMS	COUNT	T	ITERATIONS	TIME (per step)
DS	24347	10^{-14} (10^{-14})	1	41.25
IR	17318	10^{-2} (10^{-2})	3.77	16.77
IRADT	13917	10^{-1} (1)	5.54	13.00

combination SMT+LDT+IR is not an "all seasons method" and should be used carefully.

8. REFERENCES

Björck, Å. (1967) Iterative refinement of linear least squares problems, BIT 7, pp. 257-278.

Houbak, N. and Thomsen, P.G. (1979) SPARKS-a FORTRAN subroutine for the solution of large systems of stiff ODE's with sparse Jacobians, Report 79-02, Institute for Numerical Analysis, Technical University of Denmark.

Schaumburg, K., Wasniewski, J. and Zlatev, Z. (1980) The use of sparse matrix technique in the numerical integration of stiff systems of linear ordinary differential equations, Comput. Chem., 4, pp. 1-12.

Thomsen, P.G. (1979) Numerical solution of large systems of ODE's with sparse Jacobians. In "Working Papers for the 1979 SIGNUM Meeting on Numerical Ordinary Differential Equations", (R.D. Skeel, ed.), Computer Science Department, University of Illinois at Urbana-Champaign.

Wilkinson, J.H. and Reinsch, C. (1971) Handbook for automatic computations, Vol. II Linear Algebra, Springer, pp. 50-56.

Zlatev, Z. (1979) Use of iterative refinement in the solution of sparse linear systems, Report 1/79, Institute of Mathematics and Statistics, The Royal Veterinary and Agricultural University, Copenhagen.

Zlatev, Z. (1980a) On solving some large linear problems by direct methods, Report 111, Computer Science Department, Aarhus University, Denmark.

Zlatev, Z. (1980b) Modified diagonally implicit Runge-Kutta methods, Report 112, Computer Science Department, Aarhus University, Denmark.

Zlatev, Z., Barker, V.A. and Thomsen, P.G. (1978) SSLEST: a FORTRAN IV subroutine for solving sparse systems of linear equations (user's guide), Report 78-01, Institute for Numerical Analysis, Technical University of Denmark.

Zlatev, Z., Schaumburg, K. and Wasniewski, J. (1980a) Implementation of an iterative refinement option in a code for large and sparse systems, Comput. Chem., 4, pp. 87-99.

Zlatev, Z. and Thomsen, P.G. (1977) An algorithm for the solution of parabolic partial differential equations based on finite element discretization, Report 77-09, Institute for Numerical Analysis, Technical University of Denmark.

Zlatev, Z. and Thomsen, P.G. (1979) Application of backward differentiation methods to the finite element solution of time-dependent problems, Int. J. Num. Meth. Engng., 14, pp. 1051-1061.

Zlatev, Z., Wasniewski, J. and Schaumburg, K. (1980b) Comparison of two algorithms for solving large linear systems, Report 80/9, Regional Computing Centre at the University of Copenhagen (RECKU).

AUTHOR INDEX

The numbers underlined refer to the Reference pages where the references are listed in full.

SUBJECT INDEX